Smart Embedded Systems

"Smart Embedded Systems: Advances and Applications" is a comprehensive guide that demystifies the complex world of embedded technology. The book journeys through a wide range of topics from healthcare to energy management, autonomous robotics, and wireless communication, showcasing the transformative potential of intelligent embedded systems in these fields. This concise volume introduces readers to innovative techniques and their practical applications, offers a comparative analysis of wireless protocols, and provides efficient resource allocation strategies in IoT-based ecosystems. With real-world examples and in-depth case studies, it serves as an invaluable resource for students and professionals seeking to harness the power of embedded technology to shape our digital future.

Salient Features:

1. The book provides a comprehensive coverage of various aspects of smart embedded systems, exploring their design, implementation, optimization, and a range of applications. This is further enhanced by in-depth discussions on hardware and software optimizations aimed at improving overall system performance.
2. A detailed examination of machine learning techniques specifically tailored for data analysis and prediction within embedded systems. This complements the exploration of cutting-edge research on the use of AI to enhance wireless communications
3. Real-world applications of these technologies are extensively discussed, with a focus on areas such as seizure detection, noise reduction, health monitoring, diabetic care, autonomous vehicles, and communication systems. This includes a deep-dive into different wireless protocols utilized for data transfer in IoT systems.
4. This book highlights key IoT technologies and their myriad applications, extending from environmental data collection to health monitoring. This is underscored by case studies on the integration of AI and IoT in healthcare, spanning topics from anomaly detection to informed clinical decision-making. Also featured is a detailed evaluation and comparison of different system implementations and methodologies.

This book is an essential read for anyone interested in the field of embedded systems. Whether you're a student looking to broaden your knowledge base, researchers looking in-depth insights, or professionals planning to use this cutting-edge technology in real-world applications, this book offers a thorough grounding in the subject.

Smart Embedded Systems

Advances and Applications

Edited by

Arun Kumar Sinha
Abhishek Sharma
Luiz Alberto Pasini Melek
Daniele D. Caviglia

CRC Press
Taylor & Francis Group
Boca Raton London New York

CRC Press is an imprint of the
Taylor & Francis Group, an **informa** business

Image credit: metamorworks/Shutterstock

First edition published 2024
by CRC Press
2385 NW Executive Center Dr, Suite 320, Boca Raton, FL 33431

and by CRC Press
4 Park Square, Milton Park, Abingdon, Oxon, OX14 4RN

CRC Press is an imprint of Taylor & Francis Group, LLC

ISBN: 978-1-032-40417-2 (hbk)
ISBN: 978-1-032-62802-8 (pbk)
ISBN: 978-1-032-62805-9 (ebk)

DOI: 10.1201/9781032628059

Typeset in Sabon
by SPi Technologies India Pvt Ltd (Straive)

Contents

Preface

ABOUT THIS BOOK

This book, *Smart Embedded Systems: Advances and Applications*, addresses the importance of embedded systems in different sectors such as healthcare, communications, the environment, security, and more. The book provides a comprehensive guide to the current state of smart embedded systems, from basic principles to advanced applications, integrating different disciplines such as machine learning, AI, signal processing, electronics, and networking. It brings together a variety of techniques used in designing and implementing intelligent systems, ranging from hardware design and optimization to software implementation, data analysis, and machine learning. It showcases real-world applications of these techniques, with a focus on challenges and solutions in designing robust and efficient systems.

Features

- Comprehensive coverage of various facets of smart embedded systems, including design, implementation, optimization, and applications.
- Detailed analysis of machine learning techniques for data analysis and prediction in embedded systems.
- In-depth discussions on hardware and software optimizations to improve system performance.
- Exploration of real-world applications such as seizure detection, noise reduction, health monitoring, diabetic care, autonomous vehicles, and communication systems.
- Cutting-edge research on utilizing AI to enhance wireless communication and predict channel capacity.
- Discussion on important IoT technologies and their applications, from environmental data collection to health monitoring.
- Case studies on the integration of AI and IoT in healthcare, from anomaly detection to clinical decision-making.

- Illustration of different wireless protocols used for data transfer in IoT systems.
- Detailed evaluation and comparisons of different system implementations and methodologies.

CHAPTER ORGANIZATION

Chapter 1 examines the detection of epileptic seizures through the analysis of recorded electroencephalography (EEG) signals, utilizing random forest (RF) algorithms. The dual-strategy method incorporates classification, feature extraction, and RF training. **Chapter 2** presents an optimized hardware architecture for noise reduction in real electrooculogram (EOG) systems. The proposed architecture uses the differential evolution (DE) algorithm to design a noise reduction filter with fewer sign-power-of-two (SPT) terms, resulting in a more efficient hardware design. This architecture, which was implemented using Verilog HDL and tested using Altera DSP Builder, offers significant reductions in area and power consumption compared to existing architectures.

Chapter 3 explores E-health data acquisition systems for remote patient monitoring, discussing the selection of sensors and the use of neural networks for automatic health state recognition. A case study on the use of convolutional neural networks for recognizing motor imagery movement activities from EEG signals is presented. **Chapter 4** explores the impact of the digital transformation on diabetes management, highlighting how sensor technology, data analytics, and intelligent decision systems have revolutionized diabetic care.

Chapter 5 focuses on the control of a submersible autonomous robot (SAR) using an optimal control algorithm, while considering uncertain hydrodynamic parameters. A robust closed-loop system for the SAR is demonstrated, enabling it to maintain its depth effectively. **Chapter 6** highlights the design of intelligent embedded systems for future applications in various sectors. This chapter introduces a novel in-memory compute architecture that acts as a neuromorphic accelerator, compatible with a RISC-V processor core or any commercial central processing unit. A standard deep net architecture designed using resistive memories is discussed, serving multiple real-time applications with varying accuracy levels.

Chapter 7 provides insights into the role of AI in enhancing 5G and 6G wireless communication systems and demonstrates the challenges and solutions for increasing radio channel capacity, especially under conditions of vegetation obstruction. AI algorithms are employed to predict and mitigate vegetation attenuation, improving channel capacity and performance in areas with high levels of vegetation interference. The chapter underscores how AI-based systems can be instrumental in optimizing wireless communication networks. In **Chapter 8**, a smart office cabin integrated with embedded computing platforms and sensor technology is developed. The system encompasses office security, ambiance monitoring, and individual well-being monitoring. Biometrics authorize cabin entry, sensors monitor environmental variables, and image streams and biomedical sensors gage the well-being of staff.

Chapter 9 emphasizes the growth of wireless connectivity and sensor applications, and the pivotal role of wireless protocols in this surge. It examines three open-standard protocols—Sigfox, Lorawan, and Nb-IoT—detailing their specifications and making comparative analyses based on coverage area, power budget, and data transmission capacity. This evaluation offers valuable insights into selecting appropriate protocols for specific applications. **Chapter 10** looks at energy management, detailing the design and testing of a prototype chip for maximizing power extraction from a thermoelectric generator (TEG). It presents a successful integration of this prototype in a self-powered node that harvests power from thermal sources, paving the way for more autonomous, energy-efficient systems.

Chapter 11 presents a novel adaptive machine learning model, the dynamic adaptive weighted ensemble (DAWE), for handling continuous data streams in IoT-based health monitoring systems. It tackles the issue of concept drift, a phenomenon in data streams where statistical properties change over time. By employing the exponential weighted moving average (EWMA) technique, DAWE continuously refines predictive performance, achieving a significantly high accuracy rate. This reflects the importance and effectiveness of adaptive learning approaches in managing large-scale, fast-paced data streams in health monitoring. **Chapter 12** tackles the integration of IoT and electronic health records (EHR) in the healthcare sector and the challenge of efficient resource allocation. A distributed model called GraLSTM is introduced which combines graph neural networks (GNNs) with long short-term memory (LSTM). GraLSTM outperforms conventional models by efficiently capturing spatial and temporal data dependencies, offering faster computational speed and scalability to support clinical decision-making.

As editors we have worked hard to make *Smart Embedded Systems: Advances and Applications* as clear and easy to understand as possible. We hope that teachers, students, and anyone interested in this topic will find it useful. We invite you to share your thoughts on how we can make this book even better. Please email us your suggestions and feedback at: er.arunsinha.nsit@gmail.com. We value your opinion.

Editors

Dr Arun Kumar Sinha received his PhD in electronic engineering from the University of Genova, Genova, Italy, in 2013. He has 10 years' experience in industry and teaching at Agilent Technologies (Gurugram, India), Netaji Subhas Institute of Technology (New Delhi, India), Jaypee Institute of Information Technology (Noida, India), and Mekelle University (Mekelle, Ethiopia), and post-doctoral research experience at Federal University of Santa Catarina (Florianopolis, Brazil). Currently, he is working as an associate professor at VIT-AP University, Andhra Pradesh, India. Among the subjects he has taught are digital logic and microprocessors (undergraduate students, UG), fundamentals of electrical and electronics engineering (UG), principles of electrical engineering (UG), VLSI system design (UG), analog VLSI design (UG), analog IC design (postgraduate students, PG), and analog devices and circuits (UG), process dynamics and multivariable control (PG), real-time control and embedded systems (PG), electrical machines and instruments (UG), basic electronics device and circuits (UG), electronic instruments and measurement (UG), principles of electrical engineering (UG), biomedical instrumentation (UG), and electrical measurements (UG). He is the author of 35 papers in international journals/conferences, one conference proceedings as editor, and three patents. He is a member of the Hyderabad section of the IEEE, India. His research interests include intelligent algorithms, integrated circuit design, digital and analog VLSI, semiconductor devices, and mathematical modeling.

Dr Abhishek Sharma received his BE in electronics engineering from Jiwaji University, Gwalior, India, and a PhD in engineering from the University of Genoa, Italy. He is presently working as an assistant professor in the Department of Electronics and Communication Engineering at The LNM Institute of Information and Technology, Jaipur, RJ, India, where he also coordinates the ARM University partner program, Texas Instruments Lab, and Intel Intelligent Lab. He is the center lead of LNMIIT's Center of Smart Technology (L-CST). He is a member of the IEEE, the Computer Society and Consumer Electronics Society, a Lifetime Member of the Indian Society for Technical Education (ISTE), a Lifetime Member of the Advanced Computing

Society (ACS), India, and a Life Member of the Computer Society of India (CSI). He is also Membership Secretary of IEEE CTSoC, Application-specific CE for Smart Cities (SMC). He has taught microprocessors and interface, the internet of things and embedded systems at undergraduate level, and applied machine learning and system on chip at postgraduate level. His research interests are real-time systems and embedded systems.

Dr Luiz Alberto Pasini Melek received his BE in electronic engineering (cum laude) from the Federal University of Paraná (UFPR), Curitiba, Brazil, in 1998. He holds a Diploma in Telecommunications, 2001, and a Diploma in Embedded Systems Design, 2014. He received his MS degree in 2004, and PhD in 2017, both from the Federal University of Santa Catarina (UFSC), Brazil. In 2015–16 he was a professor in the Electrical Engineering Department of UFPR, where he taught classes in basic electronics, analog filters, laboratory of electronics, and laboratory of electricity. He has more than 20 years of experience in industry and since 2017 he has worked as a hardware specialist engineer at Pumatronix, Curitiba, Brazil. He has designed both hardware and firmware for applications such as embedded systems, automotives, telecommunications, metering, and agriculture. He has authored or co-authored 14 papers in international journals and conferences, and works as a volunteer. He is a reviewer for the journals *IEEE Transactions on Circuits and Systems I: Regular Papers, IEEE Transactions on Circuits and Systems II: Express Briefs, IEEE Transactions on Device and Materials Reliability*, and *World Journal of Engineering*. His main interests are ultra-low voltage digital and mixed-signal circuits for VLSI, and the development of embedded systems for a broad range of applications.

Prof. Daniele D. Caviglia graduated in electronic engineering and specialized in computer engineering at the University of Genoa, Italy, in 1980 and 1982, respectively. In 1983, he joined the Institute of Electrical Engineering, University of Genoa, as Assistant Professor of Electronics. Since 1984, he has been with the Department of Biophysical and Electronic Engineering (DIBE), as associate professor from 1992 and as full professor since 2000. From 2002 to 2008, he was the Director of DIBE. He is currently with the Department of Electrical, Electronics and Telecommunications Engineering and Naval Architecture (DITEN), University of Genoa, teaching postgraduate courses in electronic systems for telecommunications, distributed electronic systems and technologies for environmental monitoring and integrated electronics, and electronic systems laboratory undergraduate courses. He is also active in the development of innovative solutions for environmental monitoring for civil protection and security applications. His research interests include the design of electronic circuits and systems for telecommunications, electronic equipment for health and safety, and energy harvesting techniques for internet of things (IoT) applications. He is a Life Member of the IEEE.

Contributors

Swetha Annangi
School of Electronics Engineering, VIT-AP University
Near Vijayawada, Andhra Pradesh, India

Abhranil Das
Department of Electronics and Electrical Engineering, IIT Guwahati
North Guwahati, India

Soham Das
Department of Chemical Engineering, IIT Kharagpur
Kharagpur, India

Anirban Dasgupta
Department of Electronics and Electrical Engineering, IIT Guwahati
North Guwahati, India

Parishmita Deka
Department of Electronics and Communication Engineering, IIIT Guwahati, India

Sneha Edupuganti
School of Computer Science and Engineering, VIT-AP University
Near Vijayawada, Andhra Pradesh, India

Afroz Fatima
Department of Electrical and Electronics Engineering, Birla Institute of Technology and Science Pilani
Goa, India

M. Geetha Pratyusha
School of Electronics Engineering, VIT-AP University
Near Vijayawada, Andhra Pradesh, India

Kothamasu Venkata Naga Durga Sai Harshith
School of Computer Science and Engineering, VIT-AP University
Near Vijayawada, Andhra Pradesh, India

Gundugonti Kishore Kumar
Department of Electronics and Communication Engineering, V R Siddhartha Engineering College
Vijayawada, Andhra Pradesh, India

Puli Kishore Kumar
Department of Electronics and Communication Engineering, National Institute of Technology
Tadepalligudem, Andhra Pradesh, India

Sachin Kumar
Head of Technology Product Strategy AI, Samsung Research
Amaravati, Korea

Subhasish Mahapatra
School of Electronics Engineering, VIT-AP University
Near Vijayawada, Andhra Pradesh, India

Tamas Majoros
IT Systems and Networks, University of Debrecen
Debrecen, Hungary

Luiz Alberto Pasini Melek
CayennE-k Tecnologia
Curitiba, Brazil

Anupama Namburu
School of Engineering, Jawaharlal Nehru University
New Delhi, India

Stefan Oniga
Electric, Electronics and Computer Engineering, Technical University of Cluj-Napoca, North University Center of Baia Mare, Romania
and
Baia Mare, Romania
IT Systems and Networks, University of Debrecen
Debrecen, Hungary

Iuliu Alexandru Pap
Electric, Electronic and Computer Engineering, Technical University of Cluj-Napoca
North University Center of Baia Mare
Baia Mare, Romania

Abhijit Pethe
Department of Electrical and Electronics Engineering, Birla Institute of Technology and Science Pilani
Goa, India

G.V. Ramesh Babu
Sri Venkateswara University
Tirupati, India

V. H. Prasad Reddy
Department of Electronics and Communication Engineering, National Institute of Technology
Tadepalligudem, Andhra Pradesh, India

Aravapalli Rama Satish
School of Computer Science and Engineering, VIT-AP University
Near Vijayawada, Andhra Pradesh, India

T. Senthil Siva Subramanian
Head, Institute Industry Interface, Sharda Group of Institutions
Mathura, India

Kapil Sharma
Information Technology Department, Delhi Technological University (DTU)
Delhi, India

Vadapalli Siddhartha
School of Electronics Engineering, VIT-AP University
Near Vijayawada, Andhra Pradesh, India

Arun Kumar Sinha
School of Electronics Engineering, VIT-AP University
Near Vijayawada, Andhra Pradesh, India

Bhanu Teja Veeramachaneni
School of Computer Science and Engineering, VIT-AP University
Near Vijayawada, Andhra Pradesh, India

Chapter 1

A reconfigurable FPGA-based epileptic seizures detection system with 144 μs detection time

Swetha Annangi and Arun Kumar Sinha
VIT-AP University, Near Vijayawada, India

1.1 INTRODUCTION

Around 50 million people worldwide are affected by epilepsy, according to the World Health Organization (WHO) [1]. The International League Against Epilepsy (ILAE) has standardized the classification of epileptic seizures into partial and generalized. Partial seizures are restricted to the discrete area of the cerebral cortex; generalized seizures involve diffuse regions of the brain simultaneously in a bilaterally symmetrical fashion [2]. Electroencephalography (EEG) plays a major role in the analysis of epileptic seizures [3]. The availability of modern structural and functional imaging technologies has not reduced the impact of scalp EEG in the evaluation of patients with seizures [2].

The use of machine learning (ML) and deep learning (DL) techniques in healthcare has been widespread over recent decades. ML algorithms include multi-layer perceptron (MLP) [4], artificial neural networks (ANN) [5], support vector machine (SVM) [6, 7], multi-class SVM [8], random forest (RF) [8–10], XGBoost [11], K-nearest neighbor (KNN) [8, 12, 13], naïve Bayes (NB) [8, 13], and extreme learning machine (ELM) [14]. These, and DL algorithms like convolutional neural network (CNN) [15, 16], deep CNN [17], long-short-term memory (LSTM) and bi-directional LSTM (Bi-LSTM) [18], have been employed in the detection of epileptic seizures.

In recent years, many researchers have worked on field programmable gate array (FPGA)-based ML algorithms for the detection of epileptic seizures, which have advantages of reliability, flexibility and lower costs [19–25]. Lichen Feng *et al.* [21] implemented feature extraction (FE) and SVM modules using Verilog HDL. The FE module has discrete wavelet transform (DWT), mean absolute value (MAV) and variance sub-modules. In [22], Yuanfa Wang *et al.* proposed the VLSI architectures of the DWT FE module, the modified sequential minimal optimization (MSMO) training module, and one-against-one multi-class SVM classification module. ChichTsow *et al.* [23]

DOI: 10.1201/9781032628059-1

designed an FE module and NN architecture and implemented the designed modules on FPGA. Heba Elhosary [24] implemented FE and SVM classifier modules using the MATLAB® HDL coder. Rijad Saric [25] designed an ANN module using very high-speed integrated circuit hardware description language (VHDL).

This chapter describes both software and hardware implementation of a random forest (RF) algorithm for the detection of epileptic seizures. The feature extraction module, RF training and inference modules were designed using Verilog Hardware Description Language (HDL) in XILINX VIVADO 2019.2. The speed of the classification process improved with the designed hardware modules.

The rest of the paper is organized as follows. Section 1.2 gives details of the proposed methodology based on the software implementation. In Section 1.3, the hardware implementation-based epileptic seizures detection system is explained with the feature extraction, RF training and inference module architectures. The experimental results are described in Section 1.4 and Section 1.5 provides a conclusion.

1.2 SOFTWARE-BASED IMPLEMENTATION

The block diagram of the proposed software-implemented epileptic seizures detection system is shown in Figure 1.1. The software-based implementation uses the MATLAB® R2019a tool. First, the loaded raw EEG data are converted into .mat files and total data are segmented into non-overlapping one-second segments which are fed into the feature extraction module. After the features have been extracted, the total data is divided into 80% for training, and 20% for testing. The training data are given to the RF training module and test data are given to the RF inference module, which classifies it into three classes: normal, partial seizures, and generalized seizures, represented as N, P, and G, respectively.

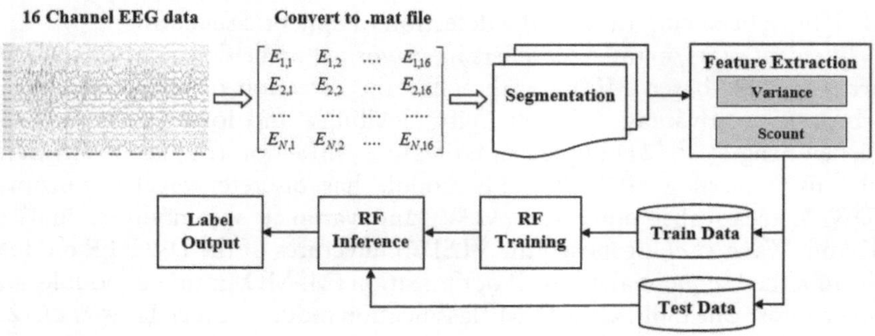

Figure 1.1 Block diagram of the proposed methodology.

1.2.1 EEG data

The raw EEG data of 40 normal and epileptic patients were collected from the Medysis Hospitals. The collected data is from both male and female patients aged 4 to 80 years. The input EEG data were collected from 16 channels with a sampling frequency of 256 Hz which are placed according to ISS 10–20. The data were analyzed by the neurologist and labeled N, P, and G, respectively. A sample of the seizure patient data acquired is shown in Figure 1.2. The raw EEG signals are converted into .xlsx files by the Allengers Virgo EEG software. The .xlsx files are given as input to MATLAB® and converted into .m files.

1.2.2 Segmentation

The loaded EEG data were segmented into non-overlapped, 1-second segments. Each segment of 1-second data has 256 samples. This study included 36,070 seconds of EEG data; thus 36,070 segments were given to the feature extraction module.

1.2.3 Feature extraction

The dimensionality of the input data can be reduced by extracting the features. The variance of each 1-second segmented EEG data is calculated using (1.1).

$$V_k = \frac{1}{N}\sum_{i=1}^{N}\left(X_{i,k} - \mu\right)^2 \tag{1.1}$$

where N indicates the number of samples in a segment. In this case, N is equal to 256. $X_{i,k}$ is the ith sample from the kth channel. μ is the mean of N samples. The calculated variances of the normal and seizure patient data are plotted in Figure 1.3, which shows that the calculated variance of the normal patient is less than that of the seizure patient. The variance value that differentiates the normal and seizure patient data is taken as threshold (τ).

The Scount is calculated by counting the number of channels with variances greater than or equal to τ. By considering all 16 channels, Scount gives the number of channels with seizure data. In this software-based design, variance and Scount are considered as features.

1.2.4 RF algorithm

Random forest (RF) [26] is an ensemble learning classifier which has a collection of weak learners. The decision tree (DT) is a weak learner in the

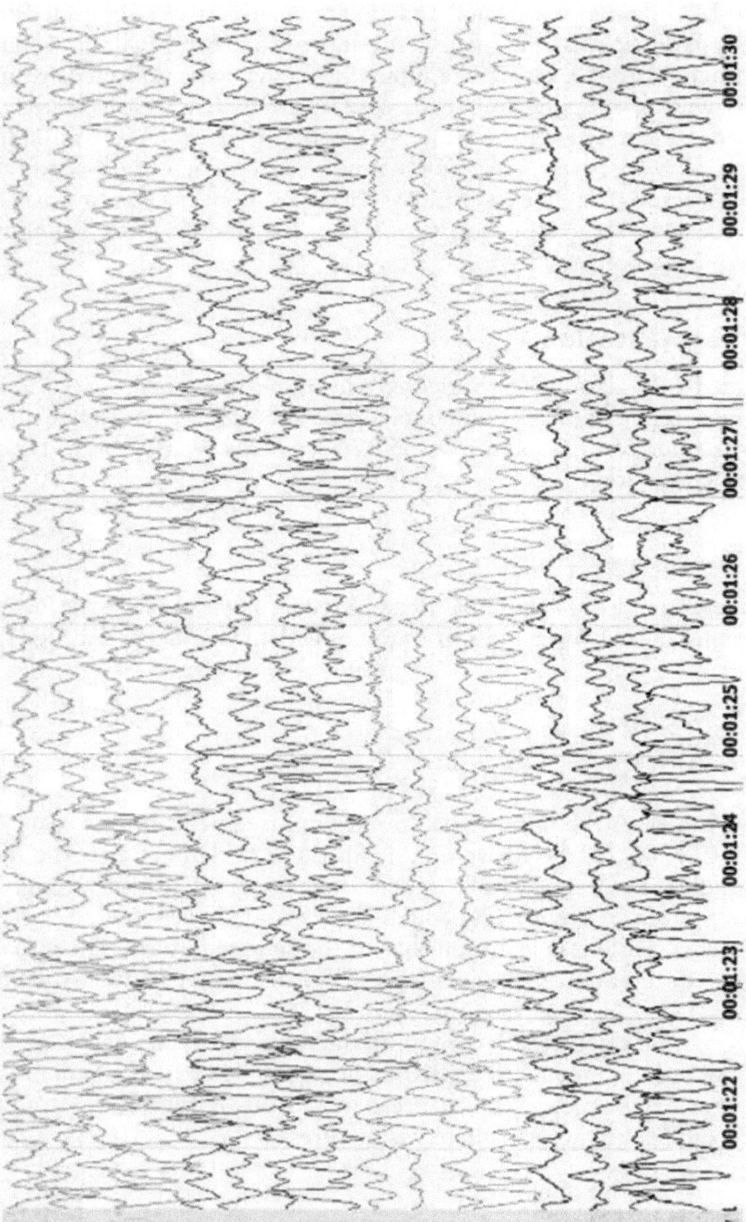

Figure 1.2 Sample EEG data of epileptic seizures patient.

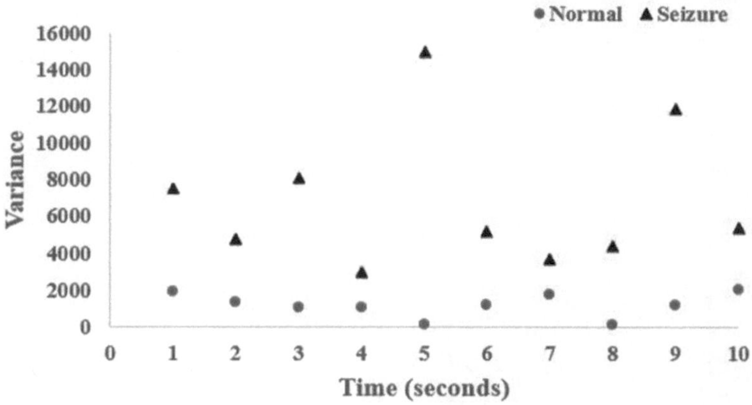

Figure 1.3 Computed variances of normal and seizure patient data.

RF classifier. Each node in the DT involves different tests by comparing the given attributes with the threshold values. There are two types of DT: univariate, in which only one attribute is considered in the test condition; and multivariate, in which two or more combinations of attributes are considered at each node for testing. The axis parallel DT is a univariate DT, the oblique and non-linear DTs are multivariate DTs [27]. In axis-parallel test function, f has the following form [28]:

$$f = A_i + a_i \tag{1.2}$$

where a_i is the threshold value of the attribute A_i. The task of the axis-parallel DT-building algorithm is to find the best values for i and a_i in every node of the tree. In oblique DT, the test function f is defined as given in (1.3).

$$f = \sum_{i=1}^{n} a_i \cdot A_i + a_{n+1} \tag{1.3}$$

where a_{n+1} represents the threshold value of the function f. In nonlinear DTs, function f is represented as the second-order polynomial of n variables as given in (1.4).

$$f_{2nd} = \sum_{i=1}^{n}\sum_{j=1}^{n} W_{i,j} \times A_i \times A_j + \sum_{i=1}^{n} W_i \times A_i + W \tag{1.4}$$

where A_i and A_j are the single attributes with new weights $W_{i,j} = a_i \times a'_j + a'_i \times a_j$ and $W_i = a_{n+1} \times a_i + a'_{n+1} \times a_{n+1}$, and new threshold $W = a_{n+1} \times a'_{n+1}$. The three commonly used DT methods, axis parallel, oblique, and non-linear, are

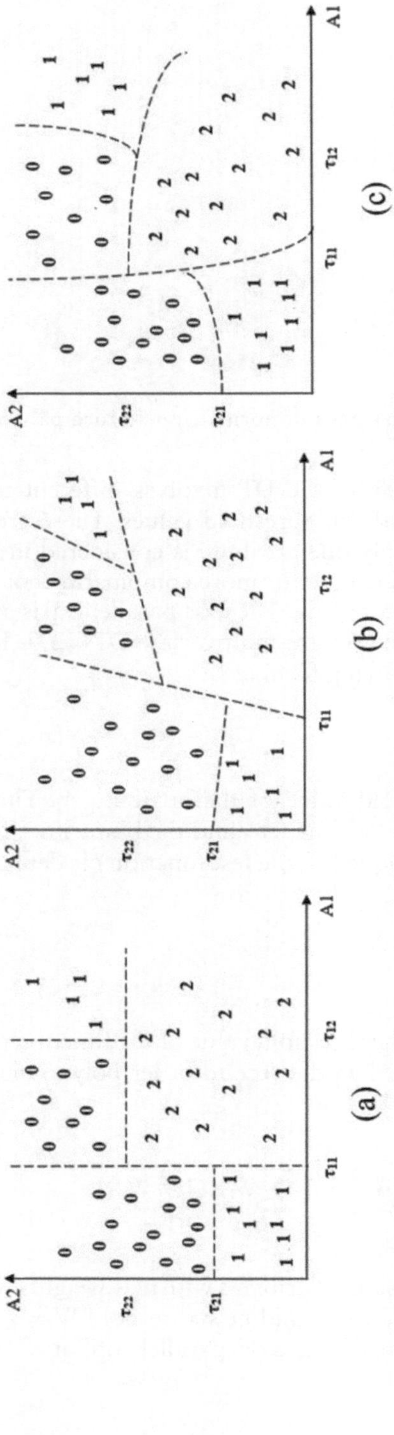

Figure 1.4 Decision tree methods: (a) Axis parallel DT, (b) Oblique DT, (c) Nonlinear DT.

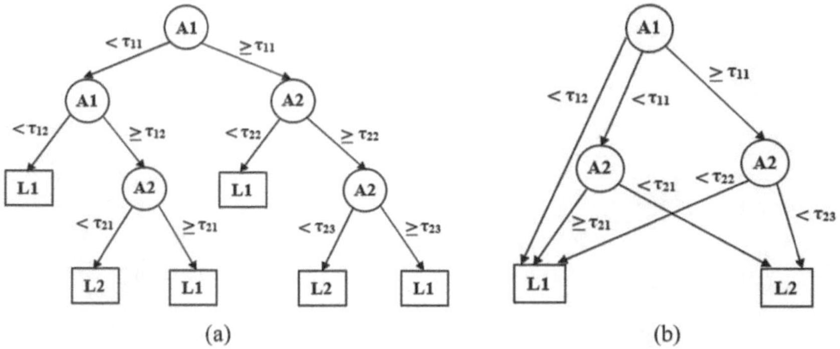

Figure 1.5 (a) Tree diagram based on BDD and (b) Tree diagram based on MDD.

differentiated as shown in Figure 1.4. The two co-ordinate axes represent A1 and A2 attributes with three classes: 0, 1, and 2. The τ_{11} and τ_{12} represent threshold values of A1, and τ_{21} and τ_{22} represent threshold values of A2. In axis parallel DT, the partition lines and attribute axes are parallel to each other. In oblique DT, the partition lines are oblique straight lines. For non-linear DT, the decision boundaries are curves.

The decision diagrams can be designed in two ways: binary-valued decision diagram (BDD) and multi-valued decision diagram (MDD) (Figure 1.5). In BDD, each terminal of a node has two outgoing edges. In MDD, there can be more than two outgoing edges [29].

1.3 HARDWARE-BASED IMPLEMENTATION

The hardware implementation of the epileptic seizures detection system is shown in the state diagram in Figure 1.6. The segmented 1-second EEG data are given to the feature extraction module when the enable (en) signal is active. After feature extraction, the training data and test data are loaded to the RF training and inference modules, respectively and the output label is displayed.

1.3.1 Feature extraction

The architecture to compute the variance of the segment from *k*th channel, V_k is shown in Figure 1.7. Rather than division, the right-shift operation is used here in the design of the variance module. Since the number of samples in a segment is 256, i.e., 2^8, the eight bits have been right shifted.

In hardware-based design, instead of taking variance values of 16 channels as features, Scount is considered as a feature, thus reducing the hardware requirement. The architecture to calculate Scount is shown in Figure 1.8.

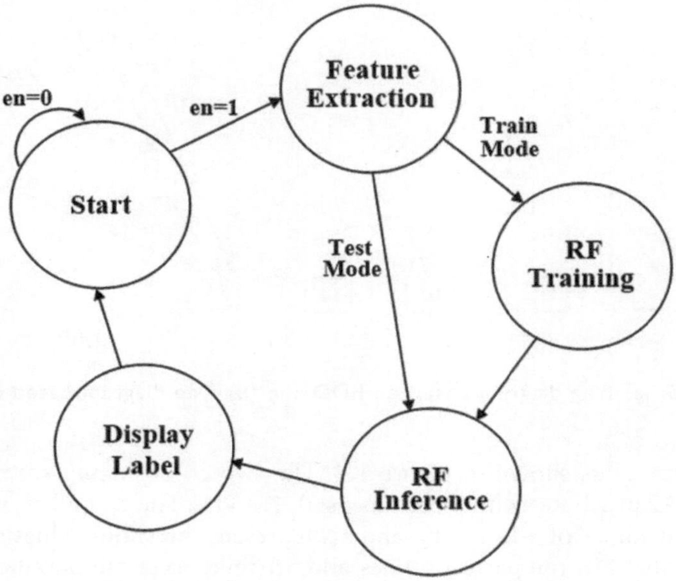

Figure 1.6 State diagram of the epileptic seizure detection system.

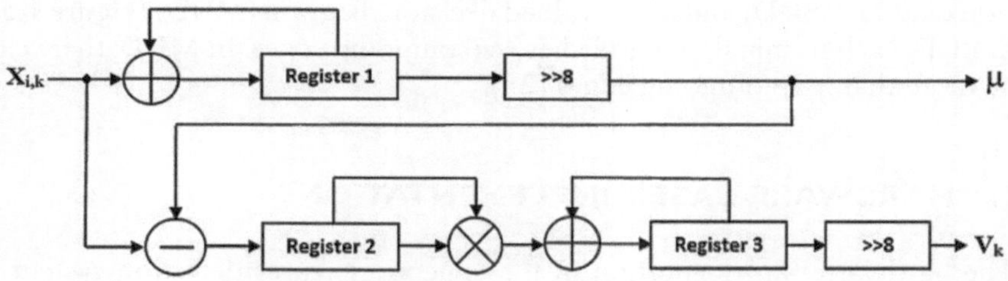

Figure 1.7 Variance calculation architecture.

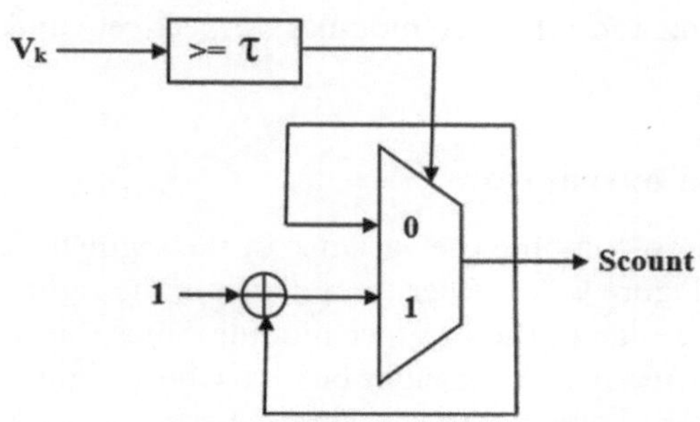

Figure 1.8 Scount calculation architecture.

Initially, Scount is zero. If the variance of the *k*th channel is greater than or equal to the τ, then Scount will be incremented. Otherwise, it remains the same.

After feature extraction, the total data are divided into 80% for training, and 20% for testing.

1.3.2 RF training module

Axis-parallel DT with MDD achieves high-speed classification. The RF algorithm has training and inference phases. In the training phase, each DT is assigned sample data randomly selected from the set of training data by a technique known as bagging, which uses bootstrap sampling. In the RF training process (Figure 1.9), the training data are randomly divided into six datasets and given to DTs. Each DT generates the thresholds by sorting the feature value at each node. Hence, RF needs comparison circuits to compare thresholds with the input value at each node.

When the training mode is selected, the training data are loaded to the RF training module, which consists of six axis-parallel DTs, each with two nodes and three leaves. After training the data, the threshold values for the N, P, and G classes are decided. These values are τ_N, τ_P, and τ_G, respectively, and are computed from the minimum and maximum values generated from the RF training module of the respective class labels (τ_{Nmin}, τ_{Nmax}, τ_{Pmin}, τ_{Pmax}, τ_{Gmin}, and τ_{Gmax}). The tree diagrams of the six DTs generated are shown in Figure 1.10. The threshold circuit in the DT design needs combinational logic suitable for VLSI implementation. The threshold function is easy to design in digital, which saves significant silicon area compared to the sigmoid or radial basis functions used in MLP [30]. The labels for N, P, and G classes are represented as 0, 1, and 2 respectively.

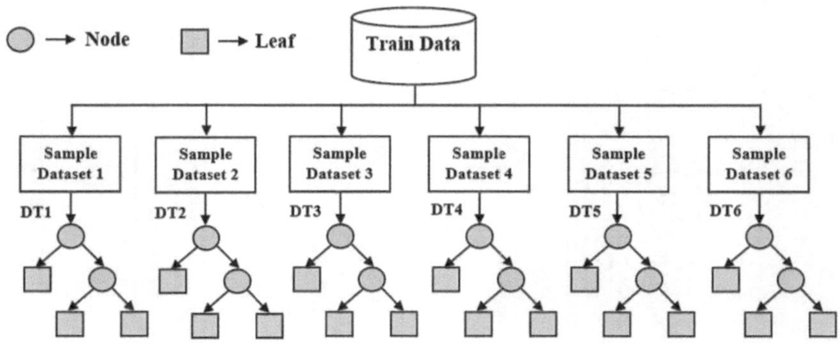

Figure 1.9 The RF training phase.

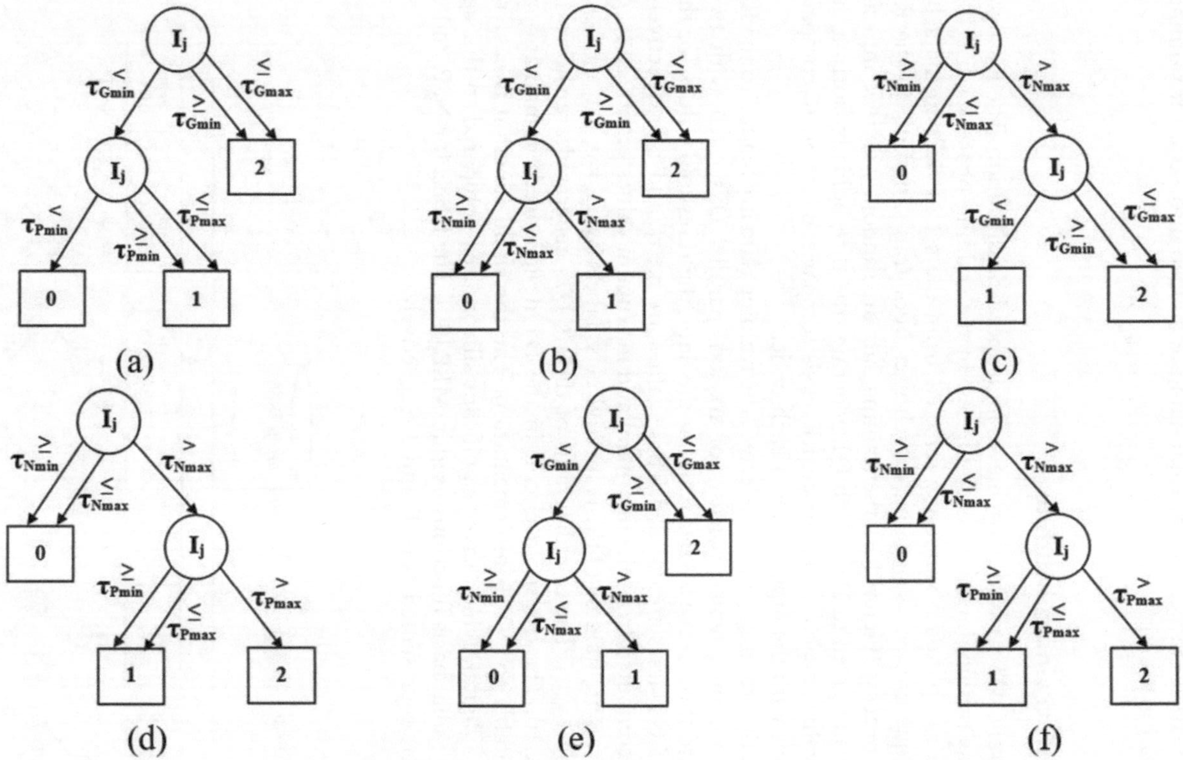

Figure 1.10 Tree diagrams of six DTs generated based on MDD: (a) DT1, (b) DT2, (c) DT3, (d) DT4, (e) DT5, (f) DT6.

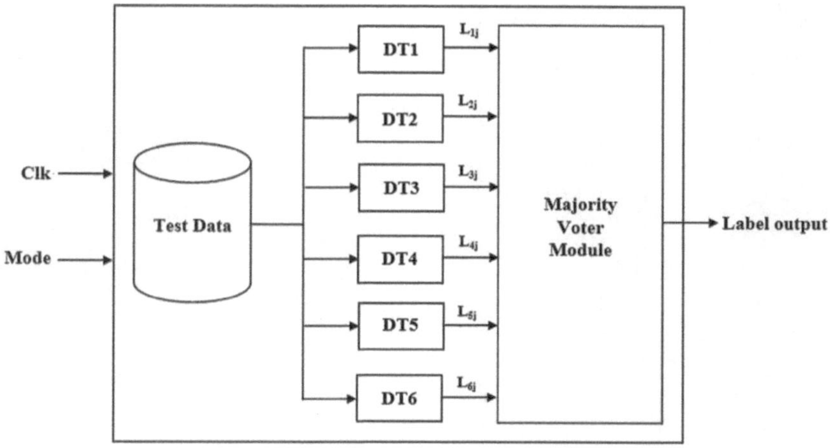

Figure 1.11 RF inference module block diagram.

1.3.3 The RF inference module

In the RF inference phase, the test data is classified into one of the classes using the designed DTs and the majority voter. The threshold values generated in the RF training phase are loaded into the inference module. When test mode is on, the test data are given to the RF inference module (Figure 1.11). In this phase, each input data is given to six designed DTs, the internal architectures of which are shown in Figure 1.12. Each DT decides the label output. The decision of the ith DT for the jth instance is labeled L_{ij}. The labelled outputs from all six DTs are given to the majority voter. The majority voter determines the most repeated class label of the jth instance as given in (1.5). The output of the majority voter is displayed. The internal architecture of the majority voter module is given in Figure 1.13. The weights for the label outputs are calculated after obtaining the output from all DTs. The weights of N, P, and G classes are represented as W0, W1, and W2, respectively. The label corresponding to the class with maximum weight will be the output of the majority voter and will be displayed at the output.

$$Labeloutput_j = \underset{i\in\{1,2,3,4,5,6\}}{Mode}\left(L_{ij}\right) \tag{1.5}$$

1.3.4 Display label

The three classes N, P, and G are labeled 0, 1, and 2, respectively. The majority voter module gives the output label for the corresponding class, which is displayed on the LCD display of the FPGA board.

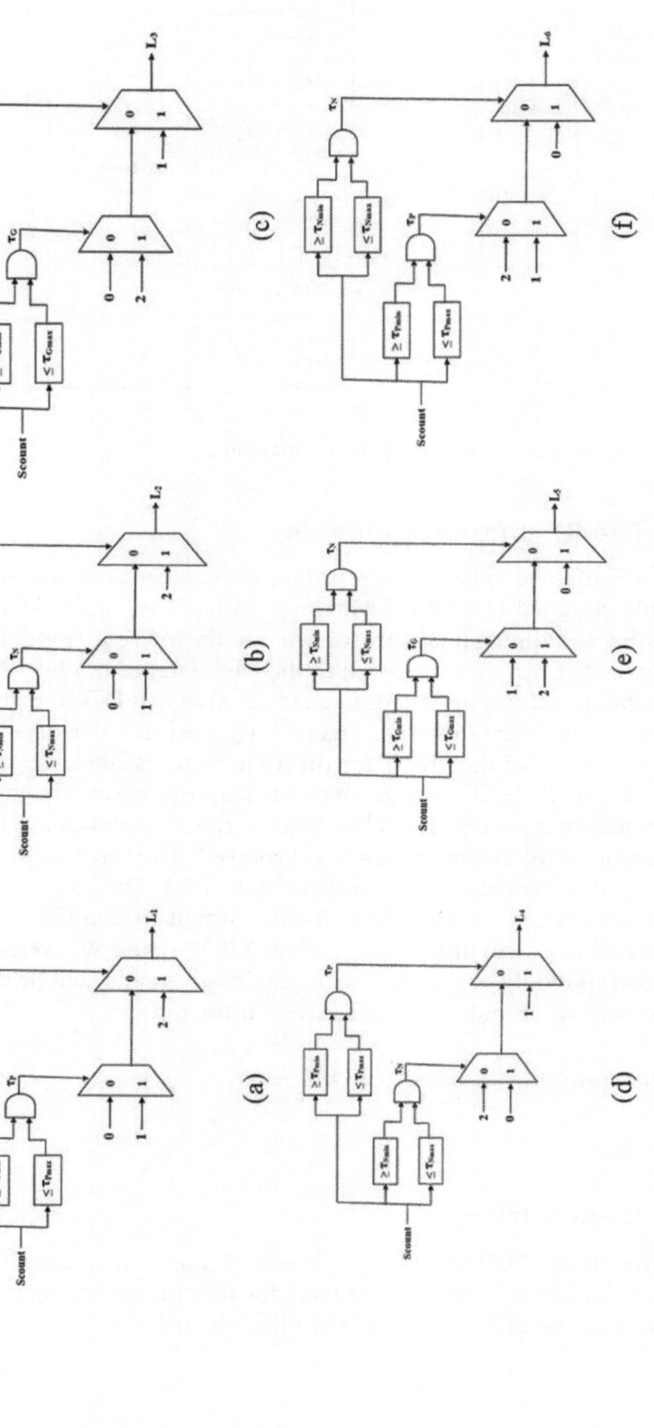

Figure 1.12 Internal architecture of six DTs: (a) DT1, (b) DT2, (c) DT3, (d) DT4, (e) DT5, (f) DT6.

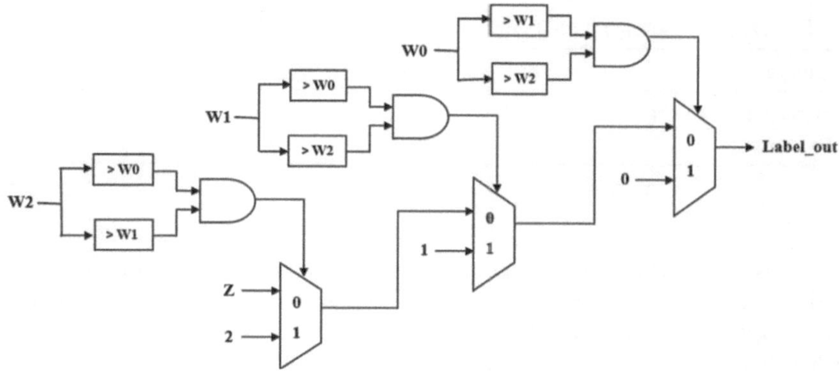

Figure 1.13 Internal architecture of majority voter module.

1.4 EXPERIMENTAL RESULTS

This section presents the experimental results obtained from both software and hardware implementations. The performance of the proposed detection system is measured using accuracy and classification time. Accuracy [31] is defined as the ratio of correctly classified classes to the total number of classes and is calculated using (1.6).

$$Accuracy = \frac{TP + TN}{TP + TN + FP + FN} \times 100 \tag{1.6}$$

where *TP* represents true positive, *TN* represents true negative, *FP* represents false positive and *FN* represents false negative.

1.4.1 Software implementation results

The EEG data from normal and epileptic seizure patients is divided into 80% for training and 20% for testing. The RF classifier is used to detect epileptic seizures and classify the input EEG signals into three classes: N, P, and G. The number of DTs considered are six. The software implementation is done using MATLAB® R2019a considering two features: variance and Scount. From the classification accuracies obtained using these two features (see Table 1.1), it is observed that Scount has given better accuracy than variance.

1.4.2 FPGA implementation results

The results obtained from MATLAB® indicate that the Scount feature classifies the seizures efficiently, compared to considering variances from 16

Table 1.1 The result from MATLAB® algorithm

Feature	*Accuracy (%)*	*Classification time (seconds)*
Variances of 16 channels	97.15	10
Scount	99.87	9

channels. In the DT design, the hardware requirement reduces in accordance with the number of features. In the case of variances from 16 channels as features, the DT needs a greater number of nodes to decide the class label. When Scount is a feature, the DT design requires only two nodes. Hence, Scount reduces the hardware requirement. The computed Scount features are given to the RF classifier. The proposed FE, RF training and RF inference modules are designed using Verilog HDL in XILINX VIVADO 2019.2. The simulation results of FE, RF training and RF inference modules are shown in Figures 1.14 to 1.16, respectively.

This study used the XILINX EDGE ZYNQ XC7Z010-1CLG400C FPGA development board. This board is integrated with 28K logic cells, 17,600 look-up tables (LUT), and 80 digital signal processor (DSP) slices and has a clock frequency of 50 MHz [32]. After the Verilog HDL modules have been designed, the following steps are performed: synthesis, implementation, bitstream generation, and programming the FPGA device. FPGA resources used after synthesis are listed in Table 1.2. After all the steps have been performed,

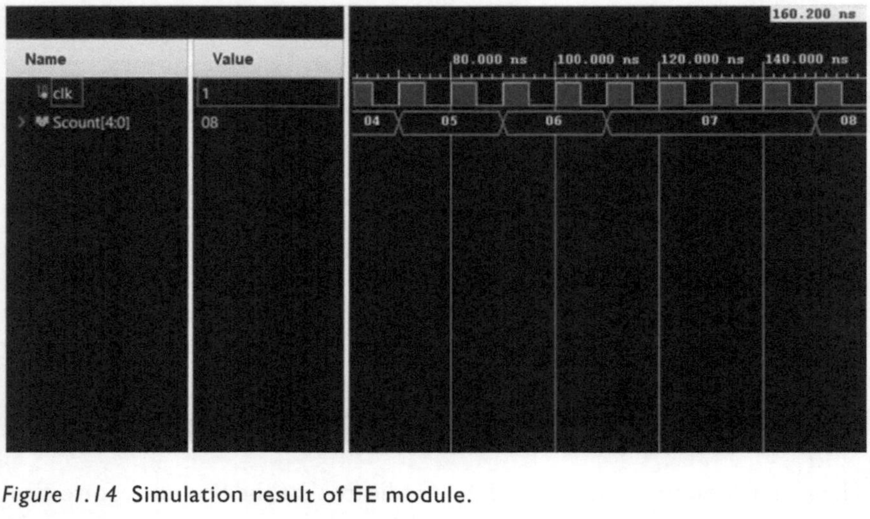

Figure 1.14 Simulation result of FE module.

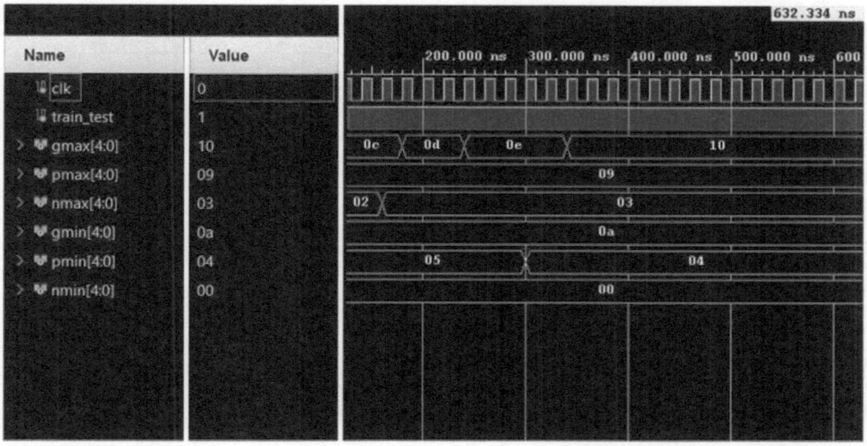

Figure 1.15 Simulation result of RF Training module.

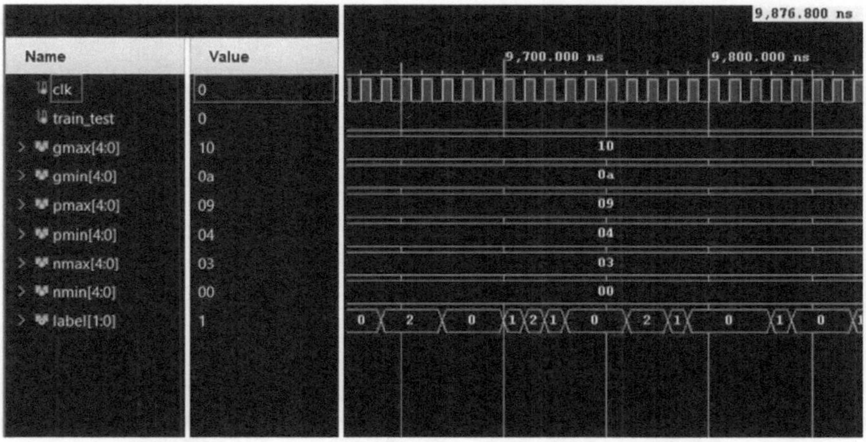

Figure 1.16 Simulation result of RF Inference module.

the label is displayed on the FPGA board. The experimental set-up displaying the type of seizure is shown in Figure 1.17.

Table 1.3 compares the proposed work with state-of-the-art works on FPGA-based epileptic seizure detection. The FPGA boards used, the modules designed, FE, classifier, operating frequency, performance metrics, and classification time needed for both software and hardware implementations are summarized in the table, which shows that the proposed design reduces the time required for efficient detection of seizures.

Table 1.2 FPGA resources utilized in this work

Resources	*Available*	*Used*	*Utilized (%)*
	FE Module		
LUTs	17600	70	0.40
Registers	35200	46	0.13
F7 Muxes	8800	4	0.05
F8 Muxes	4400	1	0.02
Bonded IOB	100	6	6
	RF Training Module		
LUTs	17600	1474	8.38
Registers	35200	174	0.49
F7 Muxes	8800	644	7.32
F8 Muxes	4400	314	7.23
Bonded IOB	100	32	32
	RF Inference Module		
LUTs	17600	312	1.77
Registers	35200	46	0.13
F7 Muxes	8800	115	1.31
F8 Muxes	4400	51	1.16
Bonded IOB	100	39	39

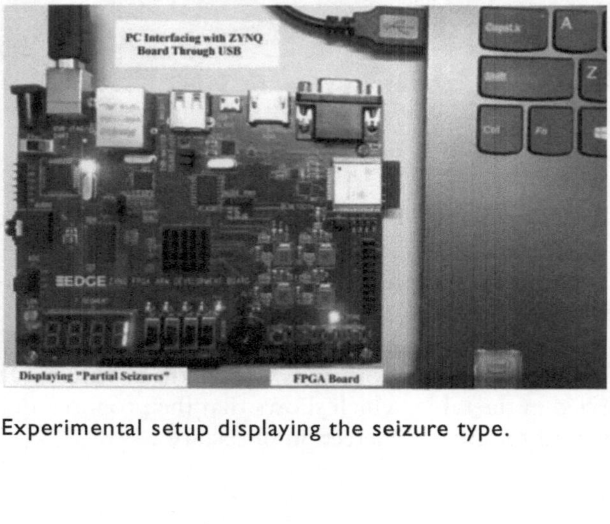

Figure 1.17 Experimental setup displaying the seizure type.

Table 1.3 Comparison of present work with the state-of-the-art works

Ref.	*FPGA board*	*Modules designed*	*F.E.*	*Classifier*	*Operating frequency (MHz)*	*Performance metrics*	*Software classification time*	*Hardware classification time*
[20]	ZYNQ 7000 XC7Z020	FE, and SVM	STFT	SVM	100	SEN: 98.4%	413 μs	313 μs
[21]	Altera Cyclone II	FE, and SVM	Time-frequency domain	NLSVM	50	DR: 96.8%	> 24.7 ms	24.7 ms
[22]	Virtex-5	FE, MSMO, and Multi class SVM	DWT, Maximum, SD	NLSVM	–	ACC: 95.1 (S/W) 93.9 (H/W)	1695 ms	686 ms
[23]	–	FE, and NN	Amplitude, Frequency, AE, and SD	NN	8	ACC: 98.76% (S/W) 89.88% (H/W)	–	160 ms
[24]	Virtex-7	SMO, FE, and SVM	FD, HE, and CL	SVM	100	ACC: 90.36% SEN: 96.7% SPE: 90.34%	–	16 ns
[25]	Altera D2-115 Cyclone IVE Series	Digital ANN	Time-frequency domain	ANN	50	ACC: 95.14%	>4.78 μs	4.78 μs
[33]	Virtex-6	Seizure detection system	WT, and Statistical	SVM	–	ACC: 96.7%	–	–
This work	**EDGE ZYNQ SoC**	**FE, RF training, and RF inference**	**Variance, Scount**	**RF**	**50**	**ACC: 99.87% (S/W) 97% (H/W)**	**9 s**	**144 μs**

NLSVM-non-linear support vector machine; DWT-discrete wavelet transform; AE-approximate entropy; SD-standard deviation; NN-neural network; ANN-artificial NN; FD-fractal dimension, HE-Hurst exponent, CL-coast line; GNS-generalized non-specific seizure; FNS-focal non-specific seizure; NS-non-specific and unknown seizure; SMO-sequential minimal optimization; ACC-accuracy; SEN-sensitivity; SPE-specificity; DR-detection rate.

1.5 CONCLUSION

In this work, FE, RF training, and RF inference modules using Verilog HDL were designed for detecting epileptic seizures. The detection system was implemented using both MATLAB® R2019a and XILINX VIVADO 2019.2. The modules were implemented on XILINX EDGE ZYNQ XC7Z010-1CLG400C FPGA development board using EEG data obtained from the Medysis Hospitals. The FPGA-based machine learning algorithm improved performance in the detection of epileptic seizures. The experimental results show that the FPGA-based system detected more rapidly than software-based implementation. This detection system can therefore be efficiently utilized by the neurologist to analyze seizure activity from the recorded long-duration EEG signals.

ACKNOWLEDGMENT

The authors would like to thank Medisys Hospitals, Hyderabad for providing raw EEG data. The authors also express their gratitude to Dr Pokala Vijaya Lakshmi, neurologist, Medisys Hospitals, Hyderabad for detailed explanations of seizure activity, and Mr Ambati Anil Kumar, Neuro Technician, Osmania General Hospital, Hyderabad for help with the analysis of EEG signals.

REFERENCES

[1] World Health Organization, Available online at: https://www.who.int/health-topics/epilepsy#tab=tab_1, Last accessed 22-04-2021.

[2] U. K. Misra, and J. Kalita, *Clinical Electroencephalography* E-Book, Elsevier, 2018.

[3] J. Engel Jr., "A practical guide for routine EEG studies in epilepsy," *Journal of Clinical Neurophysiology: Official Publication of the American Electroencephalographic Society*, vol. 1, no. 2, pp. 109–142, 1984.

[4] H. U. Amin, M. Z. Yusoff, and R. F. Ahmad, "A novel approach based on wavelet analysis and arithmetic coding for automated detection and diagnosis of epileptic seizure in EEG signals using machine learning techniques," *Biomedical Signal Processing and Control*, vol. 56, pp. 101707, 2020.

[5] L. Guo, D. Rivero, J. Dorado, J. R. Rabunal, and A. Pazos, "Automatic epileptic seizure detection in EEGs based on line length feature and artificial neural networks," *Journal of neuroscience methods*, vol. 191, no. 1, pp. 101–109, 2010.

[6] H. T. Shiao et al., "SVM-based system for prediction of epileptic seizures from iEEG signal," *IEEE Transactions on Biomedical Engineering*, vol. 64, no. 5, pp. 1011–1022, 2016.

[7] A. Temko, E. Thomas, G. Boylan, W. Marnane, and G. Lightbody, "An SVM-based system and its performance for detection of seizures in neonates," *Annual Int. Conf. of the IEEE Engineering in Medicine and Biology Society*, Minneapolis, USA, pp. 2643–2646, 2009.

[8] A. Swetha, A. K. Sinha, D. N. Jayakody, and A. Sharma, "Performance analysis of supervised machine learning algorithms applied toward epileptic seizures detection," *IEEE Int. Conf. on Information and Automation for Sustainability*, Negombo, Sri Lanka, 2021.

[9] M. Mursalin, Y. Zhang, Y. Chen, and N. V. Chawla, "Automated epileptic seizure detection using improved correlation-based feature selection with random forest classifier," *Neurocomputing*, vol. 241, pp. 204–214, 2017.

[10] T. Zhang, W. Chen, and M. Li, "AR based quadratic feature extraction in the VMD domain for the automated seizure detection of EEG using random forest classifier," *Biomedical Signal Processing and Control*, vol. 31, pp. 550–559, 2017.

[11] L. Torlay, M. P. Bertolotti, E. Thomas, and M. Baciu, "Machine learning–XGBoost analysis of language networks to classify patients with epilepsy," *Brain Informatics*, vol. 4, no. 3, pp. 159–169, 2017.

[12] J. Birjandtalab, M. B. Pouyan, and M. Nourani, "Nonlinear dimension reduction for eeg-based epileptic seizure detection," *IEEE-EMBS International Conference on Biomedical and Health Informatics (BHI)*, pp. 595–598. IEEE, 2016.

[13] A. Sharmila, and P. Geethanjali, "DWT based detection of epileptic seizure from EEG signals using naive Bayes and k-NN classifiers," *IEEE Access*, vol. 4, pp. 7716–7727, 2016.

[14] Y. Song, and P. Liò, "A new approach for epileptic seizure detection: sample entropy based feature extraction and extreme learning machine," *Journal of Biomedical Science and Engineering*, vol. 3, no. 6, pp. 556–567, 2010.

[15] M. Zhou et al., "Epileptic seizure detection based on EEG signals and CNN," *Frontiers in Neuroinformatics*, vol. 12, p. 95, 2018.

[16] X. Wei, L. Zhou, Z. Chen, L. Zhang, and Y. Zhou, "Automatic seizure detection using three-dimensional CNN based on multi-channel EEG," *BMC Medical Informatics and Decision Making*, vol. 18, no. 5, pp. 71–80, 2018.

[17] U. R. Acharya, S. L. Oh, Y. Hagiwara, J. H. Tan, and H. Adeli, "Deep convolutional neural network for the automated detection and diagnosis of seizure using EEG signals," *Computers in Biology and Medicine*, vol. 100, pp. 270–278, 2018.

[18] A. Swetha, and A. K. Sinha, "Epileptic seizures classification based on deep neural networks," in *Proc. of the International Conference on Paradigms of Computing, Communication and Data Sciences (PCCDS 2020)*, pp. 785–795. Springer Singapore, 2021.

[19] H. G. Daoud, A. M. Abdelhameed and M. Bayoumi, "FPGA implementation of high accuracy automatic epileptic seizure detection system," *IEEE 61st International Midwest Symposium on Circuits and Systems (MWSCAS)*, Windsor, Canada, pp. 407–410, 2018.

[20] H. Wang, W. Shi, and C.-S. Choy, "Hardware design of real time epileptic seizure detection based on STFT and SVM," *IEEE Access*, vol. 6, pp. 67277–67290, 2018.

[21] L. Feng, Z. Li, and Y. Wang, "VLSI design of SVM-based seizure detection system with on-chip learning capability," *IEEE Transactions on Biomedical Circuits and Systems*, vol. 12, no. 1, pp. 171–181, 2017.

[22] Y. Wang, Z. Li, L. Feng, H. Bai, and C. Wang, "Hardware design of multiclass SVM classification for epilepsy and epileptic seizure detection," *IET Circuits, Devices & Systems*, vol. 12, no. 1, pp. 108–115, 2017.

[23] C. Tsou, C.-C. Liao and S.-Y. Lee, "Epilepsy identification system with neural network hardware implementation," *IEEE International Conference on Artificial Intelligence Circuits and Systems (AICAS)*, Hsinchu, Taiwan, pp. 163–166, July 2019.

[24] H. Elhosary et al., "Low-power hardware implementation of a support vector machine training and classification for neural seizure detection," *IEEE Transactions on Biomedical Circuits and Systems*, vol. 13, no. 6, pp. 1324–1337, 2019.

[25] R. Sarić, D. Jokić, N. Beganović, L. G. Pokvić, and A. Badnjević, "FPGA-based real-time epileptic seizure classification using Artificial Neural Network," *Biomedical Signal Processing and Control*, vol. 62, pp. 1–10, 2020.

[26] L. Breiman, "Random forests," UC Berkeley TR567, 1999.

[27] R. R. Lopez, and J. C. Reich, "Construction of near-optimal axis-parallel decision trees using a differential-evolution-based approach," *IEEE Access*, vol. 6, pp. 5548–5563, 2018.

[28] Q. Li, and A. Bermak, "A low-power hardware-friendly binary decision tree classifier for gas identification," *Journal of Low Power Electronics and Applications*, vol. 1, no. 1, pp. 45–58, 2011.

[29] H. Nakahara, A. Jinguji, S. Sato, and T. Sasao, "A random forest using a multi-valued decision diagram on an FPGA," *IEEE 47th international symposium on multiple-valued logic (ISMVL)*, Novi Sad, Serbia, pp. 266–271, 2017.

[30] A. Bermak, and D. Martinez, "A compact 3D VLSI classifier using bagging threshold network ensembles," *IEEE Transactions on Neural Networks*, vol. 14, no. 5, pp. 1097–1109, 2003.

[31] Hui Lin, and Ming Li, Introduction to Data Science, 2014.

[32] Invent Logics, "EDGE ZYNQ SoC FPGA Development Board User Manual." [Online]. Available: https://allaboutfpga.com/edge-zynq-soc-fpga-development-board-user-manual/

[33] V. Geethu, and S. S. Kumar, "An efficient FPGA realization of seizure detection from EEG signal using wavelet transform and statistical features," *IETE Journal of Research*, vol. 66, no. 3, pp. 315–325, 2020.

Chapter 2

Hardware architecture for denoising of EOG signal using a differential evolution algorithm

V. H. Prasad Reddy
National Institute of Technology, Tadepalligudem, India

Gundugonti Kishore Kumar
V R Siddhartha Engineering College, Vijayawada, India

Puli Kishore Kumar
National Institute of Technology, Tadepalligudem, India

2.1 INTRODUCTION

The electrooculogram (EOG) signal used for eye movement recording measures corneo-retinal potential difference and is used in many biomedical and control device applications [1]. Human–computer interface (HCI) systems have an important role to play today in patient communication for biomedical applications. The EOG signal can be used effectively in locked-in syndrome patients who communicate with vertical eye movements and blinking [2]. The EOG signal also is also used for diagnosis of conditions such as dizziness and vertigo [3].

Digital filtering is one of the initial EOG signal filtering processes. The most important operations in digital filtering are addition and multiplication. In ASIC implementation, the shift-and-add approach is substituted for multiplication operations. The adder count in the shift-and-add approach can be minimized, based on the sign-power-of-two (SPT) terms present in each coefficient. A number of researchers have focused on multiplier-less filter implementation for real EOG systems [4, 5]. The implementations in [4, 5] are based on variable shifter and distributed arithmetic approaches with high throughput.

This study reports on an FIR filter with a differential evolution (DE) algorithm with reduced sign-power-of-two terms, which is implemented using shift-and-add operations and is designed to obtain the desired frequency response. The DE algorithm finds a suitable coefficient set suitable for implementation of the FIR filter with mutation, crossover and selection [6, 7].

DOI: 10.1201/9781032628059-2

2.2 FIR FILTER ARCHITECTURE

The difference equation of the FIR filter is given in Equation 2.1,

$$y(n) = \sum_{n=0}^{N-1} h(n) \times x(n-(N-1)) \tag{2.1}$$

In Equation 2.1, $x(n)$ and $y(n)$ represent filter input and outputs respectively. The coefficients are represented as $h(n)$ with order of the filter N. The basic building blocks of FIR filter architecture are multiplier, adder and delay. From Equation 2.1, it is clear that the FIR filter requires $N-1$ delays, N multipliers and $N-1$ adders.

There are two approaches to implementation of the FIR filter: the direct form approach (Figure 2.1) and the transposed direct form approach (Figure 2.2) [8]. The choice of filter implementation is clock frequency or maximum combinational path delay. The maximum combinational path delays of the direct form and transposed direct form approaches are given in Equation 2.2:

$$\begin{aligned} T_{direct\,form} &= T_M + T_A \times (N-1) \\ T_{transposed\,direct\,form} &= T_M + T_A \end{aligned} \tag{2.2}$$

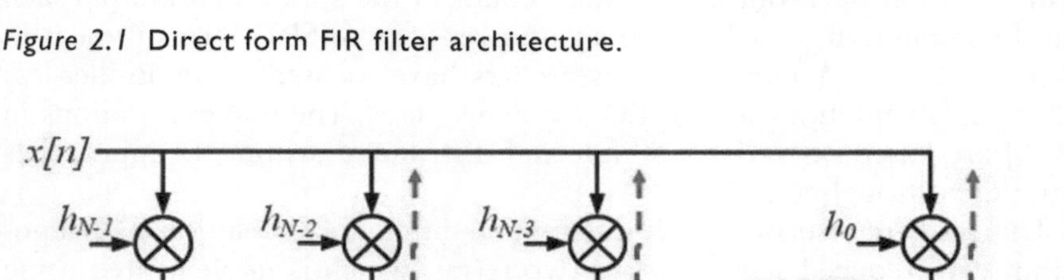

Figure 2.1 Direct form FIR filter architecture.

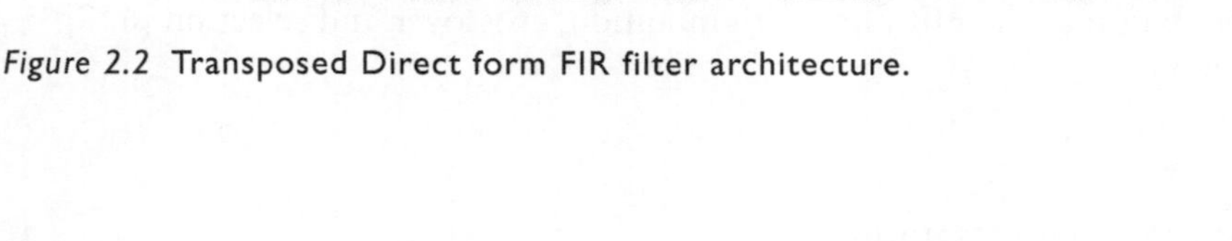

Figure 2.2 Transposed Direct form FIR filter architecture.

where T_A and T_M are adder and multiplier delays. From Equation 2.2, the direct form FIR filter architecture consists of $N - 1$ adders and one multiplier delay. The transposed direct form FIR filter implementation consists of one adder and one multiplier delay. From these architectures, it can be observed that the transposed direct form architecture has less critical path delay than direct form architecture.

2.3 FILTER DESIGN FOR DENOISING OF EOG SIGNAL USING DE ALGORITHM

The aim of this research is to design a filter with the desired frequency response with fewer sign-power-of-two terms than the conventional filter design approaches. The objective function of the proposed filter design for finding the filter coefficient set is given by Equation 2.3:

$$\text{Minimize} : E = \{\omega_c, \delta_s\}_{DE} - \{\omega_c, \delta_s\}_{desired'}$$

$$\text{subject to} : SPT_{DE} \leq SPT_{\text{Re}m} \tag{2.3}$$

In Equation 2.3, E is the error function which is determined in each iteration. SPT_{DE} and $SPT_{\text{Re}m}$ are the total number of SPT terms calculated in the DE algorithm and SPT terms of the Remez algorithm respectively. The error function E is the difference of stopband attenuation obtained in each iteration to the desired stopband attenuation. Similarly, the difference of cut-off frequency is also calculated for the obtained filter coefficient set. The flowchart of the DE-based filter design for a real EOG system is shown in Figure 2.3.

The inputs to the algorithm start with specifying filter specifications such as order of the filter (N), normalized cut-off frequency (ω_c) and stopband attenuation (δ_s). Along with the filter specification the key parameters of the DE algorithm, population size (P), maximum number of iterations ($I_{\max}$), crossover ration (C_r) and mutation factor (F) should be given to the algorithm. The modified DE-based filter design is executed as follows:

1. The coefficients are generated according to the population size with boundary conditions.
2. Each iteration of the DE algorithm consists of the following steps [6]:
   a. Mutation: A new mutant coefficient set is generated using the mutation equation.
   Crossover: The crossover trail coefficient set (either from initial population or mutant population) is selected based on C_r value.

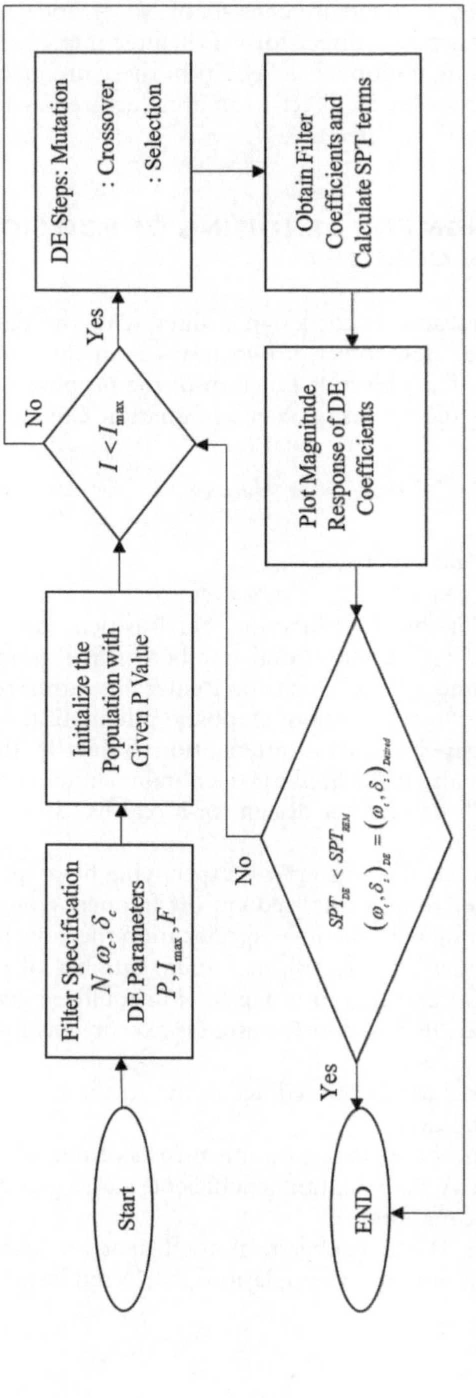

Figure 2.3 Flow chart of Denoise filter design using DE algorithm.

   b. Selection: The frequency response of the crossover coefficient set is compared with that of the initial population coefficient set. A new coefficient set is created based on the error function.
3. The SPT terms are calculated for the newly generated coefficient sets.
4. The frequency response and SPT terms are compared with the Remez coefficient set.
5. If the SPT terms are lower than the Remez coefficient set and satisfy the frequency response, the algorithm is terminated; alternatively, it goes to the next iteration.

The above modified method is followed to implement the FIR filter for denoising of the EOG signal. The implementation of the DE algorithm is discussed in the next section.

2.4 DE ALGORITHM WITH MINIMIZED COEFFICIENTS

The DE algorithm with reduced adder count is implemented using MATLAB® software. The FIR filter with an order of N = 24 and cut-off frequency of 30Hz is taken [5]. First, the FIR filter coefficients are generated using the MATLAB® function Remez. Later, these coefficients are converted into canonic signed-digit (CSD) representation. The actual Remez coefficients and their CSD representations are shown in Table 2.1, which indicates that the Remez filter coefficient set requires 66 sign-power-of-two terms to obtain the desired frequency response.

In this research work, the objective is to design the coefficient set with reduced SPT terms with desired frequency response. It was observed that the DE parameters with mutation factor of 0.7, strategy 2, and crossover 1 result in a reduction of SPT terms. The population size is 100. In order to reduce the number of SPT terms, the FIR filter coefficients need to be modified with the DE algorithm. After several iterations, a DE filter coefficient set with fewer SPT terms is obtained. The DE coefficients obtained and their CSD representations are given in Table 2.2. The frequency responses of the FIR filter with Remez function and DE algorithm with modified coefficient are shown in Figure 2.4. From the frequency response plot, it can be observed that though the filter coefficients in the DE algorithm are changed, the filter specifications are maintained as per the requirement. The number of SPT terms in the DE algorithm with minimized coefficients is also reduced to 50 from the previous 66. The modified FIR filter design using a DE algorithm with minimized coefficients resulted in a minimized filter coefficient set without compromising on the frequency response for filter implementation.

Table 2.1 FIR filter coefficients using Remez function and SPT terms

Actual	*Quantized*	*CSD*	*No. of SPTs*
−0.00586	−6	00000 $\bar{1}$010	2
−0.00879	−9	00000 $\bar{1}$00$\bar{1}$	2
−0.01074	−11	0000 $\bar{1}$0101	3
−0.01074	−11	0000 $\bar{1}$0101	3
−0.00586	−6	00000 $\bar{1}$010	2
0.00684	7	00000100 $\bar{1}$	2
0.02637	27	000100 $\bar{1}$0$\bar{1}$	3
0.05176	53	0010 $\bar{1}$0101	4
0.08008	82	001010010	3
0.10840	111	0100 $\bar{1}$000$\bar{1}$	3
0.12988	133	010000101	3
0.14160	145	010010001	3
0.14160	145	010010001	3
0.12988	133	010000101	3
0.10840	111	0100 $\bar{1}$000$\bar{1}$	3
0.08008	82	001010010	3
0.05176	53	0010 $\bar{1}$0101	4
0.02637	27	000100 $\bar{1}$0$\bar{1}$	3
0.00684	7	00000100 $\bar{1}$	2
−0.00586	−6	00000 $\bar{1}$010	2
−0.01074	−11	0000 $\bar{1}$0101	3
−0.01074	−11	0000 $\bar{1}$0101	3
−0.00879	−9	00000 $\bar{1}$00$\bar{1}$	2
−0.00586	−6	00000 $\bar{1}$010	2
	Total no. of SPT Terms		66

Table 2.2 DE with minimized coefficients and SPT terms

Actual	*Quantized*	*CSD*	*No. of SPTs*
−0.00488	−5	000000 $\bar{1}$0$\bar{1}$	2
−0.00977	−10	00000 $\bar{1}$0$\bar{1}$0	2
−0.01172	−12	0000 $\bar{1}$0100	2
−0.01172	−12	0000 $\bar{1}$0100	2
−0.00781	−8	00000 $\bar{1}$000	1
0.00391	4	000000100	1
0.02344	24	00010 $\bar{1}$000	2
0.05078	52	0010 $\bar{1}$0100	3
0.08008	82	001010010	3
0.10938	112	0100 $\bar{1}$0000	2

(*Continued*)

Table 2.2 (Continued)

Actual	*Quantized*	*CSD*	*No. of SPTs*
0.13281	136	010001000	2
0.14453	148	010010100	3
0.14453	148	010010100	3
0.13281	136	010001000	2
0.10938	112	0100$\bar{1}$0000	2
0.08008	82	001010010	3
0.05078	52	0010$\bar{1}$0100	3
0.02344	24	00010$\bar{1}$000	2
0.00391	4	000000100	1
−0.00781	−8	00000$\bar{1}$000	1
−0.01172	−12	0000$\bar{1}$0100	2
−0.01172	−12	0000$\bar{1}$0100	2
−0.00977	−10	00000$\bar{1}$0$\bar{1}$0	2
−0.00488	−5	000000$\bar{1}$0$\bar{1}$	2
	Total no. of SPT Terms		50

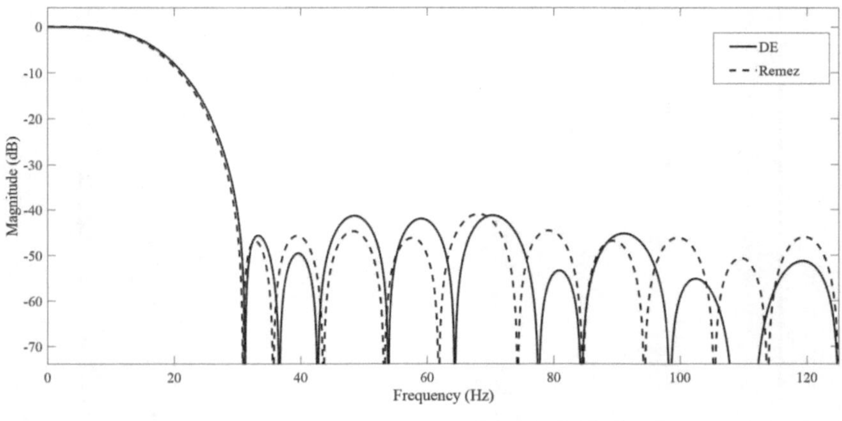

Figure 2.4 Frequency response of DE and Remez filter coefficients.

2.5 FUNCTIONAL VERIFICATION OF DE WITH MINIMIZED COEFFICIENT (DEWMC)-BASED DENOISED FIR FILTER

The denoising filter based on DEWMC is implemented using Verilog HDL. The denoising filter has one 16-bit data input, one clock input, and one 32-bit data output port. The EOG signal data is collected from the standard EOG Physionet database [9]. This EOG data is the input to the HDL filter block. The noisy and denoised EOG signals of Modelsim simulator waveform are shown in Figures 2.5 and 2.6. The mean square error (MSE)

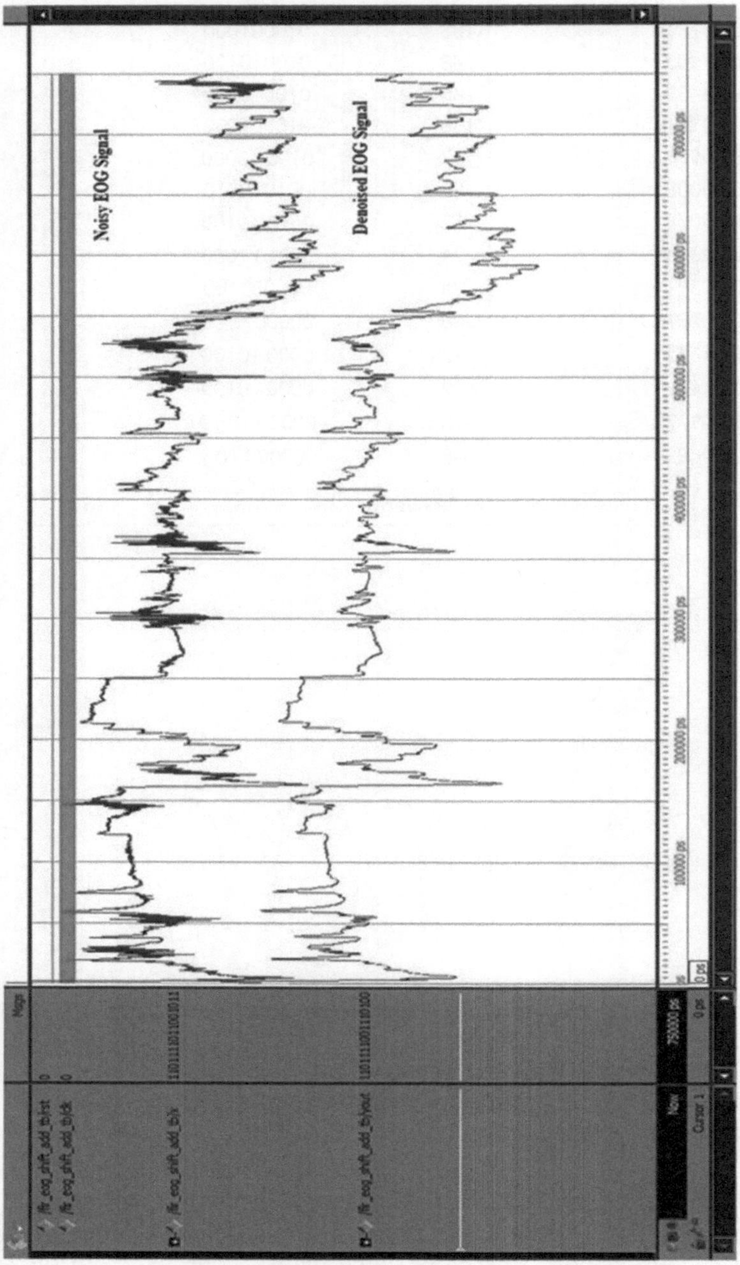

Figure 2.5 Noisy EOG signal and denoised EOG signal. The y-axis is amplitude and x-axis is time.

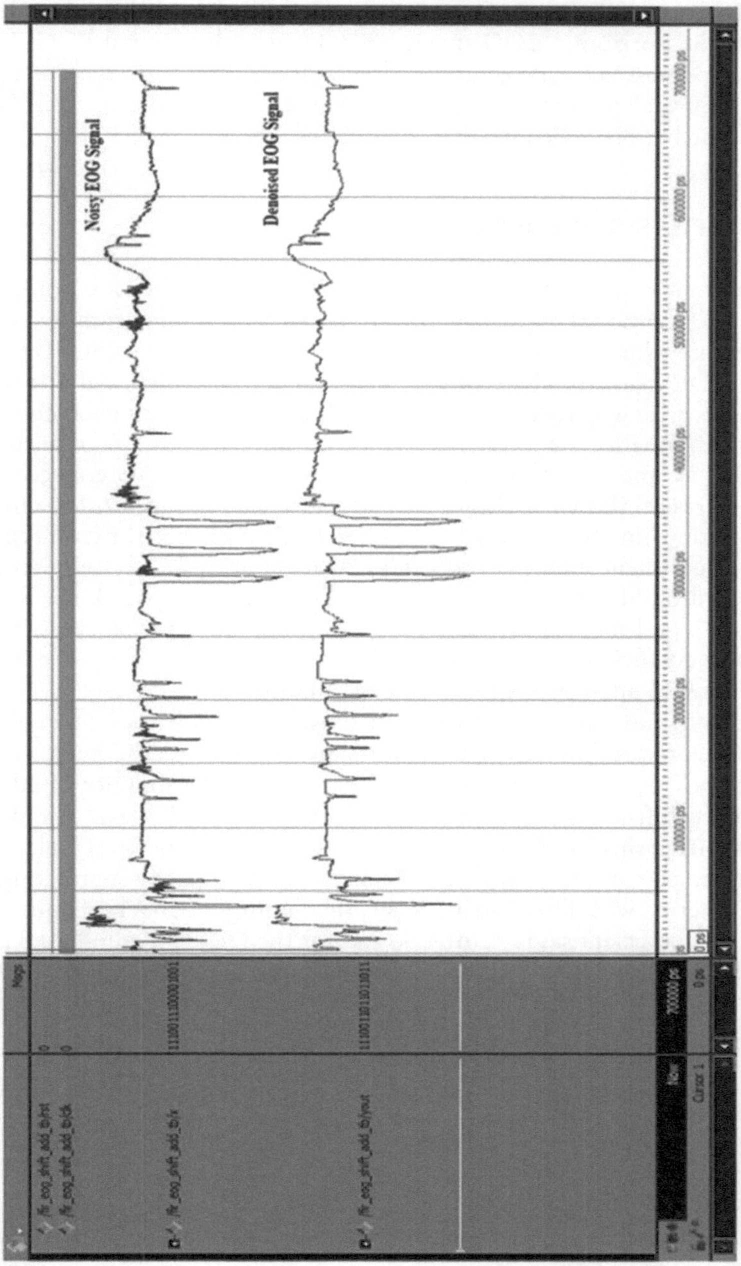

Figure 2.6 Noisy EOG signal and denoised EOG signal. The y-axis is amplitude and x-axis is time.

and signal-to-noise ratio (SNR) of the EOG signal are 0.6246 and 27.21*db* respectively. The noise in the raw EOG signal is greatly reduced by the DEWMC-based FIR filter. Figure 2.5 shows the noisy and denoised EOG signal of the horizontal EOG signal. This denoised horizontal EOG signal is further processed for detecting saccades. Figure 2.6 shows the noisy and denoised EOG signal of the vertical EOG signal. This denoised vertical EOG signal is further processed for detecting blinks.

2.6 SYNTHESIS RESULTS

The synthesis results of the DE algorithm-based FIR filter are compared with the conventional, Remez algorithm-based filter implementation along with denoise filter implemented using sum of power of two (SOPOT) [4]. Here the DE algorithm-based filter is a modified FIR filter. All filters are implemented using gate-level Verilog HDL. The filters are implemented with filter order 24 with 16-bit input and 32-bit output data samples respectively. The ASIC synthesis results are tabulated in Table 2.3. The conventional FIR filter resuts shown in Table 2.3 use radix-4 multiplier for coefficient multiplication. In the SOPOT-based filter [4], a variable shifter circuit is used for coefficient multiplication. The throughput of the shifter circuit in [4] is based on SPT terms present in a coefficient. In Table 2.3, the Remez algorithm-based filter is the CSD-based shift-and-add implementation for the Remez coefficient set. The DEWMC shown in Table 2.3 also uses CSD-based shift-and-add circuits for coefficient multiplication.

The ASIC synthesis results are obtained using the Cadence RTL compiler for UMC 90nm technology. The performance metrics for the implemented filter are power, area and delay. In order to reduce the number of adders, CSD-based shift-and-add operations are performed. In Tables 2.1 and 2.2, the total SPT terms for Remez and DE are 66 and 50 respectively. The total number of adders required for a coefficient set is obtained by subtracting the total SPT terms with filter order. Hence, the number of adders required for the Remez coefficient set is 42(66-24) and for the DE algorithm-based filter coefficient set it is 26(50-24) for a 24-order filter. However, since the order

Table 2.3 Synthesis results of modified method

Parameters	*Conventional*	*SPOT [4]*	*Remez algorithm*	*DE algorithm*
Area (mm^2)	114.92	99.86	83.92	72.58
% Reduction	–	13.11	26.98	36.85
Power (mW)	16.62	17.33	13.52	11.14
% Reduction	–	−4.27	18.65	32.97
Delay (nsec)	12.89	12.76	13.16	11.98
% Reduction	–	1.01	−2.09	7.06

of the filter has symmetrical coefficients, the actual number of adders used for Verilog HDL implementation is 21 for the Remez algorithm and 13 for the DEWMC filter.

The area and power consumption of the DEWMC can be reduced by reducing the number of adders. The ASIC synthesis results in Table 2.3 also show that the DE algorithm-based filter achieves 36.85% reduction in area and 32.97% reduction in power over the conventional FIR filter. The synthesis results also show that the DEWMC-based filter is implemented with a smaller area and lower power than the Remez algorithm and SOPOT-based filter [4].

2.7 CONCLUSION

This study has described a modified FIR filter for denoising EOG signals. The denoise filter is designed using a DE algorithm with reduced SPT terms in a coefficient set. The SPT terms play an important role in filter implementation using the shift-and-add approach. The total number of adders needed for filter implementation is calculated with the total SPT terms that exist in a filter coefficient set. Total SPT terms are reduced using the DE algorithm, which leads to the reduction in total adders needed to implement the filter architecture. The Modelsim simulator environment for EOG signal is used to check the performance of the filter. The DEWMC-based FIR filter in ASIC implementation offers 36.85% reduction in area and 32.97% reduction in power compared to conventional filter architectures.

REFERENCES

[1] A. Bulling, J. A. Ward, H. Gellersen, and G. Troster, "Eye movement analysis for activity recognition using electrooculography," *IEEE Transactions on Pattern Analysis and Machine Intelligence*, vol. 33, no. 4, pp. 741–753, 2011.

[2] K. Lee, W. Chang, S. Kim, and C. Im, "Real-time eye-writing recognition using electrooculogram," *IEEE Transactions on Neural Systems and Rehabilitation Engineering*, vol. 25, no. 1, pp. 37–48, 2016.

[3] K. Archawut, W. Tangsuksant, P. Thumwarin, M. Sangworasil, and T. Matsuura, "Realization of FIR system characterizing eye movement based on electrooculogram," in *International Conference on Biomedical Engineering (BME-HUST)*, pp. 23–26, December 2016.

[4] J. F. Wu, A. M. S. Ang, K. M. Tsui, H. C. Wu, Y. S. Hung, Y. Hu, J. N. F. Mak, S. C. Chan, and Z. G. Zhang, "Efficient implementation and design of a new single-channel electrooculography-based human–machine interface system," *IEEE Transactions on Circuits and Systems II: Express Briefs*, vol. 62, no. 2, pp. 179–183, 2015.

[5] S. Agarwal, V. Singh, A. Rani, and A. P. Mittal, "Hardware efficient denoising system for real EOG signal processing," *Journal of Intelligent & Fuzzy Systems*, vol. 32, no. 4, pp. 2857–2862, 2017.

[6] K. Price, R. M. Storn, and J. A. Lampinen, *Differential Evolution: A Practical Approach to Global Optimization*. Springer Science & Business Media, 2006.
[7] N. Srinivas, G. Pradhan, and P. K. Kumar, “FPGA implementation of zero frequency filter,” in *Conference on Information and Communication Technology (CICT)*, pp. 1–5, October 2018.
[8] B. Venkataramani, *Digital Signal Processors*, Tata McGraw-Hill Education, 2002.
[9] A. L. Goldberger et al., “PhysioBank, PhysioToolkit, and PhysioNet: Components of a new research resource for complex physiologic signals,” *Circulation*, vol. 101, no. 23, pp. 215–220, 2000.

Chapter 3

Implementation considerations for an intelligent embedded E-health system and experimental results for EEG-based activity recognition

Stefan Oniga

Technical University of Cluj-Napoca, North University Center of Baia Mare, Baia Mare, Romania

University of Debrecen, Debrecen, Hungary

Iuliu Alexandru Pap

Technical University of Cluj-Napoca, North University Center of Baia Mare, Baia Mare, Romania

Tamas Majoros

University of Debrecen, Debrecen, Hungary

3.1 INTRODUCTION

The main global causes of death are associated with cardiovascular diseases (ischemic heart diseases, stroke), and respiratory diseases (chronic obstructive pulmonary disease, lower respiratory tract infections). Prevention and treatment of these diseases is based on the daily monitoring of vital signs. There is an increasing need for inexpensive portable monitoring equipment to provide rapid and continuous remote access to physiological parameters, to improve diagnosis and treatment.

E-health system applications are becoming more widely used with improvements in many fields of healthcare such as remote patient monitoring, interactive patient care and prediction and detection of specific health conditions. Alongside other technological advances, E-health applications have made significant contributions to such issues as detection of heart failure and sleep apnea, prediction of cardiovascular disease and antibiotic resistance, nutrition assessment, precision medicine, and to reducing healthcare costs. Internet of Things (IoT) technologies have an ever-increasing role in improving E-health applications [1], by bringing remote patient monitoring solutions to healthcare and if adoption increases, even future in-home health monitoring systems.

DOI: 10.1201/9781032628059-3

Research in telemedicine, telehealth, mHealth, and all other subsets of E-health brings solutions focused on different methods of interacting with the patient, but an ever-increasing number of implementations incorporate elements of artificial intelligence (AI), one of the benefits of which is that it mimics some characteristics of human thinking, being able to analyze both structured and unstructured data. AI algorithms can be implemented through machine learning (ML) by training a model without programming it precisely, the most frequently used approaches being supervised, unsupervised, semi-supervised or reinforcement learning. One of the ML techniques which can be used to extract features from raw bio-signals using multiple layers is deep learning (DL). Implementations such as [2] seek to leverage the benefits of IoT and AI to assess stress levels with a balance between edge and cloud.

Although AI algorithms improve when employing large amounts of data, this can prove challenging for E-health embedded systems. Acquisition of biomedical data, such as electroencephalography (EEG) recordings, requires timeseries data collection to be done at sampling rates generally at the highest end of the range supported by the sensor platform. At the same time, embedded systems designed for data acquisition must also meet multiple requirements such as space constraints, power consumption and affordability that have a direct impact on the performance of the system, so every component must behave as efficiently as possible. Figure 3.1 presents a structural diagram of the flow of this chapter and highlights key sections.

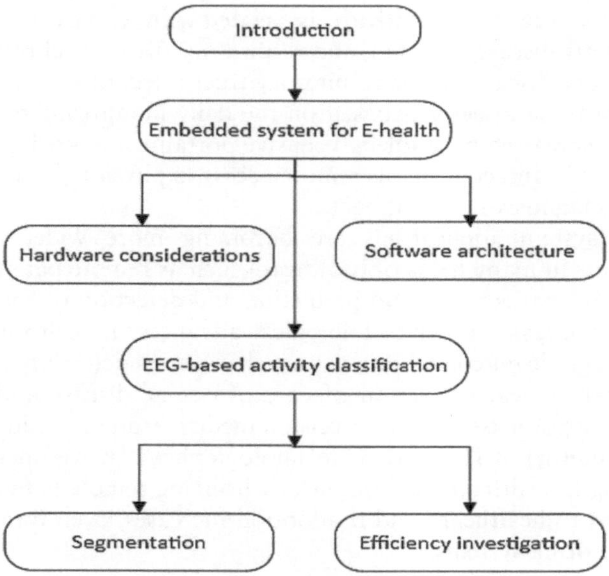

Figure 3.1 Structure of the chapter.

3.2 EMBEDDED ACQUISITION SYSTEM FOR E-HEALTH

Devices for monitoring a single parameter or a small number of parameters (blood pressure, heart rate and blood oxygen saturation, etc.) increasingly require affordable embedded systems that operate at the patient's home or the general practitioner's office. These systems must enable the connection of many types of sensors or measuring modules with various wired (UART, USB, SPI, I2C, etc.) or wireless (BT, Zigbee, ANT+, BlueRobin, SimpliciTI, SimpleLink, etc.) communication protocols. The acquired data must be able to be displayed both locally and on the web. The system must also be provided with multiple connectivity possibilities (LAN, WIFI, GSM, GPRS, etc.) in order to be monitored remotely by general practitioners.

One of the most complete E-health medical development platforms in the market connecting more than 15 sensors is MySignals [3]. The crowd-funded Crowd Supply [4] is responsible for HealthyPi, a wireless, wearable, open-source vital signs monitor.

Most of the work presented in the following sub-sections is based on the IoT-based E-health data acquisition system of [5, 6].

3.2.1 Hardware considerations, choices, and implementations

The choice of hardware will dictate how every other piece of the implementation will fit together and requires the main characteristics of the acquisition system to be considered.

The features, in order of importance, are:

1. *Sensor connectivity*. Sensors that record physiological parameters can be connected in multiple ways: wired, wireless or independent. To be able to generate and display patient data graphs, all the recorded data needs to be stored in the same place, so if independent devices such as a standalone blood pressure monitor are used, the device needs to offer synchronization capabilities. Wireless sensors such as a Bluetooth EEG headset will need a corresponding Bluetooth adapter and are easier to maneuver during data acquisition but more expensive and need battery level monitoring. From a connectivity standpoint, wired sensors provide the best reliability and communication speeds, so if the system needs to collect large amounts of raw sensor data with little to no delay, this may be the only choice.
2. *Data storage*. Recording large amounts of physiological data can be a challenge, especially on low-performance, high power-efficiency devices. First, each sensor dictates the speed at which the recording is done. Some sensors can be used at low sampling rates (e.g., temperature), while others require high sampling rates (e.g., EEG headset, air flow), so the acquisition device must be capable of processing large

amounts of data in a short time frame. Second, the storage medium must provide enough space and read/write speeds for the acquisition to be successful. Third, all the collected patient information must be saved in the same format, so the software components responsible for the graphical representation of physiological data can read and interpret any recording session. This simplifies both data graph generation and data streaming.
3. *Data visualization*. Physiological data sometimes needs to be collected in an urgent setting, so having to access a secondary device to visualize data graphs may be unacceptable. The system should therefore have at least basic data visualization capabilities, either on a dedicated screen or on a remote easily accessible device such as a smartphone or a tablet.
4. *Remote access*. If the chosen hardware can be deployed in a compact device, then the acquisition system can be portable, so the patient can use it at home or while on the road. This can be extremely helpful, especially for reducing unnecessary hospital visits. Remote access can provide essential insights for clinical staff, who can either view previously recorded patient information or remotely trigger recording sessions and view live graphs in real time – especially useful for non-mobile patients.

The many hardware choices able to run custom acquisition software that mostly satisfies these feature requirements include the following:

1. A desktop computer connected to the same network or internet is the easiest method of running any custom software, since it can use any operating system or architecture, but from a mobility standpoint, the system would be limited to a data acquisition room.
2. A laptop would provide some added mobility, but the installation would still be cumbersome for some elderly patients.
3. A tablet or smartphone, depending on the screen size, provides the highest mobility, but restricts the type of sensors that can connect to such a system, since tablets are not built with wired connectivity in mind.
4. A microcontroller development board like the Arduino system is a low-cost alternative, providing plenty of connectivity options, but the limited performance of the board does not allow the implementation of remote patient monitoring and complex data visualization capabilities.
5. The single-board computer (such as the Raspberry Pi (RPi) development board) is one of the most balanced solutions available. It is small in size, so the system can easily be transferred from one patient to another. It is compatible with Linux operating systems, allowing software to run in a reliable environment. Connectivity-wise, the RPi is

packed with features, from general input–output pins to wireless connectivity with support for the full TCP/IP stack, making it possible to build a high-compatibility remote patient monitoring web interface. The availability of dedicated sensor platforms with connectors and converters for multiple sensors is one of the main advantages.

Table 3.1 lists the most important RPi hardware features and their advantages pertaining to E-health systems.

In its final form, the RPi-based system described in this study supports measurement of the following health parameters [5, 6]:

1. Pulse rate and oxygen saturation in the blood: Contec CMS50D+ pulse oximeter
2. Systolic and diastolic blood pressure: Kodea KD-202F blood pressure monitor
3. Breathing intensity (air flow)
4. Skin conductance and resistance (galvanic skin response)
5. Body temperature
6. EEG signals from the brain: NeuroSky Mindwave or OpenBCI Ganglion headset

Table 3.1 The key Raspberry Pi hardware features

Feature	*Description*	*Advantages*
Power input	The RPi can be powered by 5 V power adapters	The E-health system can be battery powered, improving safety concerns and mobility
GPIO pins	General Purpose Input–Output pins that allow the RPi to be alternatively powered and to handle different communication methods, allowing the user to control these pins.	The E-health Sensor Platform is connected to the RPi through the GPIO Header
Storage	RPi can boot the operating system from a microSD card as well as access its contents. The RPi also supports USB storage devices.	This feature facilitates the customization of the boot process to create an immersive, kiosk-like interface that prevents the user from having to launch the application from the Linux desktop. The memory card is also used to store patient information. As an alternative, a USB drive could be used to download recordings or other patient data.
Wi-Fi and Bluetooth	The RPi can connect to wireless networks and can communicate with Bluetooth devices	EEG headsets can be connected to the RPi through Wi-Fi or Bluetooth.

The sensors connected to the RPi through the E-health Sensor Platform v2.0 for Arduino and Raspberry Pi were using an E-health shield over Raspberry Pi. The pulse oximeter is connected to one of the RPi board's USB ports. The EEG headsets used the RPI's integrated Bluetooth connectivity. To control the system locally, a RPi touch display is attached directly to the RPi board through a ribbon cable, allowing the user to interact with the interface by touch. Figure 3.2 illustrates how each component is connected to the E-health acquisition system and how Bluetooth sensors are connected

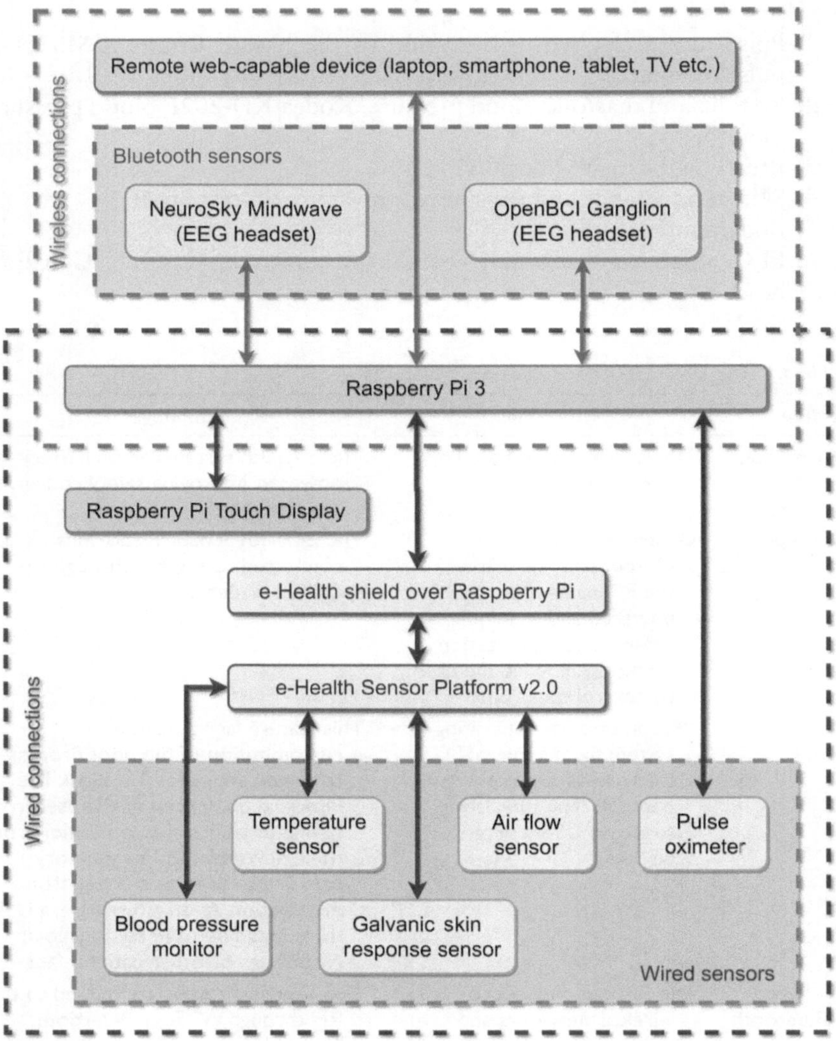

Figure 3.2 Connectivity diagram.

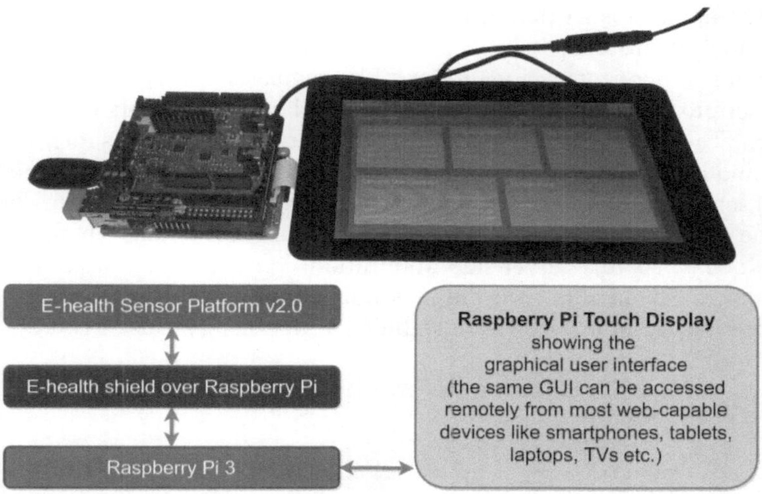

Figure 3.3 Hardware stack.

directly to the RPi (due to its internal Bluetooth adapter), while some wired sensors need additional boards to connect to the RPi. Thanks to the RPi's Wi-Fi capabilities, the same graphical user interface can be accessed from the local RPi touch display or from any remote web-capable device.

Figure 3.3 shows the E-health IoT acquisition system's hardware stack, with the RPi board connected to the sensor platform and the touch display.

The RPi development board supports microSD memory cards, so the recording sessions and other patient information can be stored on the card and easily accessed through the operating system.

3.2.2 Software architecture and implementation

For better compatibility and performance, E-health systems are often split into multiple components, at hardware, software or both levels. On one hand, hardware data acquisition components have their own constraints regarding physical connectivity, sampling rates, data transfer and power consumption. On the other hand, software components must meet the requirements set by hardware manufacturers pertaining to sampling rates and data transfer, while often the most demanding obstacle can be the availability of software libraries provided by the manufacturers for the specific devices used in the project. Software development kits provided by hardware manufacturers are sometimes written in different programming languages from that of the E-health system's core software, meaning that either the entire code has to be rewritten in a single programming language, or a modular system must be created that can use something similar to an application programming interface to communicate with the core software components, even if each component is written in a different programming language. The system presented in this chapter is implemented in such a way, running different software components written in multiple languages like C, C++, Python and JavaScript. This way, the time required to adapt new sensors to the system is significantly reduced. Figure 3.4 shows the modular system's inter-connection and how its software components communicate in order to collect data from sensors and even stream the required information to a remote device.

The software for this E-health IoT data acquisition system is mainly based on Node.js, so this server-side application controls every other software component, from data recording sessions to graphical user interface (GUI) interaction. To improve system stability, a modular approach is used, each

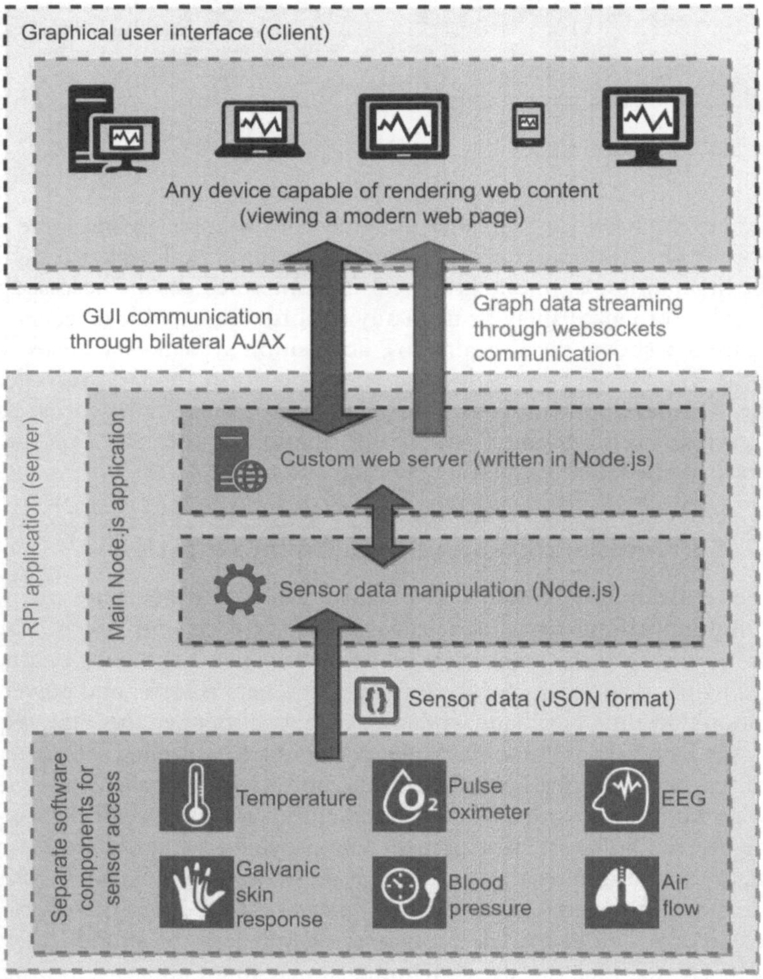

Figure 3.4 The software architecture.

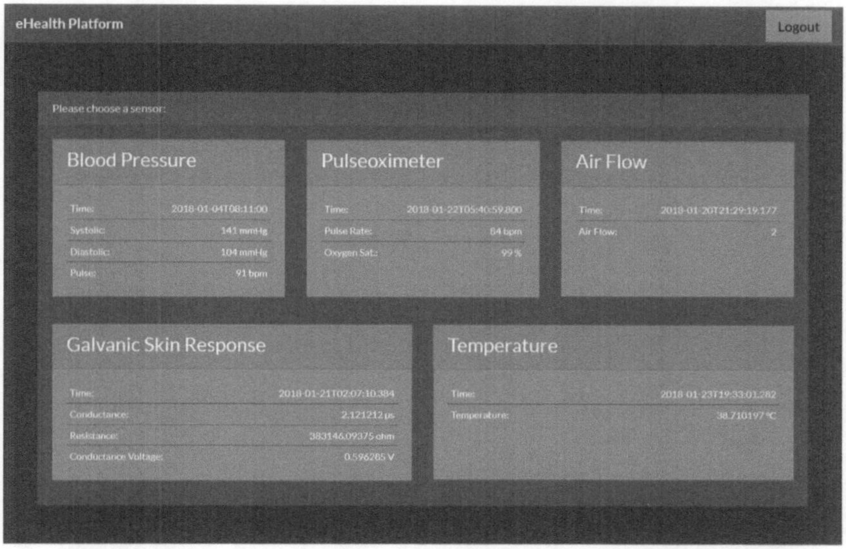

Figure 3.5 Main menu of the graphical user interface.

sensor having its own software component running independently from the main user interface. The GUI can be accessed from any web-capable device, because the entire interface is a web application. When sensor data is needed, the main Node.js app triggers specific software components that stream data directly to the browser through Websockets, so the recorded data is displayed on graphs that update in real time. The touch-friendly user interface is presented in Figure 3.5, showing the main menu offering access to all the active pages.

The main menu consists of multiple sections representing each connected sensor, showing the latest recorded data, while acting as shortcuts for the sensor pages that can retrieve and show the entire recorded data for the specific sensor. Each recording session is stored in JavaScript Object Notation (JSON) files, an accessible file format, making it easy to adapt the system to migrate information to specific database servers.

The architecture of this system has allowed it to also become a real-time remote patient monitoring system that can stream detailed graphs of biomedical data live over the internet to any web-capable device, making it possible for doctors or caregivers to remotely trigger recordings and monitor physiological parameters (body temperature, blood pressure, pulse oximeter, galvanic skin response and air flow). Such features can be extremely helpful in telehealth solutions during pandemics [7], but privacy and data security issues must be resolved for these technologies to be adopted in healthcare [8].

Figure 3.6 shows how the user is able to manipulate the graph information by using the tools provided on the screen to better interpret the systolic, diastolic and pulse fluctuations over multiple recordings.

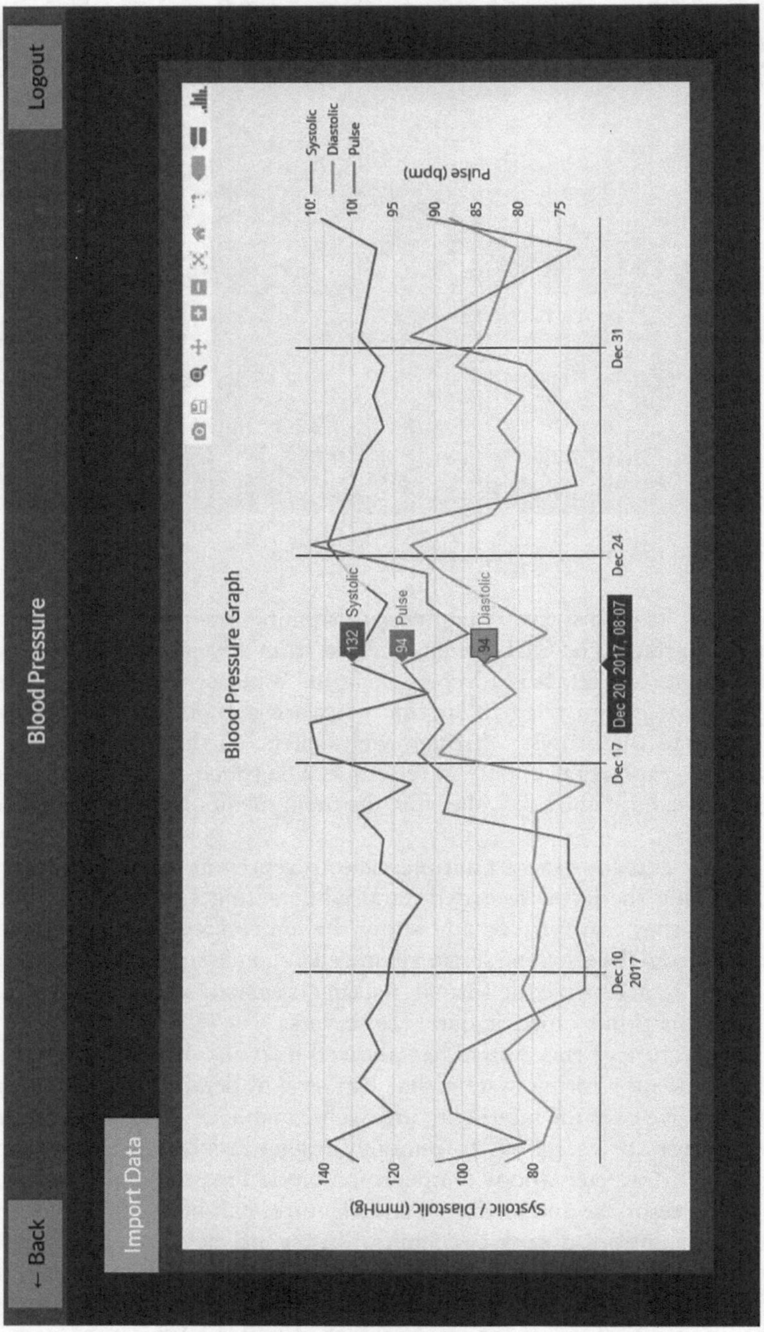

Figure 3.6 Blood pressure sensor page displaying an interactive and detailed graph.

These systems need easy access to medical data analysis to provide true value for the patient or the medical staff. In the system presented in [6], the data acquisition part of the system is separated from the machine learning data analysis part so that collection and patient interaction can be implemented in a power-efficient, compact, mobile, and relatively affordable device, while the data analysis can be done on a remote server that has the processing power to quickly pre-process, train models and classify the medical data provided. For data processing, the system uses the open-source software library TensorFlow and the Keras high-level API to create a convolution neural network (CNN) that attempts to classify eye movement by processing many EEG recordings made through a custom application specifically designed to ensure that each recording was correctly captured. One of the challenges faced at this stage is the quality and similarity of the recordings.

A production-ready E-health system sharing these characteristics can greatly benefit the healthcare sector. Especially during challenging times like the COVID-19 pandemic, collaboration by researchers, governments, and institutions can bring new innovations to light by bringing together IoT, AI, big-data and even blockchain technologies [9].

3.2.3 Related works

Table 3.2 shows IoT-related and remote patient monitoring systems additional to the system described in this chapter.

3.3 EEG-BASED CLASSIFICATION OF MOTOR IMAGERY ACTIVITIES

The system described above can even be used for activity recognition based on brainwave measurement. With proper electrodes and data acquisition systems, voltage fluctuations during brain activity can be measured and recorded to produce an electroencephalograph (EEG). Traditional interpretation of this complex EEG signal demands the expertise of a skilled professional. Nowadays, however, machine learning algorithms can extract information from EEG brain activity recordings for use in many EEG applications and research. For example, machine learning techniques are central components of many EEG-based brain–computer interfaces (BCIs) for clinical applications in communication and rehabilitation [15]. Using BCI, brain waves can be converted into actual physical movement without using muscles and, people with limited mobility can use their thoughts to perform movements that they would not be able to do on their own, significantly improving their quality of life. The robust automatic evaluation of EEG signals makes it possible to operate these systems more efficiently with reduced reliance on trained professionals.

Table 3.2 Related works

System	*Description*	*Key characteristics*
[10]	Low-cost monitoring platform based on dedicated social network	This system is also based on a Raspberry Pi 3, uses Bluetooth communications and performs bio-signal data acquisition for the following: heart rate, blood pressure, blood glucose, oxygen saturation, body temperature and weight. The system proposes the use of a dedicated social network.
[11]	IoT implementation for healthcare	A system based on the Intel Galileo Generation 2 board, running Linux, compatible with the Arduino IDE, which communicates with XBee sensor nodes to access temperature data provided by the LM35 sensor.
[12]	eBPlatform	A platform dedicated to improving the communication between patients and doctors by improving the control and treatment of patients with non-communicable diseases (NCDs). The system is based on eBox devices that use ARM9 processors and allow for ECG sampling and blood pressure measurement.
[13]	Remote monitoring for heart disease patients	A wearable solution to notify patients about their heart disease. The system uses a wearable BioHarness 3 sensor and is based on a client-server application written in Java.
[14]	Remote monitoring improvements in eHealth	A proof-of-concept solution for real-time e-health applications that aims to improve the user experience and reduce the deployment complexity. The communication between the remote user client and the cloud of gateways is done through WebRTC. The solution offers video monitoring features comparable with multiple commercial solutions.

During the application of automatic evaluation, a number of decisions need to be made: what preprocessing of the available data is needed; which features should be extracted from the data, and which of the countless machine learning algorithms is best suited to the task at hand. At this stage, other constraints, such as resource requirements and speed, must be considered. Finally, the architecture and parameters of the applied method must be decided. After training and performance evaluation further refinements can be made if necessary.

Activity recognition from EEG signals presents many challenges. One of these is the poor signal-to-noise ratio of the EEG, because in most cases noise from an external source, such as artifacts from blinking, different movements, muscle activity or the electrical network, is superimposed on the actual, useful signal. Physiological differences between individuals can also significantly affect the performance of a machine learning model: a model that performs well in recognizing the EEG signals of one individual can give very poor results for another subject.

The basic idea of neuron activity recognition from EEG signals is that different activities can be distinguished from each other based on the unique patterns characteristic of each activity that are generated by the brain. Several machine learning algorithms can be used for this purpose, including shallow and deep learning methods. Commonly used shallow machine learning methods are the support vector machine (SVM), k-nearest neighbors (kNN), the naive Bayes (NB) classifier, or the decision tree (DT). Using several independent decision trees, the random forest (RF) method can also be applied.

Artificial neural networks (ANNs) are a large subset of machine learning methods. One of the simplest and most frequently used ANNs is the multilayer perceptron (MLP), which is a feed-forward neural network. These represent a transition between shallow and deep learning techniques: models with one hidden layer are classified as shallow, while models with multiple hidden layers are classified as deep learning algorithms. Other types of ANNs are the recurrent neural network (RNN), which includes feedback, and the convolutional neural network (CNN), which supplements traditional fully connected layers with other types of layers. The combination of the latter two, the recurrent convolutional neural network (RCNN) is another possible alternative.

Many researchers have investigated the suitability of these machine learning methods in EEG-based activity recognition, but no one algorithm has been clearly shown to be more efficient than the others. For example, the authors of [16] found naive Bayes to be more effective than kNN, DT and SVM for four subjects, but for the other five individuals the NB was less effective than the other methods. The authors of [17] found CNN to be more accurate than SVM in each case. In contrast, for five of the nine subjects in the [18] study, SVM performed better than CNN. At the same time, even within a given type, researchers used networks with different architectures on data with different preprocessing, so it is not possible to draw a general conclusion. Nevertheless, in recent years the convolutional neural network has become the most widely used algorithm in this research topic [19].

The classification ability of neural networks on EEG data is demonstrated through an example where different convolutional networks were used to recognize motor imagery activities. The EEG data were obtained from the publicly available PhysioNet database [20], which was prepared using data from 109 volunteers. The sampling rate of the data is 160 Hz. The subjects participating in the study performed several different activities or imagined performing the activity without actual physical movement. In the recordings that were used, the individual imagined closing either both hands or both feet, depended on whether an object appeared randomly at the bottom or top of a screen. The target was visible for four seconds, then disappeared for another four seconds, then the subject relaxed. This eight-second period was repeated 15 times during one measurement. Three such measurements were

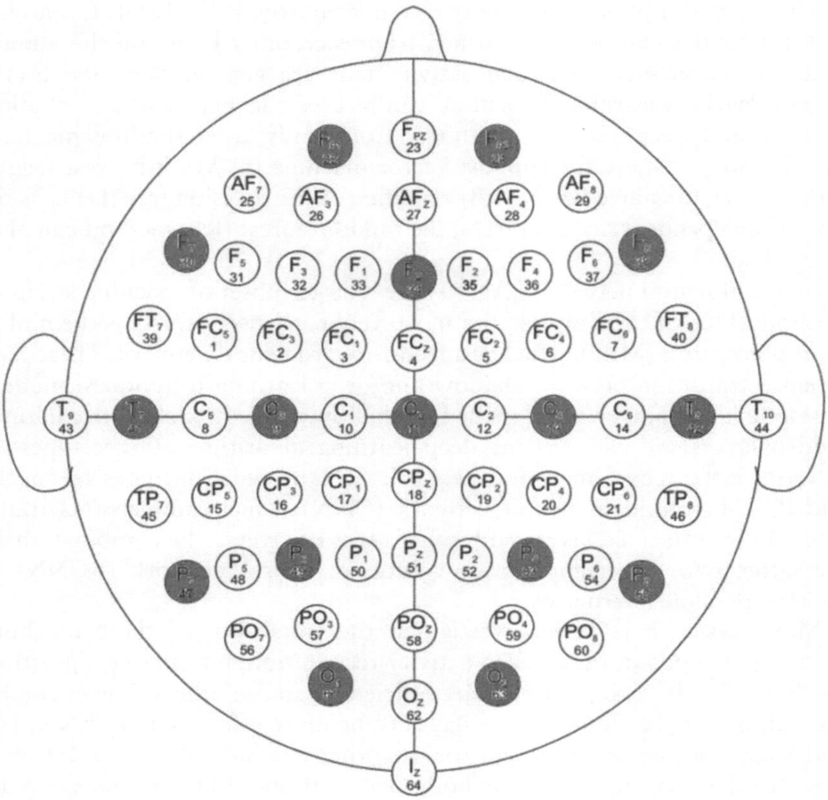

Figure 3.7 EEG electrodes used in experiment.

made per subject and the data of 10 and 20 subjects were selected, so 60 and 120 minutes of EEG recordings provided the input data for the machine learning models. The data was recorded on 64 channels, but only 16 were used, in order to be able to compare results with measurements made with a 16-channel EEG cap. These channels, marked in red in Figure 3.7, are Fp1, Fp2, F7, Fz, F8, T7, C3, Cz, C4, T8, P7, P3, P4, P8, O1 and O2, using the international 10–10 system notation.

3.3.1 Segmentation

The measured EEG signal is typically available as a long data stream, during which the subject can even perform several different activities, so splitting up the data stream, at least when the activity changes, is inevitable. Typically, however, for better performance, breaking up into smaller sections is optimal. This kind of splitting of the data stream is called segmentation, and the individual segments are called windows.

It is crucial to choose the appropriate window size because there is an optimal value which can maximize the accuracy of the machine learning task. With a window size smaller than this, i.e., with a too short time interval, it may happen that the window does not contain enough information about the performed activity, which reduces the accuracy of the classification. And a large window can contain data from different activities, especially if the activity changes are relatively frequent. Although the segments during which the activity changed can be discarded, if there are too many of them, it can cause a significant reduction in the number of available training samples. Another problem occurs during real-time activity recognition: after the change of the performed activity, the result of the classification shows up on the output with a longer delay.

There are two approaches to data segmentation: splitting up the data stream or using a sliding window. In the latter case there is an overlap between successive data segments. In this application example, the windows were almost completely overlapping, a window of size N containing the current and previous N-1 measurement points.

3.3.2 Neural network efficiency investigation

A useful property of convolutional network is that they are capable of automatic feature extraction with the help of their convolutional layers, which usually gives better results than using a static feature extraction approach.

Even with a specific network type, however, there are many choices of architecture and other parameters: how deep the network should be, what type the individual layers should be, and the correct parameterization of these layers, i.e., in the case of a convolution layer, how many kernels and what size can obtain the best performance. However, in the absence of an analytical method, network designers rely on experimentation and experience.

In the example in [21], neural networks based on two fundamentally different concepts are used as a starting point. The CNN1 network is a purely convolutional network, built from blocks containing convolution, batch normalization and ReLU activation functions. The number of kernels in the convolutional layers was 16, 32, 64, and 64 respectively, while their size was 5x5, 5x5, 3x3, 2x8. Except for the last convolution layer, no padding was used. Figure 3.8 shows the structure of the network.

The CNN2 network, on the other hand, contains blocks of convolution, ReLU activation and maximum pooling layers, and then fully connected layers with their associated activation function. The number of kernels in the convolutional layers was 8 and 16, respectively, and their size was 3x3 and 5x5. The fully connected layers consist of 64, 32 and 3 neurons, respectively. The structure of this network is presented in Figure 3.9.

Training and testing were done on a balanced data set, with Adam optimizer, initially using the data of ten people. For the first time, a small window of 32 samples, i.e., 0.2 seconds, was used, with which a recognition

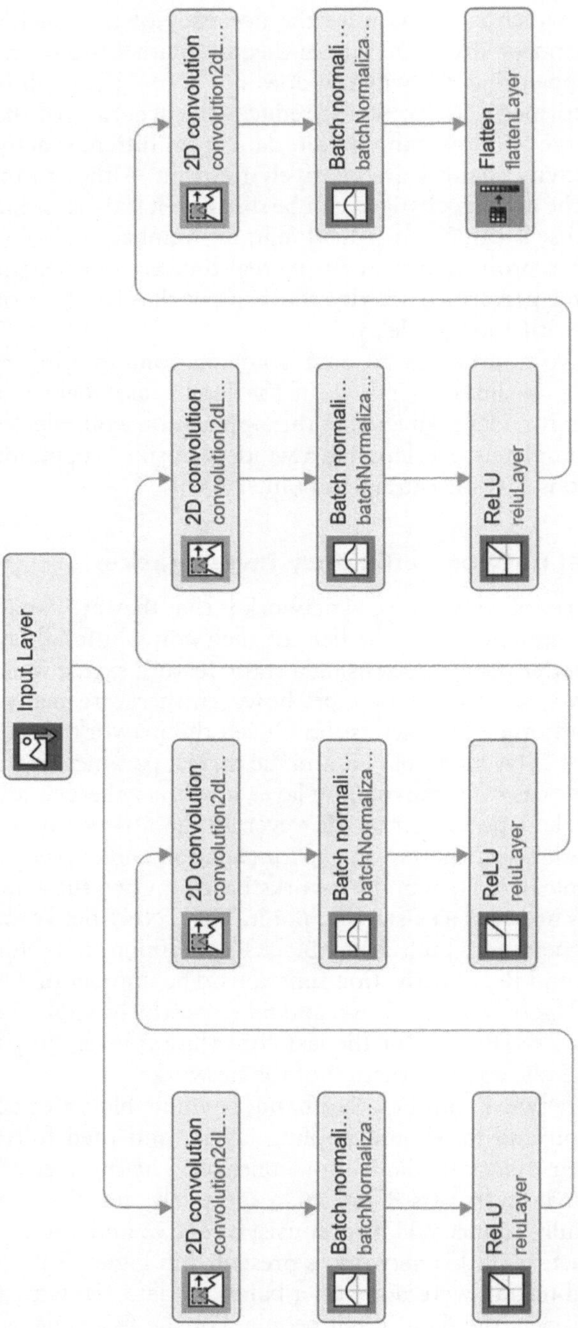

Figure 3.8 Architecture of CNN1 network.

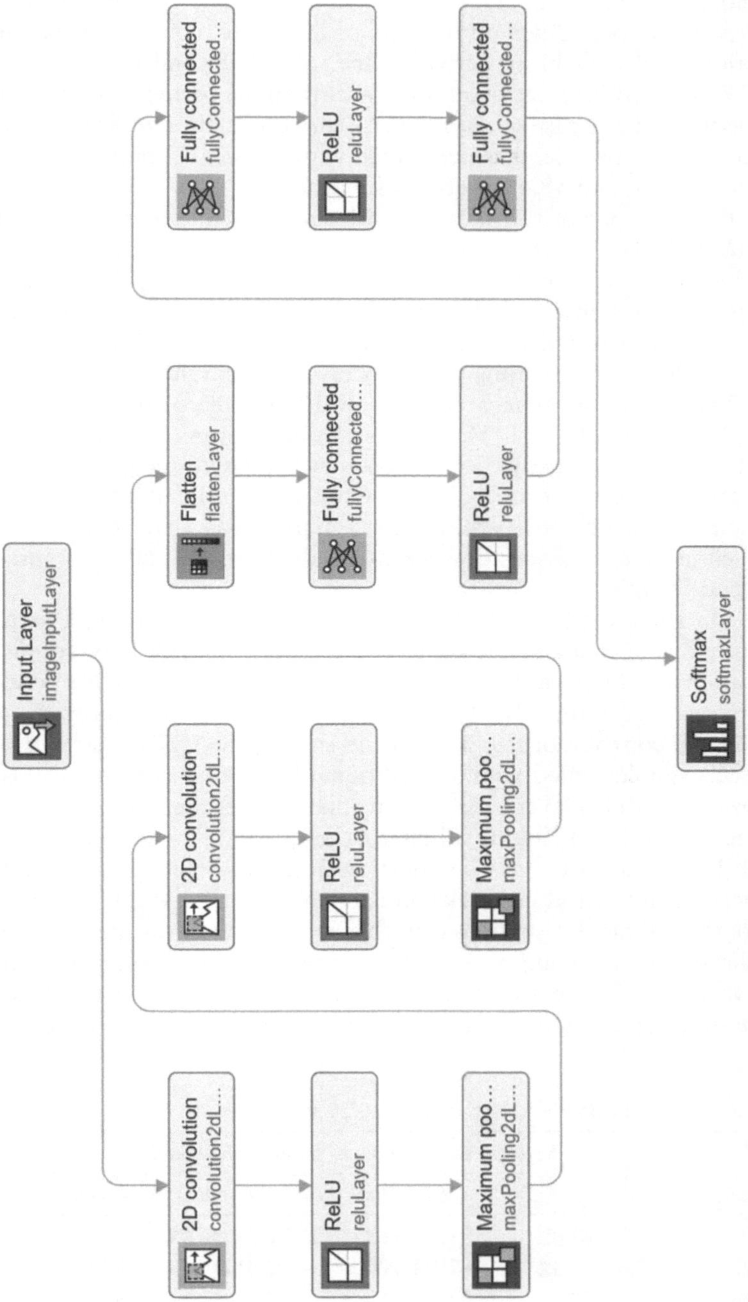

Figure 3.9 Architecture of CNN2 network.

accuracy of 79.2% was achievable for the three classes (closing fists, closing feet, relaxation) with the CNN1 network. The results obtained from the same experiment performed on the data of 20 subject confirm the previous assumption that such brain activities are subject dependent. The overall result is 71.8%, which is substantially lower than the performance provided by the network in the case of ten subjects. The results are significantly worse when the same experiments is performed with the CNN2 network: 62.5% accuracy with ten and 58.4% with 20 subjects.

Since the efficiency of the network with such a small window size is not satisfactory, the size must be increased. With segments of size 64 (0.4 seconds), the accuracy improves greatly. By further increasing the size of the segments to 128 samples (0.8 seconds), even better performance can be obtained. The last tested window size was 160 samples. Since it causes a one-second delay for real-time data processing, it should not be increased further. The classification accuracy was 99.1% for ten people, and 97.7% for 20 people, using CNN1. The results can be interpreted to mean that a segment of this size already contains enough individual-independent information for the machine learning model to find a general pattern. At the same time, the performance of CNN2 is substantially lower than that of CNN1 even when using a larger window size, with the best recognition accuracies 82.6% and 76.4%.

Based on these experiences, using the CNN1 architecture as a starting point, the effect of changing the kernel size and deepening the network on the classification accuracy was investigated. The structure of the CNN3 network is the same as that of the CNN1, but the size of the kernels has been reduced to 3x3 in the convolutional layers. In the case of the CNN4 network, the CNN1 was supplemented with an additional block of convolution – batch normalization – ReLU layers, placed immediately after the input layer. In this block the size of the convolutional filter is 5x5.

Table 3.3 summarizes the classification accuracy achieved for different window sizes and neural networks on databases of ten and 20 people.

The data in Table 3.3 show that in this use case, with this network layer order, the smaller convolution kernel has a negative effect on classification accuracy. It can also be seen that the deepening of the network caused a decrease in accuracy. A network with more convolutions can extract more

Table 3.3 Accuracy on PhysioNet data for 10/20 subjects

	Segment sze (Number of samples and duration)			
Network	*32 (0.2 s)*	*64 (0.4 s)*	*128 (0.8 s)*	*160 (1 s)*
CNN1	79.2%/71.8%	91.1%/83.3%	96.8%/94.6%	99.1%/97.7%
CNN2	62.5%/58.4%	62.1%/64%	76.4%/74.4%	82.6%/76.4%
CNN3	76.5%/68.6%	87.8%/80.2%	96.4%/91.2%	97.7%/93.6%
CNN4	76.2%/70.4%	86.1%/79.8%	96.9%/92.9%	99%/96.1%

features and improve classification performance but requires more time to train because of the increased number of parameters, and is more prone to overfitting, resulting in a less generalizable model. Overall, however, it can be concluded that, except for the CNN2 model, any presented model can be used with high efficiency for EEG-based activity recognition. In addition to the overall classification performance, it is worth examining which classes can be distinguished from each other with the greatest certainty and which ones with less. Some detailed confusion matrices are presented in Tables 3.4–3.7. In each

Table 3.4 CNN1 confusion matrix (20 subjects)

	Predicted class		
True class	*Fists*	*Feet*	*Relax*
Fists	65400	379	786
Feet	560	64773	986
Relax	940	897	64579

Table 3.5 CNN2 confusion matrix (20 subjects)

	Predicted class		
True class	*Fists*	*Feet*	*Relax*
Fists	50472	5056	11037
Feet	5649	51491	9179
Relax	8275	7801	50340

Table 3.6 CNN3 confusion matrix (20 subjects)

	Predicted class		
True class	*Fists*	*Feet*	*Relax*
Fists	60064	1478	5023
Feet	761	62370	3188
Relax	1046	1215	64155

Table 3.7 CNN4 confusion matrix (20 subjects)

	Predicted class		
True class	*Fists*	*Feet*	*Relax*
Fists	64481	766	1318
Feet	711	64549	1059
Relax	2283	1586	62547

case, the matrices show the results achieved on the data set of 20 people, with a window size of 1 second, in the case of the four different networks.

From the tables above, it can be seen that the networks confuse active activities (imagery movement of fists or feet) with each other to a much lesser extent than with the relaxation.

3.4 CONCLUSION

Design and development of an affordable embedded system capable of measuring multiple health parameters requires many considerations: type of parameters to be recorded; choice of sensor and the data transmission protocol; selection of sufficiently high-performance development board to guarantee retrieval and processing of the measured data in real time at an acceptable cost; large storage capacity to provide data storage during periods when the system is not connected to the internet; suitable local independent data display system; and redundant communication systems so that data can be viewed remotely and, if necessary, data acquisition controlled remotely. All these considerations must also take account of the strict special requirements related to medical equipment such as patient and data safety, reliability, etc. Even such a relatively simple system can be used for real-time recognition of activities based on EEG signals by implementing pre-trained convolutional neural networks on a device with higher processing power.

REFERENCES

[1] N. Y. Philip, J. J. P. C. Rodrigues, H. Wang, S. J. Fong, and J. Chen, "Internet of things for in-home health monitoring systems: current advances, challenges and future directions," *IEEE Journal on Selected Areas in Communications*, vol. 39, no. 2, pp. 300–310, 2021.

[2] S. Jiang, F. Firouzi, K. Chakrabarty, and E. Elbogen, "A resilient and hierarchical IoT-based solution for stress monitoring in everyday settings," *IEEE Internet of Things Journal*, vol. 9, no. 12, pp. 10224–10243, 2022.

[3] Libelium Comunicaciones Distribuidas, *MySignals SW eHealth and Medical IoT Development PlatformTechnical Guide*, Document versiov1.1 - 10/2016, Libelium Comunicaciones Distribuidas S.L.

[4] ProtoCentral HealthyPi v4, "Welcome to HealthyPi v4." [Online]. Available: https://healthypi.protocentral.com/

[5] I. A. Pap, S. Oniga, I. Orha and A. Alexan, "IoT-based eHealth data acquisition system," *IEEE International Conference on Automation, Quality and Testing, Robotics (AQTR)*, Cluj-Napoca, Romania, pp. 1–5, July 2018.

[6] I. A. Pap, S. Oniga and A. Alexan, "Machine learning EEG data analysis for eHealth IoT system," *IEEE International Conference on Automation, Quality and Testing, Robotics (AQTR)*, Cluj-Napoca, Romania, pp. 1–4, July 2020.

[7] D. El-Sherif, M. Abouzid, M. Elzarif, A. Ahmed, A. Albakri, and M. Alshehri, "Telehealth and artificial intelligence insights into healthcare during the COVID-19 pandemic," *Healthcare*, vol. 10, no. 2, pp. 1–15, 2022.
[8] C. Comito, D. Falcone and A. Forestiero, "Current trends and practices in smart health monitoring and clinical decision support," *IEEE International Conference on Bioinformatics and Biomedicine (BIBM)*, Seoul, Korea (South), pp. 2577–2584, December 2020.
[9] F. Firouzi et al., "Harnessing the power of smart and connected health to tackle COVID-19: IoT, AI, robotics, and blockchain for a better world," *IEEE Internet of Things Journal*, vol. 8, no. 16, pp. 12826–12846, 2021.
[10] S. P. Korres, A. Menychtas, P. Tsanakas and I. Maglogiannis, "A low-cost IoT-based health monitoring platform enriched with social networking facilities," *IEEE International Conference on Pervasive Computing and Communications Workshops (PerCom Workshops)*, Athens, Greece, pp. 173–178, October 2018.
[11] R. K. Kodali, G. Swamy and B. Lakshmi, "An implementation of IoT for healthcare," *IEEE Recent Advances in Intelligent Computational Systems (RAICS)*, Trivandrum, India, pp. 411–416, December 2015.
[12] Y. Liu, J. Niu, L. Yang and L. Shu, "eBPlatform: An IoT-based system for NCD patients homecare in China," *IEEE Global Communications Conference*, Austin, TX, USA, pp. 2448–2453, December 2014.
[13] A. Fayoumi and K. BinSalman, "Effective remote monitoring system for heart disease patients," *IEEE 20th Conference on Business Informatics (CBI)*, Vienna, Austria, pp. 114–121, July 2018.
[14] H. Moustafa, E. M. Schooler, G. Shen and S. Kamath, "Remote monitoring and medical devices control in eHealth," *2016 IEEE 12th International Conference on Wireless and Mobile Computing, Networking and Communications (WiMob)*, New York, NY, USA, pp. 1–8, October 2016.
[15] R. T. Schirrmeister et al., "Deep learning with convolutional neural networks for EEG decoding and visualization," *Human Brain Mapping*, vol. 38, no. 11, pp. 5391–5420, 2017.
[16] D. H. Krishna, I.A. Pasha, and Savithri, "Classification of EEG motor imagery multi class signals based on cross correlation," *Procedia Computer Science*, vol. 85, pp. 490–495, 2016.
[17] Z. Chen, Y. Wang, and Z. Song, "Classification of motor imagery electroencephalography signals based on image processing method," *Sensors*, vol. 21, no. 14, pp. 1–13, 2021.
[18] Y.-T. Wu, T. H. Huang, C. Yi Lin, S. J. Tsai, and P.-S. Wang, "Classification of EEG motor imagery using support vector machine and convolutional neural network," *International Automatic Control Conference (CACS)*, Taoyuan, Taiwan, pp. 1–4, November 2018.
[19] Alexander Craik, Yongtian He, and Jose L. Contreras-Vidal, "Deep learning for electroencephalogram (EEG) classification tasks: a review," *Journal of Neural Engineering*, vol. 16, no. 3, pp. 1–29, 2019.
[20] A. Goldberger et al., "PhysioBank, PhysioToolkit, and PhysioNet: Components of a new research resource for complex physiologic signals," *Circulation*, vol. 101, no. 23, pp. 215–220, 2020.
[21] T. Majoros, and S. Oniga, "Overview of the EEG-based classification of motor imagery activities using machine learning methods and inference acceleration with FPGA-based cards," *Electronics*, vol. 11, no. 15, pp. 1–14, 2022.

Chapter 4

Embedded and computational intelligence for diabetic healthcare

An overview

Anupama Namburu
Jawaharlal Nehru University, New Delhi, India

Aravapalli Rama Satish, Bhanu Teja Veeramachaneni, Sneha Edupuganti and Kothamasu Venkata Naga Durga Sai Harshith
VIT-AP University, Near Vijayawada, India

4.1 INTRODUCTION

Diabetes is a chronic condition that can have serious health effects, especially for people who struggle to follow the strict treatment plan. Self-management activities including constant insulin and prescription use, frequent blood sugar readings, rigid provisions control, and regular workout can be very difficult. Mobile apps [1] offer patients the chance to increase adherence to these behaviors, with commercial diabetes self-management applications becoming more and more readily available.

Continuous glucose monitoring (CGM) devices worn by patients are transforming type 1 diabetes care (T1D). The sensors in these devices provide the rate of change and blood glucose concentration in real time, once every 1–5 minutes [2]. These measurements are crucial for determining when exogenous insulin should be administered and for predicting upcoming adverse outcomes, such as hypo- or hyperglycemia. The development of decision support systems for the patient's use, which automatically assess the patient's data acquired with CGM sensors and other portable devices, and offer tailored suggestions for therapy modifications, is a major focus of current research in diabetes technology. Artificial intelligence (AI) approaches are being used more frequently in these decision support systems as a result of the volume and variety of data provided by people with T1D. There are invasive and noninvasive methods of making decision support systems for diabetes management. The most popular invasive technique is fingerstick blood glucose testing, which entails taking a drop of blood from the fingertip using a tiny needle or lancing tool. The glucose level is then shown on a glucose meter once the blood has been administered [3].

Non-invasive techniques monitor glucose levels using other methods. In continuous glucose monitoring (CGM), a sensor is inserted under the skin

DOI: 10.1201/9781032628059-4

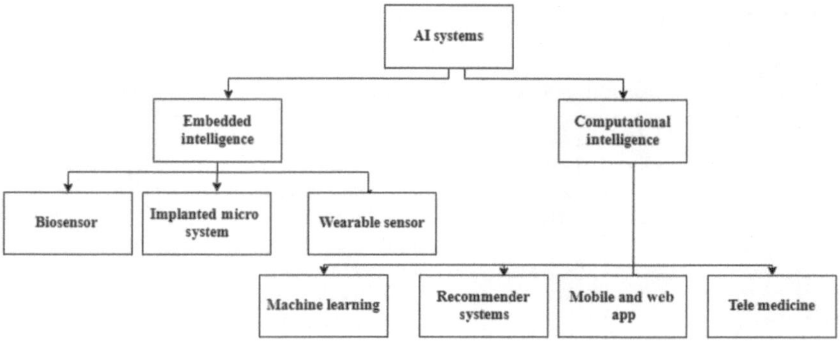

Figure 4.1 AI systems for blood glucose monitoring.

to continually detect glucose levels in interstitial fluid. Other non-invasive techniques are transcutaneous glucose monitoring (TCGM), which detects glucose levels in the tissue immediately beneath the skin, wearable devices and AI-based systems to predict glucose levels [4]. Embedded intelligence refers to the integration of intelligent systems and algorithms into physical devices and products. These systems use sensors, microprocessors, and other components to perform tasks such as data collection, analysis, and decision making. Examples of embedded intelligence include smart home devices, self-driving cars, and wearable technology.

Computational intelligence, on the other hand, is a branch of artificial intelligence that studies algorithms and techniques for creating intelligent systems. It encompasses a range of techniques, including artificial neural networks, fuzzy logic, and evolutionary algorithms. Different embedded intelligence and computational frameworks available for blood sugar monitoring are discussed in the following sub-sections and shown in Figure 4.1.

This chapter considers and analyzes the different blood glucose monitoring systems related to embedded and computational intelligence. Section 4.2 reviews biosensors, implanted sensors and wearable sensors showcasing embedded intelligence. In Section 4.3, computational intelligence and its contributing learnings are explained. Section 4.4 presents conclusions and explores the future scope of glucose monitoring systems.

4.2 EMBEDDED INTELLIGENCE GLUCOSE MONITORING

4.2.1 BioSensors

Biosensors are used for accurate and precise clinical diagnosis. A biosensor is a combined receptor-transducer device that provides specific quantification or moderately analytical data on a particular analyte such as a biosensitive material (Table 4.1). Biosensors can be enzymatic, non-enzymatic,

Table 4.1 Biosensors in diabetic health monitoring

Biosensing platform	*Transducer method*	*Substrate*	*Detection source*
Non-enzymatic [8–14]	Optical, sensing, electrochemical, amperometric, colorimetric	Hydrothermally synthesized zinc oxide (ZnO) nanorods, polydopamone immobilized concanavalin A, gold ruthenium nanoparticles, platinum nanowires/reduced zinc oxide (ZnO) nanoparticles from ocimum tenuiflorumgraphene oxide	Human serum, blood
Enzymatic [14–16]	Electrochemical	Glucose oxidase/three-dimensional porous graphene-aerogel, gold/MXene based nanocomposite, (PLDz)/(GOx)	Human serum, blood, saliva and tears
Biomarkers [7, 17, 18]	Optical, electrochemical, amperometric	sorbitol dehydrogenase, GHSA, FAO/AuPt NPs	Human serum, blood
Label-free [6, 19]	Electro-chemiluminescent	AuNPs/indium tin oxide, hydrolyzed (3-aminopropyl) trimethoxysilane/AgNPs/indium tin oxide	127L gene variation, C-peptide

biomarker and label free and can be classified into transducer units, identification elements, recognition elements, and catalytic biosensors. Biosensors based on transducers are electrodes, semi-conductors, thermal, photo and piezoelectric biosensors. Biosensors based on identification elements are enzyme, nucleic acid, microbial, cell, tissue, and immune sensors. Biosensors based on the type of recognition element are bio-affinity, metabotropic and catalytic biosensors. Various biosensors are available to detect diabetes based on human blood, serum, saliva, breath and intradermal glucose. Enzymatic diabetes biosensors containing glucosidase enzyme mounted on various materials experience inadequate stability and enzyme activity loss throughout the immobilization procedure. Numerous non-enzymatic glucose biosensors have been developed for the same purpose in order to provide interference-free and extremely sensitive sensors.

The label-free mode facilitates detection as it entails direct sensing of the signal generated via the on-site response of the transducer and analyte [5]. Zhai *et al.* [6] developed a label-free electro-chemiluminescent (ECL) biosensor based on gold nanoparticles with functionalized indium tin oxide electrodes to detect the 127L gene variant that causes type 2 diabetes.

Gessei *et al.* [7] introduced a fiber-optic biosensor based on sorbitol (a biomarker) to diagnose diabetes in people using biomarkers. The sensor made use of the more substrate-specific sorbitol dehydrogenase from the microbe *flavimonas* and assessed the fluorescence impact of the decreased form of nicotinamide adenine dinucleotide (NADH) caused by enzyme action. The NADH (light resource) was excited by an ultraviolet light-emitting diode (UV-LED), and the fluorescence of the NADH was monitored by a photomultiplier tube and a spectrometer. A Y-designed optical fiber integrated the photodetector and UV-LED. The photomultiplier tube and spectrophotometer calibration curves for the sorbitol sensor platform were 1.0–1000 mol/L and 5.0–1000 mol/L, respectively. The sensitivity and selectivity of the sorbitol sensor for the diagnosis of diabetes were thus proven.

4.2.2 Implanted micro systems

Diabetes management devices that are surgically inserted into the body are referred to as implanted micro systems. These systems include continuous glucose monitors (CGMs), insulin pumps, and other wearables with real-time glucose monitoring and control capabilities [20]. Wearable glucose monitors assess blood sugar levels in the interstitial fluid beneath the skin and display real-time glucose measurements [21]. Glucose levels may be monitored with CGMs throughout the day, providing an alert when levels are excessively high or low. Insulin pumps are wearable gadgets that can provide the body with insulin when required. The continuous glucose monitor that these pumps are attached to offers real-time glucose readings. The insulin pump can provide a tiny quantity of insulin to maintain blood sugar levels in a desired range based on the glucose measurements. In a study [22], the materializing of the data field body doubles with insulin pumps was explored. A detailed review of insulin pumps is provided in [23].

Implantable glucose sensors are gadgets that are surgically inserted beneath the skin to gauge the amount of glucose present in the interstitial fluid. In order to regulate blood sugar levels, these sensors can offer continuous, real-time glucose measurements and can be used in conjunction with an insulin pump or another wearable device. Zhang *et al.* [24] proposed an implanted automatic drug delivery system. Long-term implanted sensor monitoring for glucose is proposed in [25–27].

Artificial pancreas systems use embedded intelligence to control insulin delivery in real time based on glucose levels [28]. They use continuous glucose monitoring sensors and insulin pumps to automatically adjust insulin dosing and maintain glucose levels within a safe range. For those with diabetes, these kinds of systems have several advantages, including better glucose regulation, and improved convenience and safety. It is crucial to remember that these systems might not be appropriate for everyone and that they should only be used under the supervision of a healthcare provider. Additionally, these systems need routine maintenance.

4.2.3 Wearable sensors

In recent years wearable sensors have gained great importance due to their noninvasive or minimally invasive nature. Wearable biosensors that analyze interstitial fluid (ISF), sweat, tears and saliva are available in literature.

4.2.3.1 Wearable interstitial fluid (ISF) CGM

ISF is an extracellular fluid that surrounds cells in the body and receives its glucose from the blood via ongoing capillaries. A wearable device that monitors glucose levels and sends alerts for high or low readings is coupled to a tiny sensor inserted under the skin. ISF CGM does not require periodic fingerstick blood tests, offering real-time glucose data instead. This makes managing the disease more practical and less uncomfortable for patients with diabetes. ISF CGM systems are frequently coupled with insulin pumps to give automatic insulin dosage depending on glucose levels. They may also be linked to other wearables. The two key benefits of ISF CGM are that it is minimally intrusive and has a relatively high detection accuracy and reliability [29]. It is estimated that 44% of T1D patients have used an ISF glucose sensor. Microneedles offer a promising middle ground between invasive and noninvasive sample techniques for transdermal ISF [30, 31]. Microwave resonator-based sensors for continuous glucose monitoring via ISF were recently described by Baghelani *et al.* [32]. The closed loop, however, has a number of significant difficulties, such as the 5–10 minutes' lag between blood and ISF glucose, the time it takes for glucose to travel from plasma to ISF.

4.2.3.2 Wearable sweat CGM

Sweat contains 10-200 μM of glucose, which is roughly 1%–2% of the amount found in blood [33]. In people with diabetes, blood glucose and sweat glucose are closely connected. Other electrochemically active substances in sweat, like AA and UA, can interfere with electrochemical sweat sensors. Since sweat may easily be collected and continually monitored, sweat biosensing has recently attracted a lot of attention for noninvasive CGM. Gao *et al.* [34] created a closed-loop solution for noninvasive sweat glucose monitoring, microneedle-based point-of-care therapy, pH, temperature, and humidity measurements. A completely integrated wearable flexible sensing array for simultaneous real-time measurements of glucose, lactate, and electrolytes was demonstrated.

4.2.3.3 Wearable tear CGM

The aqueous liquid that surrounds the eye, sometimes referred to as ocular fluid, or tear, can be employed as an alternate sensing platform for continuous glucose monitoring (CGM). A smart contact lens that enables CGM and diabetic management was described by Kim *et al.* [35]. In order to monitor glucose in tears and intraocular pressure, they created a wearable multifunctional contact lens sensor that uses field-effect transistors (FETs) made with graphene-AgNW hybrid as S/D electrodes. Similar strategies are being pursued by a number of different groups, but before such CGM biosensors are commercially mature, significant obstacles like security and privacy, clinically compatible testing, etc. still need to be overcome [36].

4.2.3.4 Wearable saliva CGM

Saliva is an exocrine secretion of the salivary glands and gingival crevicular fluid that is either serous or mucous. Depending on the type of gland, the time of day, the age, and the gender, water (98%) and other compounds make up the majority of its components [37]. Studies have shown a substantial association between salivary and blood glucose levels in both diabetic and non-diabetic people [33, 38]. The concentrations of glucose in saliva are positively associated with the levels in blood plasma. However, salivary glucose is often dependent on mouthguard design, and food and drink contamination are the main sources of interference. De Castro *et al.* [39] reported a wearable sensor for saliva glucose monitoring that is based on a microfluidic paper device and printed using a craft cutter to address this issue.

4.3 COMPUTATIONAL INTELLIGENCE IN GLUCOSE MONITORING

To prevent damage to the eyes, kidneys, heart, and nervous system, people with prediabetes and diabetes need to change their lifestyles, putting an emphasis on getting enough sleep, eating well, and exercising in order to reach their blood sugar, blood pressure, fats and weight goals. It is now simpler and safer for diabetics to avoid both low and high blood sugar levels thanks to newer, safer medications that regulate blood sugar and lower the risk of developing heart disease and kidney disease. A team approach to managing diabetes puts the patient at the center and allows them to collaborate with their doctors on decisions. Figure 4.2 shows the computation intelligence approach that uses machine learning to provide better recommender systems, tele medicine, auto-administer therapy and apps.

4.3.1 Machine learning

Machine learning is a broad category of approaches, tools, and evaluations of algorithms with the goal of learning from data to discover relevant knowledge or prognostic models of a given phenomenon [40, 41].

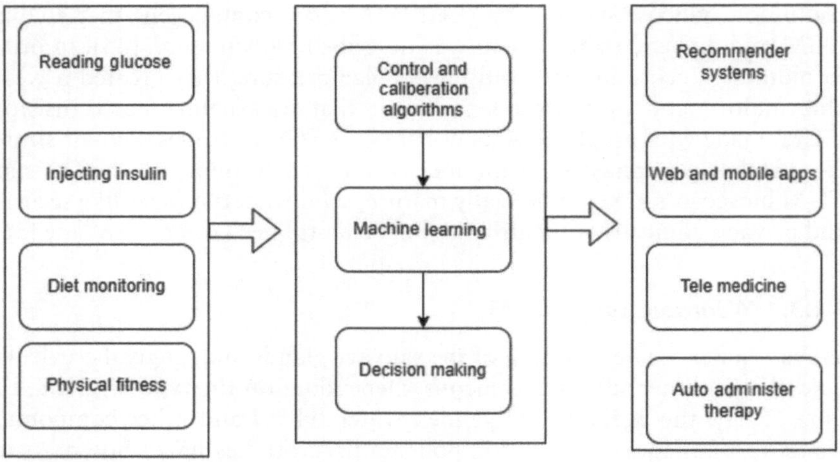

Figure 4.2 Computation intelligence for CGM.

Supervised learning and unsupervised learning are the two primary paradigms of machine learning. Support vector machines (SVMs) are supervised machine learning algorithms based on the idea of solving a two-class pattern recognition issue by locating the best separation hyperplane between two data classes. SVMs have been used to improve the analytical precision of graphene nanoelectronic-based biomolecular sensing and to assess the precision of breath-based volatile organic compound blood glucose detection [42, 43]. Regression analysis based on feature parameters was carried out for categorization of albumin protein identified from lateral flow experiments.

In their study, [44] proposed a fast and accurate artificial neural network (ANN) for diabetic recognition. Deep learning techniques are applied to predict and diagnose diabetes in a study with K-NN and ANN [45]. Principle component analysis (PCA) is an unsupervised machine learning approach for dimensionality reduction by substituting principal components for a collection of variables [46]. With the aid of SEIRA the vibrational data from glucose and fructose biosensors is extracted [47]. The k-means [46] hierarchical clustering approach known as hierarchical cluster analysis (HCA) creates a hierarchical framework to group similar objects together within a data collection. The use of decision trees (DTs) is a useful way of managing complicated behavior of papain or protein. Si surface platform is used on a biomarker mechanism named nanoscale cubic to collect free papain/protein and evaluate its behavior. Eighteen tetrapeptides were combined to form a decision tree that was used to examine papain/protein behavior in food synthesis. The use of ensemble techniques results in higher storage and computing costs despite providing the highest classification accuracy attainable. Machine learning algorithms are commercialized by businesses to provide automatic and ongoing blood glucose monitoring and propose improvements in healthcare. Model-based design is a technique used by DreaMed Diabetes [48] and Bigfoot Biomedical [49] to create automated insulin delivery devices. According to Medilyn [50], their glucose-monitoring gadget continually checks blood sugar levels using machine learning.

4.3.2 Recommender systems

A type of technology called a recommender system for diabetes monitoring can assist people with diabetes by offering them individual information, support and recommendations to better manage their condition. Recommender systems offer different approaches for managing diabetes:

Meal Suggestions: The system may assess a user's eating patterns and make healthy food suggestions that may be able to control blood sugar levels. Machine learning- and deep learning-based food recommender systems are proposed in study [51, 52], MyFitnessPal [53] and Medical ID [54]. mySugr – Diabetes Tracker Log, and Diabetes Connect are apps [55] that monitor glucose levels and make diet, exercise and medication recommendations, monitor glucose levels.

Exercise Advice: The system may assess a person's level of physical activity and provide an exercise program to enhance insulin sensitivity and blood sugar regulation. MyFitnessPal [53].

Glucose Monitoring: The program may monitor a person's blood sugar levels and offer advice. Medical ID [54], mySugr-Diabetes Tracker Log [55], Diabetes Connect.

4.3.3 Mobile and WebApp

An ophthalmoscope interface smartphone app called Portable Eye Examination Kit (PEEK) (Samsung Galaxy, Samsung Electronics Co) [56], used by Bastawrous *et al.*, was evaluated in low-income settings (Nakuru Eye Disease Cohort Study) and high-income settings in collaboration with public health technologists and physicians [57]. Results for this low-cost method are still awaited but it has significant potential influence on LMICs (low- and middle-income countries).

The D-Eye device is a handheld optical tool with a magnetic attachment that connects to a smartphone and enables fundus viewing without the use of a handheld condenser lens. A controlled review of 120 patients reported that retinal specialists viewing the fundus through dilated pupils at the point of care were able to detect diabetic retinopathy (DR) with increased gain. The authors of this study propose this technique for remote retinal DR screening [57]. Various diabetic screening devices are present in study [58–62] and are presented in Table 4.2.

Table 4.2 Smart phone applications for Diabetes Management

Device/Technology	*Company/Developer*	*Non Mydriatic*	*Price*	*Required devices/software*	*Images recorded?*	*Availability*
Smartphone and Lens	Multiple	No	$407.99	20 D Lens, Filmic Pro app	Yes	Available
Portable Eye Examination Kit (PEEK)	Peek collaboration	Yes	~$95	iPhone, Android (pending)	Yes	Available
D-Eye Portable Retinal System	D-Eye	Yes	~$400	iPhone 5/5S/6, Samsung Galaxy S4/S5	Yes	Available
iExaminer System	Welch-Allyn	No	$890.99	PanOptic Ophthalmoscope, iExaminer Pro app, iPhone 4/4s	Yes	Available
Horus Scope	JEDMED	Yes	$4400.00	Independent	Yes	Available
SmartScope Pro/ Pictor Plus	OptoMed/Volk Optical	Yes	$9995.00	Independent	Yes	Available
Ocular CellScope	UC Berkeley	Yes	N/A	iPhone 4/4S, DualAlign, i2k Retina software	Yes	Development
EyeMitra	MIT	Yes	Target $100	Independent	—	Development
EyeSelfie	MIT	Yes	N/A	Independent	—	Development

4.3.4 TeleMedicine

Telemedicine is remote medical treatment and assistance provided using technology, such as video conferencing. Telemedicine has a significant role to play in helping people with diabetes better manage their disease. Telemedicine can be used to arrange virtual consultations between diabetic patients and medical professionals. People who live in rural or isolated places or who are unable to get to a clinic or hospital may find this extremely helpful. Telemedicine became very important on the outbreak of Covid-19 [63, 64]. Telemedicine may be used to remotely check a diabetic patient's blood sugar, blood pressure, and other important health indicators. Using this data, healthcare professionals may offer patients individual advice and direction to assist them in better managing their illness. Diabetic patients can receive support and educational materials through the use of telemedicine. Healthcare professionals, for instance, can conduct training classes through video conferencing on topics like diabetes treatment strategies and healthy living [65]. Wearable technology and other diabetes management-related equipment, such as insulin pumps and continuous glucose monitoring systems, can also be remotely observed using telemedicine. These details can be used by healthcare professionals to change their practices.

4.3.5 Auto-administer therapy

The term "auto-administer therapy" refers to the application of technology to the distribution of medications for conditions like diabetes. Auto-administer treatment in the context of diabetes includes:

Insulin Pumps. Wearable gadgets that can provide the body with insulin when it is required. The continuous glucose monitor that these pumps are attached to offers real-time glucose readings. The insulin pump can provide a tiny quantity of insulin to maintain blood sugar levels in a desired range based on the glucose measurements.

Smart Insulin Pens. Track and administer insulin dosages and measure blood sugar levels in real time. Some smart insulin pens have technology that allows insulin dosages to be automatically adjusted depending on glucose measurements, assisting in maintaining target glucose levels.

Implanted devices. Insulin pumps and continuous glucose monitoring systems are examples of implantable medical equipment that may be used to automatically regulate blood glucose levels. These devices may be set up to automatically modify insulin dosages in response to glucose measurements, assisting in maintaining target glucose levels.

For diabetics, auto-administer treatment can provide a variety of advantages, including improved glucose control, convenience, and safety. However, these systems may not be suitable for everyone and should only be used under the supervision of a healthcare provider. These systems also have ongoing monitoring and maintenance requirements, as well as potential hazards and negative impacts.

4.4 CONCLUSION AND FUTURE SCOPE

Glucose levels in people with diabetes can be monitored using invasive and non-invasive methods. Invasive diabetic monitoring involves obtaining a sample of blood through a fingerstick test, which is then analyzed to determine the glucose level. This method provides an accurate measurement of glucose levels, but it can be painful and requires frequent blood tests, which can be inconvenient for patients. Non-invasive diabetic monitoring, on the other hand, does not require a blood sample and instead uses other methods to measure glucose levels. This includes continuous glucose monitoring (CGM) devices that measure glucose levels in interstitial fluid, wearable devices that use sensors to monitor glucose levels, and artificial pancreas systems that use algorithms to control insulin delivery based on glucose levels. Non-invasive diabetic monitoring has become increasingly popular in recent years due to its convenience and improved accuracy compared to invasive methods. However, non-invasive methods may not be as accurate as invasive methods and can have limitations, such as the need to calibrate the device regularly.

The use of AI in biosensors has the potential to greatly improve the management of conditions like diabetes, as it can provide more accurate and personalized glucose readings and help individuals make more informed decisions about their health. However, it is important to note that AI biosensors are still in the early stages of development and may not be suitable for everyone. Additionally, these devices should only be used under the guidance of a healthcare professional, and their accuracy and reliability may still need to be further validated.

Wearable sensors have the potential to play a major role in the future of diabetes management. The technology is rapidly advancing, and it is likely that wearable sensors will become even more sophisticated and capable in the years to come. Wearable sensors are expected to continue to improve, becoming smaller, more accurate, and more convenient to use. This will allow individuals with diabetes to continuously monitor their glucose levels, providing a more comprehensive view of their health and enabling more effective management of their condition. Wearable sensors will likely incorporate more advanced artificial intelligence and machine learning algorithms, allowing them to predict future glucose levels and alert individuals to potential hypoglycemic episodes or other health events. Wearable sensors will also likely become more integrated with insulin delivery systems, allowing for more personalized insulin dosing based on real-time glucose readings.

This will help to improve glucose control and reduce the risk of complications associated with diabetes.

Wearable sensors are expected to become more convenient and user friendly, making it easier for individuals with diabetes to monitor and manage their condition. This will likely include improvements in the design and form factor of wearable sensors, as well as more streamlined data management and reporting capabilities. Wearable sensors with embedded and computational intelligence have the potential to greatly improve the lives of individuals with diabetes, enabling them to more effectively monitor and manage their condition and live healthier, more fulfilling lives.

REFERENCES

[1] A. Thurzo, V. Kurilová, and I. Varga, "Artificial Intelligence in Orthodontic Smart Application for Treatment Coaching and Its Impact on Clinical Performance of Patients Monitored with AI-TeleHealth System," *Healthcare MDPI*, vol. 9, no. 12, pp. 1–23, 2021.

[2] D.F. Kruger, and J.E. Anderson, "Continuous glucose monitoring (CGM) is a tool, not a reward: Unjustified insurance coverage criteria limit access to CGM," *Diabetes Technology & Therapeutics*, vol. 23, no. S3, pp. S45–S55, 2021.

[3] A. Ahmed, S. Aziz, U. Qidwai, and A. Abd-Alrazaq, "Performance of artificial intelligence models in estimating blood glucose level among diabetic patients using non-invasive wearable device data," *Computer Methods and Programs in Biomedicine Update*, vol. 3, pp. 1–7, 2023.

[4] A. Piersanti, F. Giurato, C. Göbl, L. Burattini, A. Tura, and M. Morettini, "Software packages and tools for the analysis of continuous glucose monitoring data," *Diabetes Technology & Therapeutics*, vol. 25, no. 1, pp. 69–85, 2023.

[5] N. Khansili, G. Rattu, and P.M. Krishna, "Label-free optical biosensors for food and biological sensor applications," *Sensors and Actuators B: Chemical*, vol. 265, pp. 35–49, 2018.

[6] S. Zhai, C. Fang, J. Yan, Q. Zhao, and Y. Tu, "A label-free genetic biosensor for diabetes based on AuNPs decorated ITO with electrochemiluminescent signaling," *Analytica Chimica Acta*, vol. 982, pp. 62–71, 2017.

[7] T. Gessei, T. Arakawa, H. Kudo, and K. Mitsubayashi, "A fiber-optic sorbitol biosensor based on NADH fluorescence detection toward rapid diagnosis of diabetic complications," *Analyst*, vol. 140, no. 18, pp. 6335–6342, 2015.

[8] S.N. Sarangi, S. Nozaki, and S.N. Sahu, "ZnO nanorod-based non-enzymatic optical glucose biosensor," *Journal of Biomedical Nanotechnology*, vol. 11, no. 6, pp. 988–996, 2015.

[9] M. Lobry, D. Lahem, M. Loyez, M. Debliquy, K. Chah, M. David, and C. Caucheteur, "Non-enzymatic D-glucose plasmonic optical fiber grating biosensor," *Biosensors and Bioelectronics*, vol. 142, pp. 1–22, 2019.

[10] T. N. Nguyen, X. Jin, J. K. Nolan, J. Xu, K. V. H. Le, S. Lam, Y. Wang, M. A. Alam, and H. Lee, "Printable nonenzymatic glucose biosensors using carbon

nanotube-PtNP nanocomposites modified with AuRu for improved selectivity," *ACS Biomaterials Science & Engineering*, vol. 6, no. 9, pp. 5315–5325, 2020.
[11] X. Duan, K. Liu, Y. Xu, M. Yuan, T. Gao, and J. Wang, "Nonenzymatic electrochemical glucose biosensor constructed by NiCo2O4@ Ppy nanowires on nickel foam substrate," *Sensors and Actuators B: Chemical*, vol. 292, pp. 121–128, 2019.
[12] X. Fu, Z. Chen, S. Shen, L. Xu, and Z. Luo, "Highly sensitive nonenzymatic glucose sensor based on reduced graphene oxide/ultrasmall Pt nanowire nanocomposites," *Int. J. Electrochem. Sci*, vol. 13, no. 5, pp. 4817–4826, 2018.
[13] T. Dayakar, et al., "Novel synthesis and structural analysis of zinc oxide nanoparticles for the non enzymatic glucose biosensor," *Materials Science and Engineering: C*, vol. 75, pp. 1472–1479, 2017.
[14] S. Cinti, R. Cusenza, D. Moscone, and F. Arduini, "Based synthesis of Prussian Blue Nanoparticles for the development of whole blood glucose electrochemical biosensor," *Talanta*, vol. 187, pp. 59–64, 2018.
[15] R.B. Rakhi, P. Nayak, C. Xia, and H.N. Alshareef, "Novel amperometric glucose biosensor based on MXene nanocomposite," *Scientific Reports*, vol. 6, no. 1, pp. 1–10, 2016.
[16] C. Liu, Y. Sheng et al., "A glucose oxidase-coupled DNAzyme sensor for glucose detection in tears and saliva," *Biosensors and Bioelectronics*, vol. 70, pp. 455–461, 2015.
[17] C. Apiwat, P. Luksirikul, et al., "Graphene based aptasensor for glycated albumin in diabetes mellitus diagnosis and monitoring," *Biosensors and Bioelectronics*, vol. 82, pp. 140–145, 2016.
[18] U. Jain, S. Gupta, and N. Chauhan, "Construction of an amperometric glycated hemoglobin biosensor based on Au–Pt bimetallic nanoparticles and poly (indole-5-carboxylic acid) modified Au electrode," *International Journal of Biological Macromolecules*, vol. 105, pp. 549–555, 2017.
[19] X. Liu, C. Fang, J. Yan, H. Li, and Y. Tu, "A sensitive electrochemiluminescent biosensor based on AuNP-functionalized ITO for a label-free immunoassay of C-peptide," *Bioelectrochemistry*, vol. 123, pp. 211–218, 2018.
[20] B. Purohit, A. Kumar, K. Mahato, and P. Chandra, "sContinuous Glucose Monitoring for Diabetes Management Based on Miniaturized Biosensors", in *Miniaturized Biosensing Devices: Fabrication and Applications*, Singapore: Springer, pp. 149–175, 2022.
[21] Y. Yao et al., "Integration of interstitial fluid extraction and glucose detection in one device for wearable non-invasive blood glucose sensors," *Biosensors and Bioelectronics*, vol. 179, pp. 1–7, 2021.
[22] S Horrocks, "Materializing datafied body doubles: Insulin pumps, blood glucose testing, and the production of usable bodies," *Catalyst: Feminism, Theory, Technoscience*, vol. 5, no. 1, pp. 1–26, 2019.
[23] N.D. Sora, F. Shashpal, E.A. Bond, and A.J. Jenkins, "Insulin pumps: review of technological advancement in diabetes management," *The American Journal of the Medical Sciences*, vol. 358, no. 5, pp. 326–331, 2019.
[24] J. Zhang et al., "Wearable glucose monitoring and implantable drug delivery systems for diabetes management," *Advanced Healthcare Materials*, vol. 10, no. 17, pp. 1–23, 2021.

[25] J.J. Joseph, "Review of the long-term implantable senseonics continuous glucose monitoring system and other continuous glucose monitoring systems," *Journal of Diabetes Science and Technology*, vol. 15, no. 1, pp. 167–173, 2021.
[26] P. Sanchez, S. G. Dastidar, K.S. Tweden and F.R. Kaufman, "Real-world data from the first US commercial users of an implantable continuous glucose sensor," *Diabetes Technology & Therapeutics*, vol. 21, no. 12, pp. 677–681, 2019.
[27] H.K. Akturk, and S. Brackett, "A novel and easy method to locate and remove first approved long-term implantable glucose sensors," *Diabetes Technology & Therapeutics*, vol. 22, no. 7, pp. 538–540, 2020.
[28] M. Nwokolo, and R. Hovorka, "The artificial pancreas and type 1 diabetes," *The Journal of Clinical Endocrinology & Metabolism*, vol. 108, no. 7, pp. 1614–1623, 2023.
[29] Z. Xie, X. Zhang, G. Chen, J. Che, and Zhang, D, "Wearable microneedle-integrated sensors for household health monitoring," *Engineered Regeneration*, vol. 3, no. 4, pp. 420–426, 2022.
[30] H. C. Ates, et al., "End-to-end design of wearable sensors," *Nature Reviews Materials*, vol. 7, no. 11, pp. 887–907, 2022.
[31] M. Dervisevic, "Wearable microneedle array-based sensor for transdermal monitoring of pH levels in interstitial fluid," *Biosensors and Bioelectronics*, vol. 222, pp.114955(1-10), 2023.
[32] M. Baghelani, Z. Abbasi, M. Daneshmand and P. E. Light, "Non-Invasive lactate monitoring system using wearable chipless microwave sensors with enhanced sensitivity and zero power consumption," *IEEE Transactions on Biomedical Engineering*, vol. 69, no. 10, pp. 3175–3182, 2022.
[33] J. Heikenfeld, A. Jajack et al., "Accessing analytes in biofluids for peripheral biochemical monitoring," *Nature Biotechnology*, vol. 37, no. 4, pp. 407–419, 2019.
[34] H. Lee, C. Song, Hong, YS, et al., "Wearable/disposable sweat-based glucose monitoring device with multistage transdermal drug delivery module," *Science Advances*, vol. 3, no. 3, pp. 1–9, 2017.
[35] D.H. Keum, S.K. Kim et al., "Wireless smart contact lens for diabetic diagnosis and therapy," *Science Advances*, vol. 6, no. 17, pp. 1–13, 2020.
[36] R. Zou, S. Shan, L. Huang et al., "High-performance intraocular biosensors from chitosan functionalized nitrogen-containing graphene for the detection of glucose," *ACS Biomaterials Science & Engineering*, vol. 6, no.1, pp. 673–679, 2020.
[37] K.M. Jaedicke, P.M. Preshaw, and J.J. Taylor, "Salivary cytokines as biomarkers of periodontal diseases," *Periodontology 2000*, vol. 70, no. 1, pp. 164–183, 2016.
[38] J. Liu, S. Sun, H. Shang, J. Lai, and L. Zhang, "Electrochemical biosensor based on bienzyme and carbon nanotubes incorporated into an Os-complex thin film for continuous glucose detection in human saliva," *Electroanalysis*, vol. 28, no. 9, pp. 2016–2021, 2016.
[39] L.F. De Castro, S.V. de Freitas et al., "Salivary diagnostics on paper microfluidic devices and their use as wearable sensors for glucose monitoring," *Analytical and Bioanalytical Chemistry*, vol. 411, no. 19, pp. 4919–4928, 2019.
[40] M. Revathi, A.B. Godbin, S.N. Bushra and S.A. Sibi, "Application of ANN, SVM and KNN in the Prediction of Diabetes Mellitus," *International Conference on*

Electronic Systems and Intelligent Computing (ICESIC), Chennai, India, pp. 179–184, June 2022.

[41] R. Krishnamoorthi, S. Joshi et al., "A novel diabetes healthcare disease prediction framework using machine learning techniques," *Journal of Healthcare Engineering*, vol. 2022, pp. 1–10, 2022.

[42] Z. Saberi, B. Rezaei, P. Rezaei, and A.A. Ensafi, "Design a fluorometric aptasensor based on CoOOH nanosheets and carbon dots for simultaneous detection of lysozyme and adenosine triphosphate," *Spectrochimica Acta Part A: Molecular and Biomolecular Spectroscopy*, vol. 233, pp. 1–26, 2020.

[43] R. Patil, S. Tamane, S.A. Rawandale, and K. Patil, "A modified mayfly-SVM approach for early detection of type 2 diabetes mellitus," *International Journal of Electrical and Computer Engineering*, vol. 12, no. 1, pp. 1–10, 2022.

[44] R. Jader, and S. Aminifar, "Fast and accurate artificial neural network model for diabetes recognition," *NeuroQuantology*, vol. 20, no. 10, pp. 2187–2196, 2022.

[45] N. Arora, A. Singh, M. Z. N. Al-Dabagh, and S.K. Maitra, "A novel architecture for diabetes patients' prediction using K-Means clustering and SVM," *Mathematical Problems in Engineering*, vol. 2022, 2022.

[46] W. Oujidi, R. Sekhsoukh, Y. Harrar, and M.A. Mehdi, "Study of variation glycated hemoglobin in diabetic patients using PCA method. Case study: Moroccan eastern region," *Annals of Medicine and Surgery*, vol. 81, pp. 1–6, 2022.

[47] D. Pfezer, J. Karst, M. Hentschel, and H. Giessen, "Predicting concentrations of mixed sugar solutions with a combination of resonant plasmon-enhanced seira and principal component analysis," *Sensors*, vol. 22, no. 15, pp. 1–16, 2022.

[48] Endo.digital, "DreaMed," *dreamed-diabetes.com.* [Online]. Available: https://dreamed-diabetes.com/

[49] Bigfootbiomedical, "A simple way to manage diabetes," *bigfootbiomedical.com.* [Online]. Available: https://www.bigfootbiomedical.com/

[50] Medilync, "Welcome to our website," *medilync.com.* [Online]. Available: https://medilync.com/

[51] V.C. Silva, B. Gorgulho, D.M. Marchioni, and S.M. Alvim, "Recommender system based on collaborative filtering for personalized dietary advice: A cross-sectional analysis of the elsa-brasil study," *International Journal of Environmental Research and Public Health*, vol. 1, no. 22, pp. 1–12, 2022.

[52] M. Rostami, M. Oussalah and V. Farrahi, "A novel time-aware food recommender-system based on deep learning and graph clustering," *IEEE Access*, vol. 10, pp. 52508–52524, 2022.

[53] S. L. Hahn et al., "Impacts of dietary self-monitoring via MyFitnessPal to undergraduate women: A qualitative study," *Body Image*, vol. 39, pp. 221–226, 2021.

[54] C. Kaczmarek et al., "Medical ID and emergency apps: A useful tool in emergency situations or a waste of time?," *Medizinische Klinik, Intensivmedizin und Notfallmedizin*, vol. 116, no. 4, pp. 339–344, 2020.

[55] A. Maharaj, D. Lim, R. Murphy, and A. Serlachius, "Comparing two commercially available diabetes apps to explore challenges in user engagement: randomized controlled feasibility study," *JMIR Formative Research*, vol. 5, no. 6, 2021.

[56] Portable Eye Examination Kit (PEEK), "We make the invisible, visible," *peekvision.org.* [Online]. Available: https://peekvision.org/

[57] M. E. Giardini et al., "A smartphone based ophthalmoscope," *36th Annual International Conference of the IEEE Engineering in Medicine and Biology Society*, Chicago, IL, USA, pp. 2177–2180, 2014.

[58] H. Palta, and M. Karakaya, "Image quality assessment of smartphone-based retinal imaging systems," in *Proc. Smart Biomedical and Physiological Sensor Technology XIV*, Orlando, Florida, vol. 12123, 2022.

[59] B. J. Palermo, S. L. D'Amico, B.Y. Kim, and C. J. Brady, "Sensitivity and specificity of handheld fundus cameras for eye disease: a systematic review and pooled analysis," *Survey of Ophthalmology*, vol. 67, no.5, pp. 1531–1539, 2022.

[60] R. Rajalakshmi, V. Prathiba, S. Arulmalar, and M. Usha, "Review of retinal cameras for global coverage of diabetic retinopathy screening," *Eye*, vol. 35, no.1, pp. 161–172, 2020.

[61] A. Pugalendhi, and R. Ranganathan, "Development of 3D printed smartphone-based multi-purpose fundus camera (multiscope) for retinopathy of prematurity," *Annals of Biomedical Engineering*, vol. 49, no.12, pp. 3323–3338, 2021.

[62] A. Bahl, and S. Rao, "Diabetic retinopathy screening in rural India with portable fundus camera and artificial intelligence using eye mitra opticians from Essilor India," *Eye*, vol. 36, no. 1, pp. 230–231, 2020.

[63] L.M. Quinn, M.J. Davies, and M. Hadjiconstantinou, "Virtual consultations and the role of technology during the COVID-19 pandemic for people with type 2 diabetes: the UK perspective," *Journal of Medical Internet Research*, vol. 22, no.8, 2020.

[64] L. Shin, F.L. Bowling, D.G. Armstrong, and A.J. Boulton, "Saving the diabetic foot during the COVID-19 pandemic: a tale of two cities," *Diabetes Care*, vol. 43, no. 8, pp. 1704–1709, 2020.

[65] S.R. Jain, Y Sui, CH Ng, Z.X. Chen, L.H. Goh, and S. Shorey, "Patients' and healthcare professionals' perspectives towards technology-assisted diabetes self-management education. A qualitative systematic review," *PLOS one*, vol. 15, no.8, pp. 1–20, 2020.

Chapter 5

A semi-definite programming-based design of a robust depth control for a submersible autonomous robot through state feedback control

Vadapalli Siddhartha and Subhasish Mahapatra
VIT-AP University, Near Vijayawada, India

5.1 INTRODUCTION

The deep exploration of oceans/seas by human beings is a challenging task. Control scientists in various continents have invented robots that can replace humans going deep into the oceans for exploration. Many researchers have designed control algorithms to maneuver SARs. Among the important applications of SARs that require their control at the desired depth to fulfill specific tasks are bathymetry mapping, oil, and natural gas detection, inspection, and defense applications. Over the years, many diving algorithms have been designed to address various problems encountered in the applications, as mentioned earlier. To control the depth of an autonomous underwater vehicle (AUV), the authors in Zhou *et al.* [1] made use of the dynamic sliding mode control method based on multiple-model switching laws (DSMC-MMSL). It consists of two parts, diving and pitch control, to avoid significant pitch angle variations. An active disturbance rejection controller (ADRC) is designed in the diving plane of an AUV, as reported in Shen [2]. Three subsystems are proposed for speed control, diving control, and steering control of the AUV. The controller uses the enhanced tracking differentiator and ADRC. The authors in Mousavian [3] addressed the issue of identification-based robust AUV motion control. The unknown system parameters are estimated using an adaptive parameter identifier, and the gains are further optimized using the particle swarm optimization (PSO) algorithm. Wu [4] describes the tracking of desired depth trajectories using a model-free reinforcement learning technique based on the approximation of neural networks and the deterministic policy gradient theorem. The authors in Ma [5] explored robust position tracking in the vertical plane of an underwater vehicle system. A Lyapunov-based backstepping controller is designed for the nonlinear coupled input system by combining neural networks with parameter estimation. A sliding mode controller to reduce chattering is designed for AUV in the discrete-time domain in Sarif [6]. Combining the reaching law approach and discrete terminal sliding mode control reduces the chattering effect due to the quasi-sliding mode (QSM). The adaptive depth tracking control for an AUV involving hydrodynamic

 DOI: 10.1201/9781032628059-5

uncertainty, dead zone, and saturation is explored in Yu [7]. The inherent input saturation is rejected by including the gradient protection approach. In Nerkar [8], the design of the super twisting algorithm is described as a sliding mode controller based on disturbance observers (STA-SMC) to address the uncertainty issues in an oceanic environment. As discussed in the earlier literature, many controllers were designed for AUV by addressing robustness and hydrodynamic uncertainty issues through DSMC-MMSL, ADRC, SMC, etc. However, a novel design is used in this work to address the robust behavior of the closed-loop system by choosing selected uncertain hydrodynamic parameters of an SAR system. Some earlier state-of-the-art polytopic systems in various applications are also discussed to ensure the effectiveness of the proposed control law.

In Xie [9], the authors investigated stability analysis for the Roesser-type two-dimensional state-space digital filter with polytopic-type uncertainty. An efficient slack matrix variable approach is developed to examine the stability analysis. The authors in Al-Jiboory [10] discussed the robust input covariance constraint (ICC) control problem through polytopic uncertainty. A convex optimization with linear matrix inequality (LMI) approach is used to solve the ICC control problem. Morais [11] explores a digital redesign methodology to discuss the difficulties of H_2 and H_∞ discrete time stabilizing controllers. These techniques consider a predetermined continuous-time controller to determine a stabilized digital gain that deduces performance criteria to minimize the difference between continuous-time and discrete-time state trajectories. A nonlinear H_∞ control is designed in Mahapatra [12–14] to control the AUV in diving and steering planes. Taylor's series approach and nonlinear matrix inequality approach are used in the diving and steering control of AUV. An AUV in the vertical plane used the nonlinear H_∞ control design in Mahapatra [15, 16] by analyzing the state and output feedback control schemes. An estimator design for the periodic neural networks through polytopic uncertain connection weight matrices is addressed in Rao [17]. The sensor nonlinearities which occur randomly are characterized by employing Bernoulli processes. The estimator performance is improved through a Lyapunov function. Polytopic and time-inhomogeneous-based Markov jump linear systems are exploited in Lun [18]. The authors considered a transition probability matrix arbitrary in a polytopic set of stochastic matrices and varied with time. Coupled algebraic Riccati equations are used to obtain the optimal controller. In Dinh [19], delayed fractional-order linear systems through convex polytopic uncertainties are used to investigate robust stability and stabilization problems. The fractional Razumikhin stability theorem is used in terms of LMIs to derive sufficient conditions for the problems. Linear systems that include polytopic and additive uncertainties are explored in Jhu [20], by considering the design of dynamic output feedback model predictive control (DOFMPC). The formation of the bilinear matrix inequality problem is discussed. A new convex stability condition for a relaxation formula and linear matrix inequalities is proposed to trace the optimization problem. The LMI-based adequate conditions for a generalized robust state feedback control allow for solving the linear continuous-time polytopic systems problem, as addressed in Hu [21]. The suggested new LMI conditions are compared to existing parameterized LMI-based conditions. To choose uncertain parameters effectively, this work demonstrates the development of a robust and stable optimal control law by considering a polytopic SAR system. Considering the uncertainty of the selected hydrodynamic parameters of the SAR, the optimal control problem is solved using the LMI approach. The YALMIP tool in a MATLAB®/Simulink environment is used to model the control problem, discussed in Zhang [22].

The following are the main contributions of this chapter:

- The LMI approach is used to construct a state feedback strategy for an SAR that dives in the vertical plane.
- A new efficient method of selecting the hydrodynamic parameters for a polytopic SAR system is proposed to construct a robust optimal control.
- Different ranges of uncertain hydrodynamics parameters are taken into account to demonstrate the effectiveness of the suggested control algorithm in ensuring robust behavior.

The chapter is organized as follows: Section 5.2 addresses the nonlinear form of the SAR along with the problem statement for modeling the SAR in a diving plane. In Section 5.3, the design of the SAR algorithm for robust optimal control is presented. Simulation results for depth tracking under various uncertain conditions are discussed in Section 5.4. The final section, Section 5.5, highlights the conclusion and discusses the significance of the control algorithm. The organization of the chapter in flow chart form is shown in Figure 5.1.

5.2 MODELING OF SAR IN DEPTH PLANE

The mathematical modeling of SAR in depth plane is provided by considering the kinematics and dynamics of the SAR system. Figure 5.2 shows the schematic diagram of SAR based on the North-East Down (NED) frame. It describes the various motions of SAR in terms of kinematics and dynamics.

The various motions of SAR in the depth plane are as follows:

Surge Motion

$$m\dot{u}_g = C_{X_{uu}}u_g^2 + C_{X_{ww}}w_g^2 + C_{X_{qq}}q_g^2 + u_g^2 C_{X_{\delta s \delta s}}\delta_s^2 + C_{X_{\dot{u}}}\dot{u}_g + T_g \tag{5.1}$$

$$\dot{x}_g = u_g \cos\theta_g + w_g \sin\theta_g \tag{5.2}$$

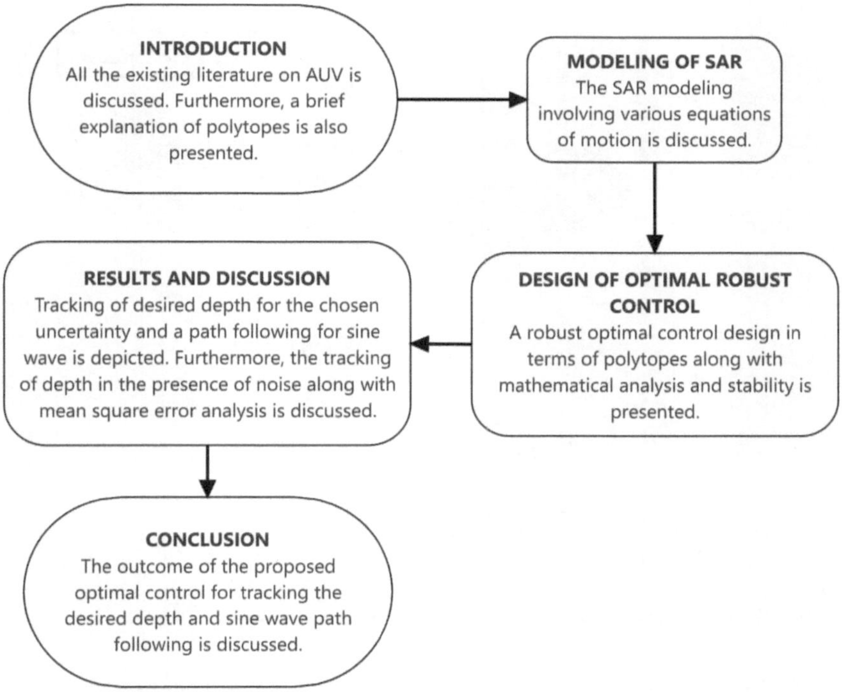

Figure 5.1 Flowchart of this chapter.

Heave Motion:

$$m(\dot{w}_g - u_g q_g) = (W - B)\cos\theta_g + C_{Z_w} u_g w_g + C_{Z_q} u_g q_g + C_{Z_{\delta s}} u_{g^2} \delta_s + C_{Z_{\dot{w}}} \dot{w}_g \tag{5.3}$$

$$\dot{z}_g = -u_g \sin\theta_g + w_g \cos\theta_g \tag{5.4}$$

Pitch Motion:

$$I_y \dot{q}_g = z_B B \sin\theta_g + C_{M_w} u_g w_g + C_{M_q} u_g q_g + C_{M_{\delta s}} u_g^2 \delta_s + C_{M_{\dot{q}}} \dot{q}_g \tag{5.5}$$

$$\dot{\theta}_g = q_g \tag{5.6}$$

The values of hydrodynamic parameters $C_{(*)}$ are considered from Silvestre [23]. The nonlinear structure of SAR can be represented by considering a constant forward velocity, i.e., $\dot{u}_g = 0$. Referring to Figure 5.2, the state vector is considered as

$$x_{g_j} = \left[w_g, q_g, z_g, \theta_g\right]^T = \left[x_{g,1}, x_{g,2}, x_{g,3}, x_{g,4}\right]^T. \tag{5.7}$$

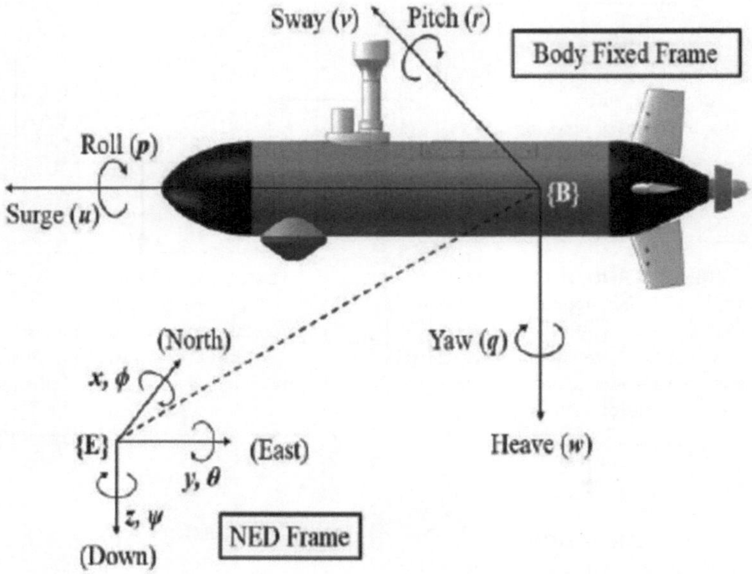

Figure 5.2 Structure of SAR showing the reference frames.

The nonlinear structure of SAR in the diving plane is considered as

$$x_{g,j} = f_g\left(x_{g,j}\right) + g_u\left(x_{g,j}\right)u_{g,j}. \tag{5.8}$$

where

$$f_g\left(x_{g,j}\right) = \left[f_{g1}\left(x_{g,j}\right), f_{g2}\left(x_{g,j}\right), f_{g3}\left(x_{g,j}\right), f_{g4}\left(x_{g,j}\right)\right]^T$$

$$f_{g1}\left(x_{g,j}\right) = C_w\left[(W - B)\cos x_{g,4} + C_{Z_w}u_g x_{g,1} + C_{Z_q}u_g x_{g,2} + m u_g x_{g,3}\right]$$

$$f_{g2}\left(x_{g,j}\right) = C_q\left[Z_B B\sin x_{g,4} + C_{M_w}u_g x_{g,1} + C_{M_q}u_g x_{g,2}\right]$$

$$f_{g3}\left(x_{g,j}\right) = -u_g \sin x_{g,4} + x_{g,1}\cos x_{g,4}$$

$$f_{g4}\left(x_{g,j}\right) = x_{g,2}$$

$$g_u\left(x_{g,j}\right) = \begin{bmatrix} C_w C_{Z_{\delta s}} \\ C_q C_{M_{\delta s}} \\ 0 \\ 0 \end{bmatrix}$$

Referring to Equation (5.8), $u_{g,\,j} = \delta_s$. Further, $C_w = \left(m - C_{z_{\dot{w}}}\right)^{-1}$ and $C_q = \left(m - C_{m_{\dot{q}}}\right)^{-1}$. This work focuses on tracking the desired depth. Hence, linear velocity in the x direction x_g in an earth-fixed frame is ignored. A generalized linear system is shown below to design the control algorithm for the nonlinear SAR model.

$$\begin{aligned} \dot{x} &= \mathcal{H}x + \mathcal{G}u \\ y &= \mathcal{C}x \end{aligned} \tag{5.9}$$

where $\mathcal{H} \in R^{nxn}$, $\mathcal{G} \in R^{nxm}$, $\mathcal{C} \in R^{rxn}$.

5.2.1 Problem statement

The state feedback optimal control algorithm is developed from consideration of Equation (5.9). The following objective needs to be achieved.

Desired depth z_D can be achieved by minimizing the depth error, i.e.,

$$\lim_{t \to 0} z_e(t) = 0 \tag{5.10}$$

where $z_e = z_g - z_D$

5.3 DESIGN OF ROBUST OPTIMAL CONTROL

This section exploits the algorithm used for robust optimal control of the diving motion of SAR shown in Figure 5.3. A robust optimal controller employing a polytopic SAR system is proposed.

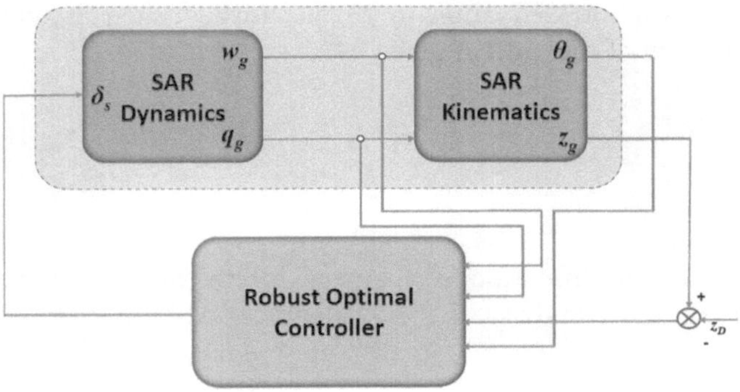

Figure 5.3 Robust LMI state feedback control for diving plane.

5.3.1 Robust optimal state feedback for polytopic SAR system

The robust behavior of the closed-loop SAR system based on LMI is studied in this section.

5.3.1.1 Preliminaries: LMI-based LQR controller

An LMI-based control algorithm is obtained by considering a linearized system of SAR, as reported in Equation (5.9). A constant gain matrix $\mathcal{K}_{\mathcal{G}} \in \mathbb{R}^{nxn}$ is formulated such that the closed-loop system is asymptotically stable for

$$u = -\mathcal{K}_{\mathcal{G}} x \tag{5.11}$$

Problem 1:

The closed-loop system is described using the linearized system Equation (5.9) and the control law Equation (5.11), which can be given as

$$\begin{aligned} \dot{x} &= \mathcal{H}x + \mathcal{G}\left(-\mathcal{K}_{\mathcal{G}} x\right) \\ \dot{x} &= \left(\mathcal{H} - \mathcal{G}\mathcal{K}_{\mathcal{G}}\right) x \end{aligned} \tag{5.12}$$

Theorem 1:

The solution to Problem 1 exists such that the necessary and sufficient condition depends on the existence of matrices $\mathcal{P}$ and $\mathcal{R}$ where $\mathcal{P} = \mathcal{P}^T$, $\mathcal{P} \in \mathbb{R}^{nxn}$,$R \in \mathbb{R}^{nxn}$ such that

$$\begin{aligned} &-\mathcal{H}^T\mathcal{P} - \mathcal{P}\mathcal{H} + \mathcal{P}\mathcal{G}\left(\mathcal{R}^{-T} + \mathcal{R}^{-1}\right)\mathcal{G}^T\mathcal{P} \leq 0 \\ &P \geq 0 \end{aligned} \tag{5.13}$$

If the above condition equation (5.13) is satisfied, then a state feedback matrix that solves Problem 1 is given by

$$\mathcal{K} = \mathcal{R}^{-1}\mathcal{G}^T\mathcal{P} \tag{5.14}$$

Proof:

The stability of the closed-loop system is exhibited by considering where $\mathcal{P}$ is a positive definite matrix, i.e., $\mathcal{P} > 0$. The closed-loop system (5.12) has to be stable, then the energy should decrease with time, i.e., $\dot{V} \leq 0$. Then, $\dot{V}$ is presented as

$$\dot{V}(x) = \dot{x}^T \mathcal{P} x + x^T \mathcal{P} \dot{x} \tag{5.15}$$

From equation (5.9) and the basic control law equation, i.e., $u = -\mathcal{K}x$

$$\dot{V}(x) = \left[(\mathcal{H} - \mathcal{G}\mathcal{K})x\right]^T \mathcal{P}x + x^T \mathcal{P}\left[(\mathcal{H} - \mathcal{G}\mathcal{K})x\right] \tag{5.16}$$

From equation (5.14), the above expression can be written as

$$\begin{aligned} \dot{V}(x) &= x^T \left[\left(\mathcal{H}^T - \left(\mathcal{R}^{-1}\mathcal{G}^T\mathcal{P}\right)^T \mathcal{G}^T\right)\mathcal{P} + \mathcal{P}\left(\mathcal{H} - \mathcal{G}\left(\mathcal{R}^{-1}\mathcal{G}^T\mathcal{P}\right)\right]x \\ \dot{V}(x) &= x^T \left[\left(\mathcal{H}^T\mathcal{P} + \mathcal{P}\mathcal{H} - \mathcal{P}\mathcal{G}\mathcal{R}^{-T}\mathcal{G}^T\mathcal{P} - \mathcal{P}\mathcal{G}\mathcal{R}^{-1}\mathcal{G}^T\mathcal{P}\right)\right]x \end{aligned} \tag{5.17}$$

Rearranging the above equation,

$$\dot{V}(x) = -x^T \left[\mathcal{P}\mathcal{G}\left(\mathcal{R}^{-T} + \mathcal{R}^{-1}\right)\mathcal{G}^T\mathcal{P} - H^T\mathcal{P} - \mathcal{P}\mathcal{H}\right]x$$

$$\dot{V}(x) = -x^T \left[\mathcal{P}\mathcal{G}\left(\left(\left(\mathcal{R}^{-T} + \mathcal{R}^{-1}\right)\right)^{-1}\right)^{-1} \mathcal{G}^T\mathcal{P} - \mathcal{H}^T\mathcal{P} - \mathcal{P}\mathcal{H}\right]x$$

$$\dot{V}(x) = -x^T \left[\mathcal{P}\mathcal{G}\left(Y^{-1}\right)^{-1} \mathcal{G}^T\mathcal{P} - \mathcal{G}^T\mathcal{P} - \mathcal{P}\mathcal{G}\right]x$$

where $Y = ((\mathcal{R}^{-T} + \mathcal{R}^{-1}))^{-1}$

Now, the following condition must be satisfied for the system to be stable $\dot{V} \leq 0$

$$\begin{aligned} &-\mathcal{H}^T\mathcal{P} - \mathcal{P}\mathcal{H} - (-\mathcal{P}\mathcal{G})Y^{-1}\mathcal{G}^T\mathcal{P} \leq 0 \\ &-\mathcal{H}^T\mathcal{P} - \mathcal{P}\mathcal{H} + \mathcal{P}\mathcal{G}\left(\mathcal{R}^{-T} + \mathcal{R}^{-1}\right)^{-1} \mathcal{G}^T\mathcal{P} \leq 0 \end{aligned} \tag{5.18}$$

The above condition is represented in LMI form, as shown below.

$$\begin{pmatrix} -\mathcal{H}^T\mathcal{P} - \mathcal{P}\mathcal{H} & -\mathcal{P}\mathcal{G} \\ \mathcal{G}^T\mathcal{P} & Y \end{pmatrix} \leq 0$$

5.3.1.2 Robust LMI-based optimal controller

Considering the state space representation as discussed in Equation (5.9), a convex combination of polytopic vertices is given as

$$\dot{x}(t) = \sum_{j=1}^{n} \alpha_j (\mathcal{H}_j x + \mathcal{G}_j u) = \mathcal{H}(\alpha)x + \mathcal{G}(\alpha)u \tag{5.19}$$

where n represent the polytope vertices.

The parameters α_j, $j = 1, 2, \ldots n$ are taken as constant and unknown real numbers. They belong to unitary simplex U_G, which is given by

$$U_G = \sum_{j=1}^{n} \alpha_j = 1, \alpha_j \geq 0, j = 1, 2, \ldots n \tag{5.20}$$

Lemma 1:

For the closed loop system to be asymptotically stable, a constant gain matrix $\mathcal{K}_{\mathcal{GR}} \in \mathcal{R}^{nxn}$ such that

$$u = -\mathcal{K}_{\mathcal{GR}} x \tag{5.21}$$

Problem 2:

Based on an uncertain system (5.19) and robust control law equation (5.21), the closed loop system is represented as

$$\begin{aligned} \dot{x} &= \mathcal{H}(\alpha)x + \mathcal{G}(\alpha)(-\mathcal{K}_{GR} x) \\ \dot{x} &= (\mathcal{H}(\alpha) - \mathcal{G}(\alpha)\mathcal{K}_{GR})x \end{aligned} \tag{5.22}$$

A similar procedure is followed to solve Problem 2 by defining the following theorem.

Theorem 2:

The solution to Problem 2 exists such that the necessary and sufficient condition depends on the existence of matrices $\mathcal{P} = \mathcal{P}^T$ and $\mathcal{R}$ where $\mathcal{P} \in \mathcal{R}^{nxn}$, $\mathcal{R} \in \mathcal{R}^{nxn}$ such that

$$-\mathcal{H}(\alpha)^T \mathcal{P} - \mathcal{P}\mathcal{H}(\alpha) + \mathcal{P}\mathcal{G}(\alpha)\left(\mathcal{R}^{-T} + \mathcal{R}^{-1}\right)\mathcal{G}(\alpha)^T \mathcal{P} \leq 0$$
$$\mathcal{P} \geq 0 \tag{5.23}$$

When equation (5.23) is satisfied, then a state feedback matrix that solves Problem 2 is given by

$$\mathcal{K}_{\mathcal{G}\mathcal{R}} = \mathcal{R}^{-1}\mathcal{G}^T\mathcal{P} \tag{5.24}$$

Proof:

To achieve the stability condition through LMI, it is intended to consider the LQR problem concerning Algebraic Riccati Equation (ARE). This problem is solved using LMIs. A similar procedure is followed to demonstrate the state feedback control in terms of LMIs, as reported in Theorem 1. Thus, the above condition (5.23) is represented in LMI form, as shown below.

$$\begin{pmatrix} -\mathcal{H}(\alpha)^T \mathcal{P} - \mathcal{P}\mathcal{H}(\alpha) & -\mathcal{P}\mathcal{G}(\alpha) \\ \mathcal{G}(\alpha)^T \mathcal{P} & Y \end{pmatrix} \leq 0 \tag{5.25}$$

where $Y = ((\mathcal{R}^{-T} + \mathcal{R}^{-1}))^{-1}$.

For the system to be stable, based on Theorem 2, $\dot{V} \leq 0$ which indicates that $\mathcal{P}\mathcal{G}(\alpha)\left(Y^{-1}\right)^{-1}\mathcal{G}(\alpha)^T \mathcal{P} - \left(\mathcal{H}(\alpha)^T \mathcal{P} + \mathcal{P}\mathcal{H}(\alpha)\right)$ it must be positive definite. From the above, it is inferred that the condition (5.23) exists for the system (5.19) where $\mathcal{H}(\alpha) = \alpha_1\mathcal{H}_1 + \alpha_2\mathcal{H}_2 + ... + \alpha_n\mathcal{H}_n$, $\mathcal{G}(\alpha) = \alpha_1\mathcal{G}_1 + \alpha_2\mathcal{G}_2 + ...$ $+ \alpha_n\mathcal{G}_n$. Finally, concerning Theorem 2, the matrix $\mathcal{P} = \mathcal{P}^T$ exists such that equation (5.23) is sufficient for solving Problem 2. Furthermore, the above condition (5.25) is represented by considering the polytopic vertices as

$$\begin{pmatrix} -\mathcal{H}_i^T\mathcal{P} - \mathcal{P}\mathcal{H}_i & -\mathcal{P}\mathcal{G}_i \\ \mathcal{G}_i^T\mathcal{P} & Y \end{pmatrix} \leq 0 \tag{5.26}$$

where $i = 1, 2, 3,...n$ are the number of polytopic vertices.

5.4 RESULTS AND DISCUSSION

The numerical analysis and results of the closed-loop system are presented in this section. The equilibrium points for the SAR nonlinear structure described in Equation (5.9) are considered as $[w_g, q_g, z_g, \theta_g] = [0,0,0,0]$.

Table 5.1 Range of $C_{Z_{\dot{w}}}$ and $C_{M_{\dot{q}}}$ based on uncertainty percentage

Hydrodynamic parameters	*Uncertainty percentage*	*Range obtained*
$C_{Z_{\dot{w}}}$	±10%	$-5079.25 \le C_{Z_{\dot{w}}} \le -4155.76$
$C_{M_{\dot{q}}}$	±10%	$-1861.64 \le C_{M_{\dot{q}}} -1523.16$

Thus, the state matrix, input matrix, and output matrix are obtained, as shown below.

$$\mathcal{H} = \begin{pmatrix} -0.876 & 1.416 & 0 & 0 \\ 1.245 & -3.716 & 0 & -0.264 \\ 1.000 & 0 & 0 & -2.000 \\ 0 & 1.000 & 0 & 0 \end{pmatrix}, \mathcal{G} = \begin{pmatrix} -0.402 \\ -0.933 \\ 0 \\ 0 \end{pmatrix}, \mathcal{C} = \begin{pmatrix} 0 & 0 & 1 & 0 \end{pmatrix}$$

The developed algorithm is simulated by considering a constant longitudinal velocity of $u_g = 2$ m/s. From the nonlinear structure of SAR, as reported in equation (5.9), it is observed that the hydrodynamic parameters $C_{Z_{\dot{w}}}$ and $C_{M_{\dot{q}}}$ associated with heave and pitch motion, respectively, affect each term of the dynamics equations. Hence, the uncertain polytopic SAR system is obtained by considering ±10% change in $C_{Z_{\dot{w}}}$ and $C_{M_{\dot{q}}}$ parameters. Table 5.1 depicts the ranges obtained for the parameters for ±10% range of uncertainties.

The proposed controller is simulated in the presence of the uncertainties for the SAR model and the chosen parameter uncertainty, ±10% as obtained in Table 5.1. Here, ±10% parameter uncertainties are considered for both hydrodynamic parameters $C_{Z_{\dot{w}}}$ and $C_{M_{\dot{q}}}$ to generate the polytope vertices.

$$\mathcal{H}_1 = \begin{pmatrix} -0.820 & 1.327 & 0 & 0 \\ 1.310 & -3.911 & 0 & -0.278 \\ 1.000 & 0 & 0 & -2.000 \\ 0 & 1.000 & 0 & 0 \end{pmatrix},$$

$$\mathcal{H}_2 = \begin{pmatrix} -0.820 & 1.327 & 0 & 0 \\ 1.185 & -3.539 & 0 & -0.252 \\ 1.000 & 0 & 0 & -2.000 \\ 0 & 1.000 & 0 & 0 \end{pmatrix}$$

$$\mathcal{H}_3 = \begin{pmatrix} -0.939 & 1.518 & 0 & 0 \\ 1.310 & -3.911 & 0 & -0.278 \\ 1.000 & 0 & 0 & -2.000 \\ 0 & 1.000 & 0 & 0 \end{pmatrix},$$

$$\mathcal{H}_4 = \begin{pmatrix} -0.939 & 1.518 & 0 & 0 \\ 1.185 & -3.539 & 0 & -0.252 \\ 1.000 & 0 & 0 & -2.000 \\ 0 & 1.000 & 0 & 0 \end{pmatrix}$$

and

$$\mathcal{G}_1 = \begin{pmatrix} -0.377 \\ -0.982 \\ 0 \\ 0 \end{pmatrix}, \mathcal{G}_2 = \begin{pmatrix} -0.377 \\ -0.888 \\ 0 \\ 0 \end{pmatrix}, \mathcal{G}_3 = \begin{pmatrix} -0.431 \\ -0.982 \\ 0 \\ 0 \end{pmatrix}, \mathcal{G}_4 = \begin{pmatrix} -0.377 \\ -0.982 \\ 0 \\ 0 \end{pmatrix}$$

Using semi-definite variable programming, the positive definite matrix $\mathcal{P}, \mathcal{Q}$, and $\mathcal{R}$ is obtained by solving Problem 2 using the YALMIP tool and given as

$$\mathcal{P} = 1 * 10^{-9} * \begin{pmatrix} 0.2932 & -0.2546 & -0.0284 & -0.1730 \\ -0.2546 & 0.3213 & 0.0172 & 0.1175 \\ -0.0284 & 0.0172 & 0.0085 & 0.0323 \\ -0.1730 & 0.1175 & 0.0323 & 0.1522 \end{pmatrix}$$

$$\mathcal{Q} = 1 * 10^{3} * \begin{pmatrix} 128.2535 & 0 & 0 & 0 \\ 0 & 128.2535 & 0 & 0 \\ 0 & 0 & 128.2535 & 0 \\ 0 & 0 & 0 & 128.2535 \end{pmatrix},$$

$$\mathcal{R} = \left(5.3877 \times 10^{-12}\right)$$

Once the $\mathcal{P}$ matrix is obtained, the gain matrix $\mathcal{K}_{GR}$ can be generated using Equation (5.24), as shown below.

$$\mathcal{K}_{GR} = \left[22.1850 - 36.6106 - 0.8541 - 7.4099\right]$$

The tracking of the desired depth using a robust LQR controller is shown in Figure 5.4a. From the figure, it is observed that the desired depth monitoring is effective despite uncertainty. Figure 5.4b ensures the smooth variation of the control signal in the presence of uncertainty. Subsequently, Figure 5.4c and 5.4d represent the variation of heave velocity and the pitch rate of SAR, respectively, in the presence of ±10% uncertainty.

Limitations of PID Control

- The tuning is limited to only three parameters, which will be challenging to generate the gains.
- The algorithms involved in tuning the parameters are time-consuming.
- The existence of uncertainties and disturbances may affect the PID control design if the tuning of parameters is not accurate.

However, the proposed algorithm uses the YALMIP tool to generate the gains. Subsequently, a path following control in the vertical plane for a sine wave is depicted in Figure 5.5a. As shown in the figure, a sine wave of 4m depth is considered a desired path and is tracked accordingly. Figure 5.5b shows the control input during path tracking. Similarly, the pitch rate is shown in Figure 5.5c, which settles to zero faster. The error signal for tracking the sine wave is observed in Figure 5.5d, and it is seen that the error signal is almost zero for the proposed control. Furthermore, Table 5.2 displays the mean square error (MSE) analysis for the proposed control and the FOPID control [24]. The table shows that the MSE value for the proposed optimal control is less than for the FOPID control. This indicates that the desired sine path is accurately tracked for the case of the proposed optimal control.

Effect of Noise in Tracking the Desired Depth

An external band-limited white noise with a noise power of 0.001 is added to the model to verify its effectiveness in tracking the desired depth. Different time instants are used to plot the desired depth, as shown in Figure 5.6a, and illustrate the depth tracking in the presence of noise. Despite the noise, the desired depth is tracked except for a slight overshoot and discrepancies. Figure 5.6b describes the energy consumed in the presence of noise. Similarly, the heave velocity and pitch rate are illustrated in Figure 5.6c and 5.6d, respectively.

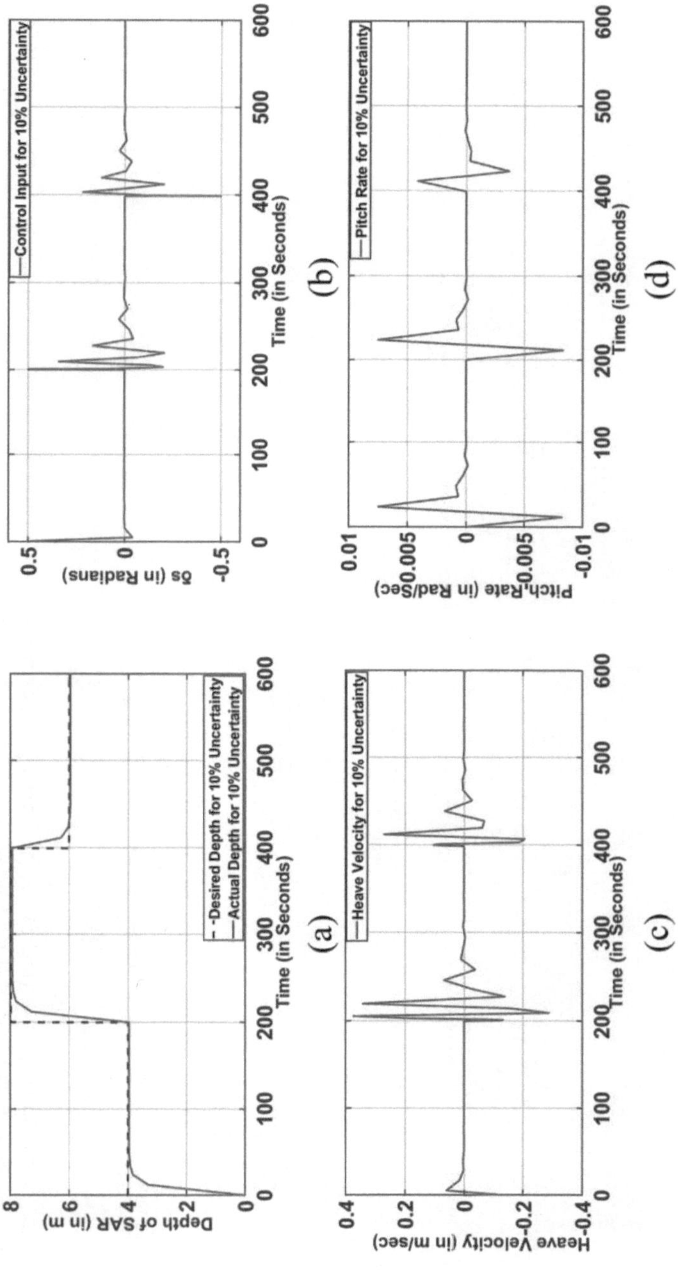

Figure 5.4 Control of depth for ±10% uncertainty: (a) Tracking of desired depth, (b) Control input, (c) heave velocity, (d) pitch rate.

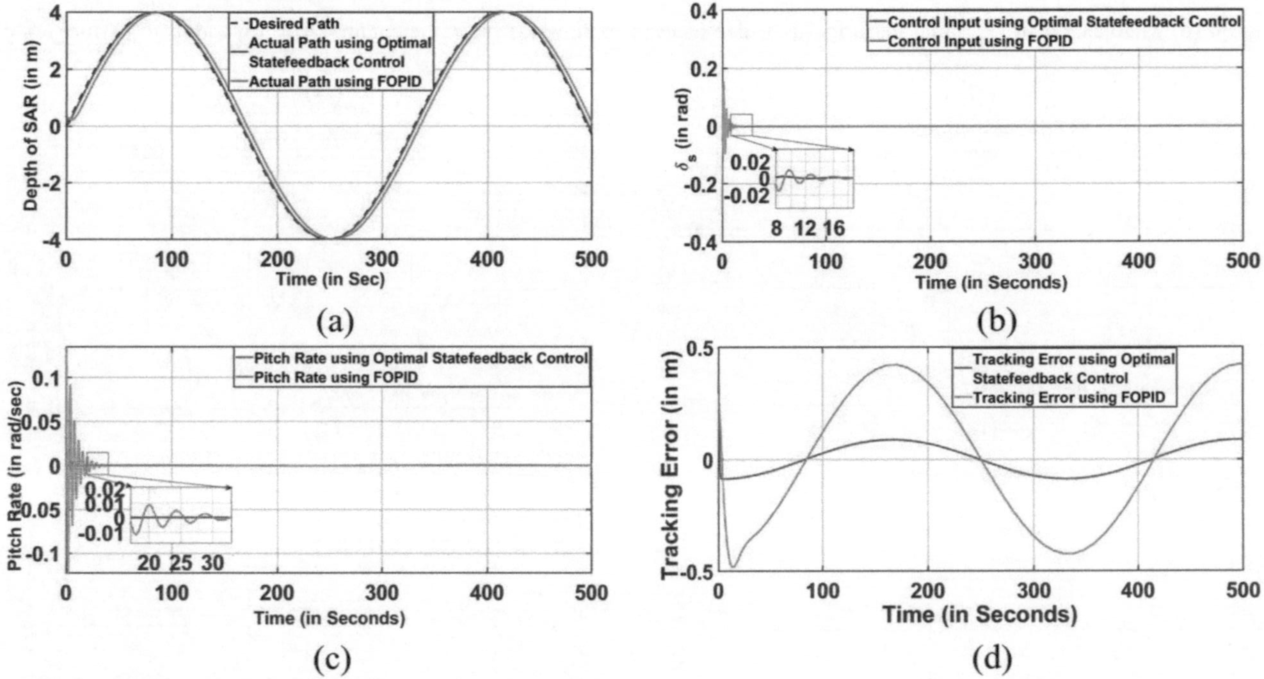

Figure 5.5 Path following in vertical plane: (a) Tracking of desired sine wave, (b) control input for a sine wave, (c) pitch rate for a sine wave, (d) error signal for a sine wave.

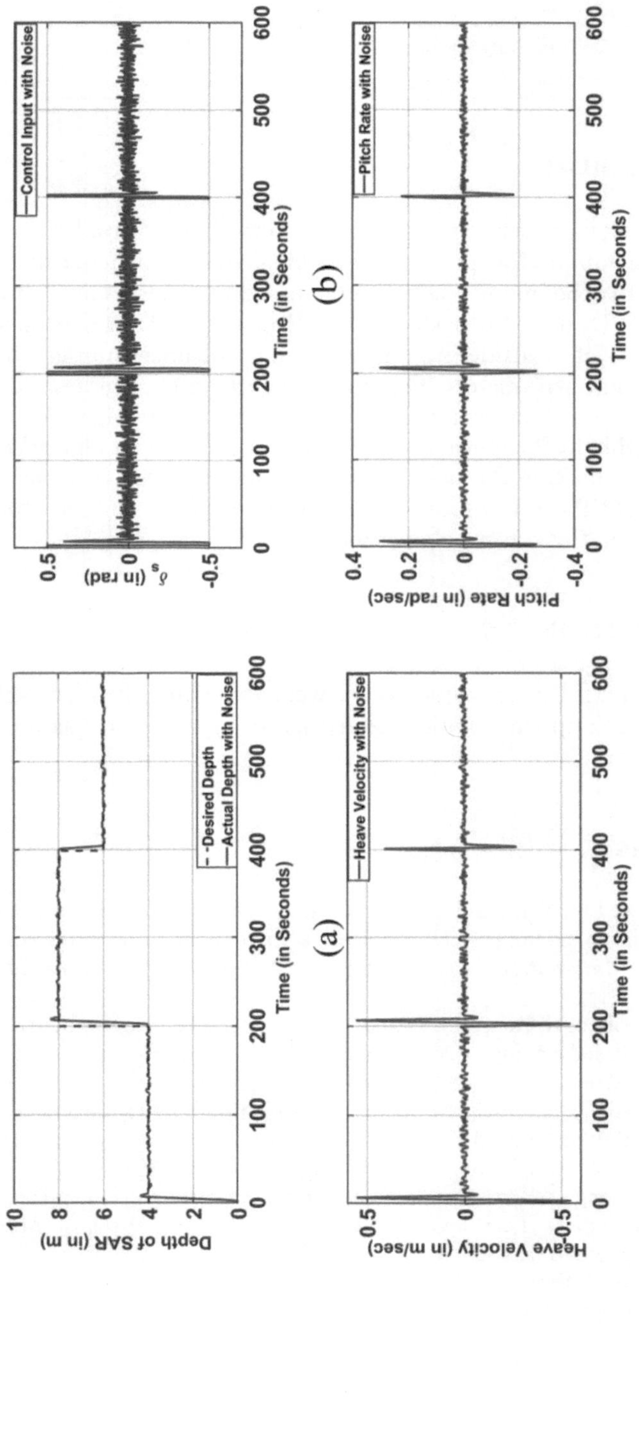

Figure 5.6 Tracking of depth in the presence of noise: (a) Tracking of desired depth, (b) control input, (c) heave velocity, (d) pitch rate.

Table 5.2 MSE between the proposed control and FOPID control

Name of control	*MSE for sine path*
Robust Optimal Statefeedback Control	0.0390
FOPID Control	0.1133

5.5 CONCLUSION

This work develops a novel state feedback control algorithm based on the robust LQR technique for an uncertain SAR system. Further, a distinctive collection of hydrodynamic parameters introduces uncertainty into the system. The polytopic SAR system is developed to design a robust control algorithm while considering multiple ranges of uncertainties on selective hydrodynamic parameters. The controller can track the desired depth despite the presence of uncertainties. Besides, a path-following task has been achieved to highlight the efficacious behavior of the control algorithm. The effectiveness of tracking the desired depth in the presence of noise is also discussed. YALMIP optimization tool in MATLAB®/Simulink environment is used to develop the control algorithm.

ACKNOWLEDGMENTS

We are very thankful to anonymous reviewers and editors for their valuable suggestions and comments, which helped us to improve the quality of the chapter.

REFERENCES

[1] H. Zhou, K. Liu, Y. Li, and S. Ren, "Dynamic sliding mode control based on multi-model switching laws for the depth control of an autonomous underwater vehicle," *International Journal of Advanced Robotic Systems*, vol.12, no. 7, pp. 1–10, 2015.

[2] Y. Shen, K. Shao, W. Ren, and Y. Liu, "Diving control of autonomous underwater vehicle based on improved active disturbance rejection control approach," *Neurocomputing*, vol. 173, no. 3, pp. 1377–1385, 2016.

[3] S. H. Mousavian, and H. R. Koofigar, "Identification-based robust motion control of an AUV: optimized by particle swarm optimization algorithm," *Journal of Intelligent & Robotic Systems*, vol. 85, pp. 331–352, 2016.

[4] H. Wu, S. Song, K. You and C. Wu, "Depth Control of Model-Free AUVs via Reinforcement Learning," *IEEE Transactions on Systems, Man, and Cybernetics: Systems*, vol. 49, no. 12, pp. 2499–2510, 2019.

[5] Z. Ma, J. Hu, J. Feng and A. Liu, "Diving Adaptive Position Tracking Control for Underwater Vehicles," in *IEEE Access*, vol. 7, pp. 24602–24610, 2019.

[6] N.M. Sarif, R. Ngadengon, H.A. Kadir, and M.H. Jalil, "Terminal sliding mode control on autonomous underwater vehicle in diving motion control," *Indonesian Journal of Electrical Engineering and Computer Science*, vol. 20, no. 2, pp. 798–804, 2020.

[7] C. Yu, Y. Zhong, L. Lian, X. Xiang, "An experimental study of adaptive bounded depth control for underwater vehicles subject to thruster's dead-zone and saturation," *Applied Ocean Research*, vol. 117, pp. 102947(1-8), 2021.

[8] S.S. Nerkar, P.S. Londhe, and B.M. Patre, "Design of super twisting disturbance observer based control for autonomous underwater vehicle," *International Journal of Dynamics and Control*, vol. 10, no. 1, pp. 306–322, 2022.

[9] X.P. Xie, S.L. Hu, Q.Y. Sun, "Less conservative global asymptotic stability of 2-D state-space digital filter described by Roesser model with polytopic-type uncertainty," *Signal Processing*, vol. 115, pp. 157–163, 2015.

[10] A.K. Al-Jiboory, G. Zhu, "Robust input covariance constraint control for uncertain polytopic systems," *Asian Journal of Control*, vol. 18, no. 4, pp. 1489–1500, 2016.

[11] C.F. Morais, M.F. Braga, E.S. Tognetti, R.C. Oliveira, and P.L. Peres, "H_2 and H_∞ digital redesign of analog controllers for continuous-time polytopic systems," *IFAC-PapersOnLine*, vol. 50, no. 1, pp. 6691–6696, 2017.

[12] S. Mahapatra, and B. Subudhi, "Design of a steering control law for an autonomous underwater vehicle using nonlinear H_∞ state feedback technique," *Nonlinear Dynamics*, vol. 90, no. 2, pp. 837–854, 2017.

[13] S. Mahapatra, and B. Subudhi, "Design and experimental realization of a backstepping nonlinear H_∞ control for an autonomous underwater vehicle using a nonlinear matrix inequality approach," *Transactions of the Institute of Measurement and Control*, vol. 40, no. 11, pp. 3390–3403, 2018.

[14] S. Mahapatra, and B. Subudhi, "Nonlinear matrix inequality approach based heading control for an autonomous underwater vehicle with experimental realization," *IFAC Journal of Systems and Control*, vol. 16, pp. 1–11, 2021.

[15] S. Mahapatra, B. Subudhi, R. Rout, and B.K. Kumar, "Nonlinear H_∞ control for an autonomous underwater vehicle in the vertical plane," *IFAC-PapersOnLine*, vol. 49, no. 1, pp. 391–395, 2016.

[16] S. Mahapatra, and B. Subudhi, "Nonlinear H_∞ state and output feedback control schemes for an autonomous underwater vehicle in the dive plane," *Transactions of the Institute of Measurement and Control*, vol. 40, no. 6, pp. 2024–2038, 2018.

[17] H.X. Rao, Y. Xu, B. Zhang, and D. Yao, "Robust estimator design for periodic neural networks with polytopic uncertain weight matrices and randomly occurred sensor nonlinearities," *IET Control Theory & Applications*, vol. 12, no. 9, pp. 1299–1305, 2018.

[18] Y. Z. Lun, A. Abate and A. D'Innocenzo, "Linear quadratic regulation of polytopic time-inhomogeneous Markov jump linear systems," *18th European Control Conference (ECC)*, Naples, Italy, pp. 4094–4099, 2019.

[19] C.H. Dinh, V.T. Mai, and T.H. Duong, "New results on stability and stabilization of delayed Caputo fractional order systems with convex polytopic uncertainties," *Journal of Systems Science and Complexity*, vol. 33, no. 3, pp. 563–583, 2020.

[20] J. Hu, "Dynamic Output Feedback MPC of Polytopic Uncertain Systems: Efficient LMI Conditions," in *IEEE Transactions on Circuits and Systems II: Express Briefs*, vol. 68, no. 7, pp. 2568–2572, 2021.

[21] C. Hu, and I.M. Jaimoukha, "New iterative linear matrix inequality based procedure for H_2 and H_∞ state feedback control of continuous-time polytopic systems," *International Journal of Robust and Nonlinear Control*, vol. 31, no. 1, pp. 51–68, 2021.
[22] J. Zhang, A.K. Swain, and S.K. Nguang, *Robust observer-based fault diagnosis for nonlinear systems using MATLAB®*, Berlin: Springer, 2016.
[23] C. Silvestre, and A. Pascoal, "Control of the INFANTE AUV using gain scheduled static output feedback," *Control Engineering Practice*, vol. 12, no. 12, pp. 1501–1509, 2004.
[24] L. Liu, L. Zhang, G. Pan, and S. Zhang, "Robust yaw control of autonomous underwater vehicle based on fractional order PID controller," *Ocean Engineering*, vol. 257, pp. 111493, 2022.

Chapter 6

Embedded system with in-memory compute neuromorphic accelerator for multiple applications

Afroz Fatima and Abhijit Pethe
Birla Institute of Technology and Science Pilani, Pilani, India

6.1 INTRODUCTION

An embedded system, as the name suggests, is the combination of a processor (single/multi-core), memory and several peripheral devices that define/execute a task within a larger entity/system, often electronic or mechanical. Recent advances in the area of neuromorphic embedded systems have introduced many parallel core architectures. A recently reported research example is the spiking neural network architecture, often referred to as SpiNNaker, designed with 57,600 processing nodes, each with 18 ARM968 processors. Total cores were 1,036,800 with 7 terabytes of memory [1, 2]. Technische Universität München developed a robotic platform called SpOmnibot [3, 4] that hosts the integrated SpiNNaker neuromorphic computing board and enables nearly one million simple neurons to be simulated in real time. It also hosts embedded dynamic vision sensors (eDVSs) and has a 9 DOF (degrees of freedom) inertial measurement unit. One of its use cases is autonomous simultaneous localization mapping.

To meet the real-time processing requirements for the large amount of sensory data arriving from various on-board sources of an embedded system, an efficient neuromorphic accelerator is needed. In-memory computation plays a significant role in building these high-performance neuro-based accelerators [5, 6]. There has been a tremendous change in the design process of in-memory computing accelerators across analog, digital and mixed-mode signal architectures using memristive technology [7, 8]. Creating a universal design that deals with industry-specific applications has motivated novel designs that use potent technology to improve memory bandwidth and memory latency and to provide an optimized solution to higher leakage currents, and that deal with sneak paths within crossbar arrays, high power consumption, high operating voltage, lower accelerator response speed, etc. [9, 10]. There have been many attempts to create application-specific accelerators using a combination of CPU (central processing unit), GPU (graphics processing unit), FPGA (field programmable gate array), ASIC (application-specific integrated circuits) and other efficient hardware methods, in particular

DOI: 10.1201/9781032628059-6

in the areas of artificial intelligence (AI) and machine learning (ML) [11]. Application design also involves decisions on tasks such as classification, regression, predictions and recommendation [11]. Creating a novel hardware-centric approach for industry-specific needs is always challenging and demands newer approaches. Among the popular ML approaches are back-propagation techniques and stochastic gradient-based approaches [12]. Design methodologies for different applications such as aerospace, big-data analytics, internet of things (IoT), computer vision, speech, language translation, and autonomous driving are discussed in [11, 13, 14] and are implemented using the novel memory technology (memristors) which to some extent has already captured the memory market [11]. This chapter proposes a novel design solution to optimize the power and speed of the analog accelerator for multiple applications using resistive memories. The proposed architecture has been designed, implemented and verified on ST Microelectronics 28nm FD-SOI (fully depleted silicon-on-insulator) technology which makes it power efficient at high computational speed. Combining memristive technology and FD-SOI technology makes our design much more efficient, optimized and impactful for a variety of applications.

The proposed architecture comprises a crossbar grid that has resistive memory devices at the intersection of each word line (WL) and bit line (BL), which helps in computing the input signal (or information) which is being propagated and also stores the computed output results to further process this signal to the next WL-BL in sequence. The resistive memories, in conjunction with NMOS (N-channel metal oxide semiconductor) transistors are used to efficiently switch the device on and off and also to process information accurately. The crossbar array structure computes the synaptic operation which is to store and process the information. The information obtained at the end of the synaptic (or crossbar) columns is linear in nature, unlike the natural neural activity where the nature of information is non-linear, computing complex information in a few milli-seconds and consuming a few watts of energy [15–17]. To map to the natural neural behavior in the analog domain, a non-linearity function is introduced in conjunction with the synapse grid so that the computed information from the crossbar and the non-linear function approximately mimics the neural activity of the brain [18]. Thus, the neural net has a compact pattern of neuron-synapse-neuron replicating the complex structure of the brain. This pattern of neuron-synapse-neuron (in biological perspective) is created from analog counterparts such as resistive crossbar memory devices along with activation circuits formulated as a core or processing element (PE) of a neuromorphic chip in addition to elements such as the AXI interconnects, routers, etc. that are required for the analog mixed-signal type of chip implementations [19–21].

The proposed accelerator uses crossbar arrays having resistive memories (RRAM) along with the swish activation function to introduce non-linearity for the denser neural net [22, 23]. The swish circuit used in this work [24, 25] is designed using five pairs of CMOS (complementary metal oxide semiconductor) transistors, two RRAM (resistive random access memory) devices and a multiplier circuit [26]. The swish circuit captures the wide range of characteristics between negative and positive voltage values and it has one-sided boundedness at zero which makes it better than ReLU (rectified linear unit) and sigmoid functions [23]. The design of analog architectures for inference and training has been demonstrated in [24, 27, 28], where forward-propagation and backward-propagation techniques are used to implement such neuro-memristive-based accelerators. This design has also been verified on different applications such as speech-based emotion recognition and classification to prove the significance of the architectures [29].

In this chapter, a standard neural architecture having deep nets that fits multiple applications is implemented. The following are this chapter's main contributions:

- A standard (one-stop) in-memory compute architecture having 32 input layer neurons, 40 hidden layer neurons, 10 output layer neurons has been implemented for training using resistive memories and swish function, the first of its kind on a 28nm FD-SOI process design kit (PDK).
- A standard (one-stop) in-memory compute architecture having 32 input layer neurons, 40 hidden layer neurons, and 10 output layer neurons has been implemented for inference using resistive memories and swish function, the first of its kind on a 28nm FD-SOI PDK.
- One-stop and individual architectural (inference and training) implementations on different industry-specific applications such as climate technology, social sciences, medical sciences, finance technology and gaming technology have been targeted for implementation using resistive memories and swish function, the first of its kind on a 28nm FD-SOI PDK.
- The inference, training and parametric analysis of resistive memory and swish function, together with the benefits of the one-stop neural accelerator, as opposed to individual circuit implementations, is also discussed. The overall power consumption of the proposed in-memory compute accelerator is 2280 μW and the accuracy is between 62% and 96% for all five applications.

This chapter is organized as follows. Section 6.2 details the circuit implementation and benefits of the one-stop in-memory compute accelerator. Section 6.3 discusses the individual circuit implementation of the five applications. Section 6.4 highlights the results obtained from the one-stop and individual applications of the accelerator. Conclusions are drawn and presented in the final Section 6.5.

6.2 ACCELERATOR FOR IN-MEMORY COMPUTE APPLICATIONS

6.2.1 Background

The in-memory compute accelerator designed in this chapter is based on gradient-descent techniques [12]. The model created has a deep net of 32 input neurons, 40 hidden neurons and 10 output neurons. The analog circuit for this type of network has been created by performing a number of steps like forward propagation, backward propagation and their individual verifications. The architectural implementation related to forward propagation (*inference*), and how the information (*signal*) is computed and propagated in the forward direction (*left-to-right*) attained by a simple step of dot-product operation based on Ohm's law and Kirchhoff's law is described in [24]. The inference circuit uses a one transistor–one resistive memory structure at the intersection of the WL and BL as the crossbar structure to compute the operation. The output obtained is in the form of potentiation and depression signals captured as spike events (*the action potential*) over a period of time. These spike events from the artificial neural networks are the activated signals which approximate the spike activity of the natural brain [24].

As regards backward propagation (*training*), the chapter explains how the network can be trained to compute a dedicated task. After performing the inference steps, the backward propagation (*right-to-left*) is computed by comparing the obtained outputs against a set of target signals in order to adjust and tune the weighted signals on the network through the error accomplished during computation. An error circuit comprising a difference amplifier, a multiplier and a switch are used for this operation. The error signals obtained are backpropagated to the crossbar of the previous layer and this process iterates until the error reaches a minimum value. Additional sub-circuits used in the backpropagation architecture include error derivative circuits at hidden and output layers which consists of difference amplifier, inverting amplifier, a multiplier and a switch to perform the desired task of computing error derivatives, a derivative circuit for swish function and weight update, and weight sign control circuits in order to perform the training in the neural network. Once the error in the network reaches a minimum value, the training is stopped and the network is analyzed for its learned behavior which can be observed by capturing the weight update signals in the network [27, 28].

6.2.2 Design and implementation of standard 32×40×10 in-memory compute architecture

A neural network having 32 input layer neurons, 40 hidden layer neurons and 10 output layer neurons is shown in Figure 6.1. The architecture uses

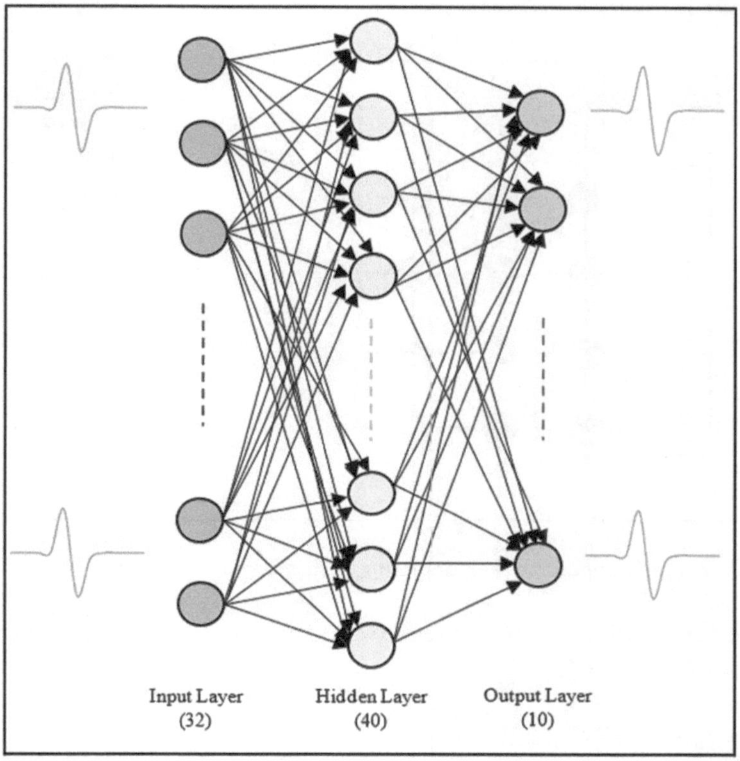

Figure 6.1 Artificial neural network (32×40×10).

crossbars in the form of one transistor and one resistive memory at the WL–BL (*row–column*) intersection of the grid (Figure 6.2). The resistive memory elements used to store and also compute the dot-product operation in the crossbar array (synapse) and in the swish activation function (neuron) have the benefits of low power usage (μW or less) and fast switching action (ns), and are non-volatile, reliable and resilient in nature [9]. The standard in-memory compute architecture having 32×40×10 structure is designed, implemented and verified on 28nm FD-SOI technology PDK in various industrial applications (see Section 6.3).

6.2.2.1 Working principle

The architecture mainly consists of a resistive memory device (RRAM), a two-terminal metal-insulator-metal stack device with bipolar switching characteristics which is calibrated using IMEC's TiN/Hf/HfO_x/TiN as defined in Equation (6.1) [22].

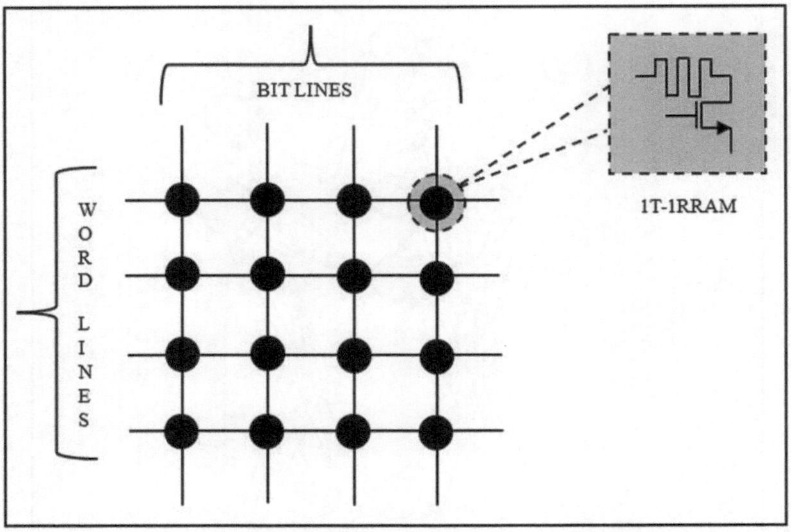

Figure 6.2 Standalone crossbar depicting resistive memories.

The I–V relationship of the resistive memory model is based on the following equation [22]:

$$I = I_o \exp\left(-\frac{g_m}{g_o}\right) \times \sinh\left(\frac{V}{V_o}\right) \tag{6.1}$$

where g_m is the gap distance between the top electrode (TE) and conductive filament (CF) tip of the device and I_0, g_0, and V_0 are the fitting parameters.

The RRAM devices used store information effectively and switch their conductance state fast from high resistance (HRS) to low resistance (LRS) and vice versa on application of input voltage or current. The device has been modeled for the ideal I–V characteristics. The parametric analysis is also performed by varying the DC input voltage between −2V and 2V. The device results can be seen in Section 6.4. Additionally, the in-memory compute architecture relies heavily on the principles of Ohm's and Kirchhoff's law (Equations (6.2) and (6.3)) to compute the inference phenomena. In general, the neuron function (N_z) can be represented as

$$N_Z = f\left(w_i x_i + \text{bias}\right) \tag{6.2}$$

where f is the swish activation function [23] given as, $f(x) = x/1 + e^{-x}$; w_i are the weights and x_i are the inputs (for i = 1 to n). From the electronics perspective, Equation (6.2) can be expressed as

$$I_Z = \Sigma\left(G_{ij} \times V_i\right) \tag{6.3}$$

where I_z is the columnar current drawn from the crossbar grid; G is the conductance of RRAM device; V is the input voltage; i and j = 1, 2, 3, ...n. The training phenomena for the in-memory compute architecture are derived from the relations and their gradients described in Equations (6.4) to (6.7). The error in the network is calculated as

$$E_N = (1/2)\sum_{i=1}^{n}(\text{Target} - \text{Output})^2 \tag{6.4}$$

where n is the number of neurons in the layer.

The error derivative at hidden layer is

$$\frac{\partial E_{Nh}}{\partial w_{ij}} = X' \cdot \delta_{Nh} = X' \cdot \left(e_{Nh} \cdot \frac{\partial N_{oh}}{\partial w_{ij}} \right) \tag{6.5}$$

The error derivative at output layer is

$$\frac{\partial E_{No}}{\partial w_{jk}} = N_{oO} \cdot \delta_{NO} = N_{oO} \cdot \left(e_{NO} \cdot \frac{\partial N_{oO}}{\partial w_{jk}} \right) \tag{6.6}$$

The first order derivative of swish activation function (f′) is given as

$$f'(x) = \frac{df(x)}{dx} = \sigma(x) + x \cdot \sigma(x) \cdot (1 - \sigma(x)) \tag{6.7}$$

where the sigmoid function is $\sigma(x) = 1/(1 + e^{-x})$.

In order to perform the forward-propagation and backward-propagation operation based on Equations (6.2) to (6.7), the in-memory compute architecture (Figure 6.3) consists of a neural compute unit comprising two memory compute units (MCU-1 and MCU-2) constructed using two crossbar arrays arranged in the form of one transistor and one resistive memory cell (1T-1RRAM), swish activation and its derivative circuit, error and its derivative circuit and the weight update circuit.

The MCU-1 and MCU-2 are the processing elements that drive the RRAM memory cell in the crossbar array. When the input voltage or current is applied to the WL, BL of the crossbar, combinatory RRAM devices are programmed through their select transistors, changing the conductance state from HRS to LRS and LRS to HRS for a short period of time. As the crossbar grid is arranged in mesh format and based on Equations (6.2) and (6.3), the columnar current obtained is applied non-linearly so that the complete structure behaves as a neural unit. Therefore, for this purpose a swish activation function is applied to create non-linear current in the path of crossbar 1 and crossbar 2. The swish activation circuit uses five pairs of

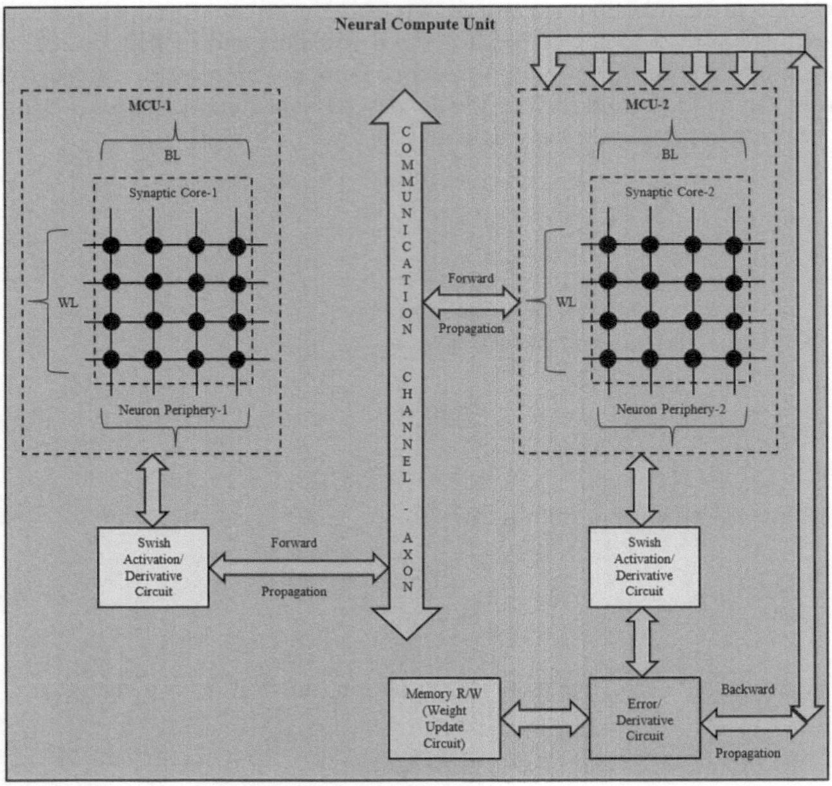

Figure 6.3 Top-level in-memory compute architecture.

CMOS transistors, two resistive memory devices and a multiplier circuit to produce the firing potential of the neuron and is described in [24]. This circuit helps invoke non-linearity in the top-level in-memory compute architecture to further build complex learning tasks in the network. The same process of forward propagation is repeated for crossbar 2, crossbar 3 … crossbar N, depending upon the network. The non-linear current obtained at the end of the first crossbar is forward propagated to the next crossbar in the circuit. The final output (voltage or current) obtained at the end of the crossbar is compared against a target value (voltage or current) through an error circuit as defined in Equation (6.4) and also elaborated in the implementation [28].

The aim here is to understand the loss in the network in order to effectively train the neurons to perform learning. This computed error is backpropagated if it is high. Iteratively this process will reduce error in the inference. Meanwhile, the gradients of error at hidden and output layer along with the gradient of swish circuit that are being trained on the 32×40×10 network are computed and thereby the layer weights are continuously updated in the weight update circuit. The weight update circuit is basically the memory read (R)/write (W) unit that stores and also updates the information computed by the error unit. This process of computing error, their gradients and updating of weights iterates until the error in the network falls to a low level.

Once the error in the network is minimal, the training process is stopped, which signifies that the network has learned the inputs which have been trained. The 32×40×10 network has been trained on five different industry-based applications and consumes less power (2280 μW) than [30] from the literature. The overall performance of the architecture is better due to the memory candidate RRAM, the swish circuit, the FD-SOI technology and its usage as an accelerator across multiple domains in real time. The in-memory compute architecture designed has MCU-1 and MCU-2 which are the main cores that drive the neuromorphic chip in addition to external peripheral units and interconnect as shown in Figure 6.4. The design, implementation and verification of the in-memory compute architecture for all five applications was performed using Cadence® simulations on ST Microelectronics

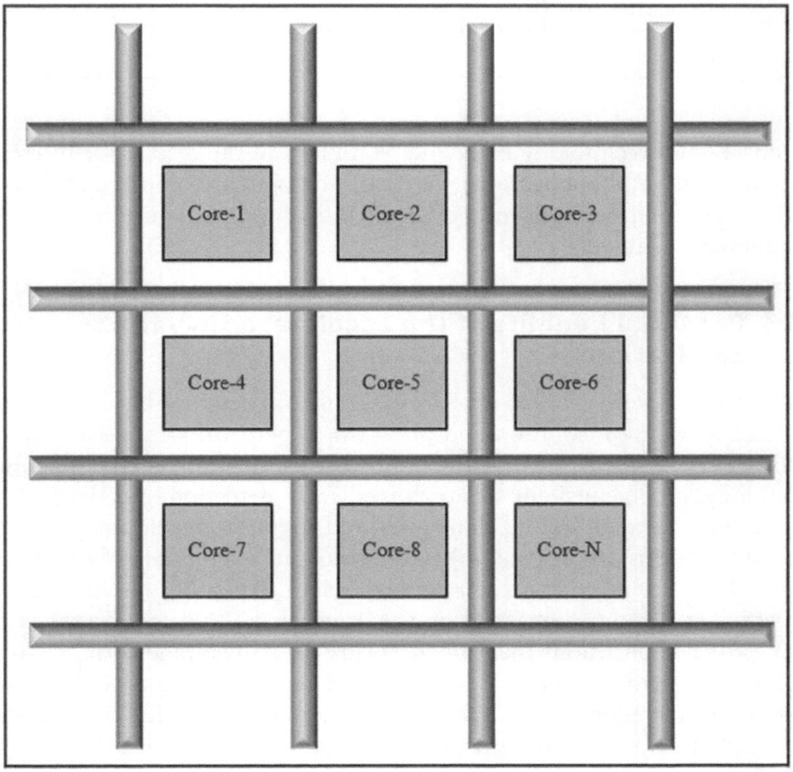

Figure 6.4 Top-level architecture of the neural compute unit.

Table 6.1 Resource utilization of in-memory compute architecture for 32×40×10 neural network

Design parameters	*Inference*	*Training*
Deep Net Structure	32×40×10	32×40×10
Hardware Details	1T-1RRAM based cores (1780 RRAMs) on 28 nm FD-SOI	1T-1RRAM based cores (1794 RRAMs) on 28 nm FD-SOI
Transistor: PFET/NFET	L = 30 nm, W = 80 nm	L = 30 nm, W = 80 nm
Memory Cell	1T-1RRAM	1T-1RRAM
Resistive Memories in Neuron Circuit	100	108
Resistive Memories in Synapse Circuit	1680	1680
Additional Resistive Memories used in the Architecture	–	6
Total Resistive Memories in the Architecture	1780	1794
Operating Voltage	1.7V	2V
Resistors	1 kΩ/2 kΩ	1 kΩ/2 kΩ
Temperature	27°C	27°C

28 nm FD-SOI technology PDK and Matlab®. All the results for inference, training circuit and applications were captured and are reported in detail in Section 6.4. The circuit inventory for the in-memory compute accelerator is highlighted in Table 6.1.

6.2.3 Potential benefits of the accelerator in various applications

The standard accelerator having 32×40×10 neural network has been curated to create a common solution and platform to perform a range of different applications. Problems related, for example, to simulation crashes in climate technology, involvement in social causes, early detection and diagnosis of cancer, financial analysis and challenges, and cognitive development through gaming platforms can be promptly addressed and acted upon. Therefore, the 32×40×10 accelerator can cater to major industry solutions in real time. The efficient response time and power usage make this accelerator circuit ideal for the applications mentioned. Its greatest advantage is the ability of the resistive memory candidate to compute and store the processed information and to retain it even when the power is shut off [22]. In addition, the crossbar array structure helps compute the functions at a better speed and allows the signals to be propagated among different layers in the structure. Also, this accelerator circuit uses a non-linear function (swish) which is much faster and more power efficient, and the overall performance is better, than the traditional functions (ReLU, sigmoid, GeLU, etc.) [23, 24, 27]. Due to the ability of resistive memory in these non-linear functions, the overall computation is fast and effective, utilizing fewer micro-watts of power than traditional memories like FLASH and DRAM [8]. The FD-SOI technology is reliable and highly robust when used with memristive devices, unlike the planar CMOS process [31, 32]. It has the advantages of low energy consumption, very low operating voltages, low noise, lower gate capacitance, and faster operation.

6.3 IN-MEMORY COMPUTE ACCELERATOR: AN EMBEDDED SYSTEM PERSPECTIVE FOR MULTIPLE APPLICATIONS

The following sub-sections discuss five different industry applications of the in-memory compute accelerator implemented on a standard network having 32×40×10 neural network. When used together with the necessary sensor, actuation, processing, memory and peripheral requirements as part of an embedded system, the in-memory compute accelerator can cater for useful real-time applications in day-to-day tasks or social life.

6.3.1 Climate technology

Identification of the success and failure of different climate data (or information) expressed through simulation technology has been demonstrated on the in-memory compute accelerator. The dataset used is the climate model simulation crashes dataset by Lawrence Livermore National Laboratory, USA [33, 34]. This dataset has 540 instances in total based on 20 parameters, and comprises records of crashed signals that occurred in an uncertainty quantification (UQ) study to observe the simulation effects on one of their models, called an ocean model parameter. The Parallel Ocean Program (POP2) in particular observed multiple crashes of the climate data. The simulation crashes are categorized as success (1) and failure (0).

The in-memory compute architecture built using the standard network having 32 input layer neurons, 40 hidden layer neurons and 10 output layer neurons performed inference and training with the climate technology dataset. The 20 parameters of the climate technology data were fed to the in-memory compute architecture, whereas the additional 12 neurons at the input layer and 8 neurons at the output layer of the crossbar element are the drop-out neurons which were temporarily turned off to satisfy the climate technology application. Therefore, the 20 parameters of the climate data were fed as inputs to the MCU-1, in which the computation begins with forward propagation through the neuro-synaptic cores having RRAM-based crossbars and swish circuits; at MCU-2, backward propagation is

Table 6.2 Performance results of in-memory compute architecture for 32×40×10 neural network

Design parameters	*Inference*	*Training*
Resistive Memory Switching	Ron: 20 kΩ, Roff: 80 kΩ	Ron: 70 kΩ, Roff: 40 MΩ
Resistive Memory Switching Time	7 ns	30 ns
Resistive Memory Conductance	4 mhos	18 mhos
Total Power	958 μW	2280 μW

computed through the neuro-synaptic cores having RRAM-based crossbars and swish circuits. The error circuit, and the derivatives at hidden, output layer and memory update units are utilized for training the climate technology. The training was performed on 70% of the samples, with validation and testing each on 15% of samples from a total of 540 samples. The training concluded after 34 iterations in which the error observed was minimum; thus the 32×40×10 neural network successfully learnt the trained climate technology data. The climate technology application observed 91.5% of the successful classifications. See Section 6.4 for discussion of the output characteristics such as the activated output signals, error minimization and memory R/W information. See Tables 6.2 and 6.3 for the output parameters observed in the in-memory circuit inventory during simulation.

6.3.2 Social sciences

The in-memory compute accelerator was used in an application to identify the percentage of women editors and their editing practices in Spanish Wikipedia. The dataset used, the Gender Gap in Spanish Wikipedia, was developed by the University of Catalonia [35, 36] and comprised 4746 instances and 20 different parameters. The contribution to Spanish Wikipedia through content making and different editing practices observed among different genders was categorized as unknown (0), male (1), female (2).

The in-memory compute architecture designed using the standard network having 32 input layer neurons, 40 hidden layer neurons and 10 output layer neurons was used to perform the inference and training with the Spanish gender gap Wikipedia dataset. The 20 parameters of the Spanish Wikipedia data were fed to the in-memory compute architecture, while the additional 12 neurons at the input layer and 7 neurons at the output layer of the crossbar element are the drop-out neurons which were temporarily turned off to satisfy this social sciences application. Therefore, the 20 parameters of the Spanish Wikipedia data were fed as inputs to the MCU-1, in which computation begins with the forward propagation through the neuro-synaptic cores having RRAM-based crossbars and swish circuits; at MCU-2 the backward propagation is computed through the neuro-synaptic cores having RRAM-based crossbars and swish circuits. The error circuit, the

Table 6.3 Performance results of in-memory compute architecture for 32×40×10 neural network on multiple applications

Target industry	*Climate technology*	*Social sciences*	*Medical sciences*	*Finance technology*	*Gaming technology*
Dataset	Climate Model Simulation Crashes	Gender Gap in Spanish Wikipedia	Wisconsin Breast Cancer	Wholesale Customers	Skill Craft Master Table
Total Attributes	20	21	32	8	20
Total Instances	540	4746	569	440	3395
Training Data	378 (70%)	3322 (70%)	399 (70%)	308 (70%)	2377 (70%)
Validation Data	81 (15%)	712 (15%)	85 (15%)	66 (15%)	509 (15%)
Testing Data	81 (15%)	712 (15%)	85 (15%)	66 (15%)	509 (15%)
Total Epochs	34 iterations	38 iterations	19 iterations	15 iterations	166 iterations
Training Confusion	0.095	0.362	0.043	0.123	0.611
Validation Confusion	0.049	0.336	0.035	0.167	0.619
Test Confusion	0.074	0.381	0.059	0.152	0.678
Learning Rate	0.01	0.01	0.01	0.01	0.01
Weight Update Frequency (i.e., interval)	1 epoch	1 epoch	1 epoch	1 epoch	1 epoch
Gradient	0.0161	0.0286	0.0195	0.0320	0.0289
Training Time	11.95 µs	117.39 µs	7 µs	4.29 µs	366.96 µs
Accuracy of the In-Memory Compute Accelerator	91.5%	63.9%	95.6%	86.6%	62.2%

derivatives at hidden, output layer and memory update units were utilized for the training. The training was performed on 70% of the samples, with validation and testing each having 15% of samples from the total of 4746 samples. The training concluded after 38 iterations where the error observed was minimum and the 32×40×10 neural network had successfully learnt the trained Spanish Wikipedia data. The social sciences application then observed 63.9% of the successful classifications. See Section 6.4 for discussion of the output characteristics such as the activated output signals, the error minimization and the memory R/W information. See Table 6.2 and 6.3 for the output parameters observed in the in-memory circuit inventory during simulation.

6.3.3 Medical sciences

In this study, the in-memory compute accelerator demonstrated understanding of various levels of tumors in cancer patients, using the Wisconsin Breast Cancer Dataset developed by the University of Wisconsin, USA [37, 38]. The dataset consists of a total of 569 instances with 30 parameters. The tumors are identified by fine needle aspiration (FNA) which displays an image of characteristics of cell nuclei. Cancers are generally identified and categorized as malignant or benign.

The in-memory compute architecture built using the standard network having 32 input layer neurons, 40 hidden layer neurons and 10 output layer neurons was used to perform the inference and training with the Wisconsin Breast Cancer Dataset. The 30 parameters of the medical sciences data were fed to the in-memory compute architecture, while the additional 2 neurons at the input layer and 8 neurons at the output layer of the crossbar element are the drop-out neurons which were temporarily turned off. The 30 parameters were fed as inputs to the MCU-1, in which the computation begins with forward propagation through the neuro-synaptic cores having RRAM-based crossbars and swish circuits; at MCU-2 the backward propagation is computed through the neuro-synaptic cores having RRAM-based crossbars and swish circuits. The error circuit, the derivatives at hidden, output layer and memory update units are utilized for the training of the medical sciences application. The training was performed on 70% of the samples, with validation and testing each having 15% of samples from a total of 569 samples. The training concluded after 19 iterations when the error observed was minimum and the 32×40×10 neural network had successfully learnt the trained Wisconsin breast cancer data. Subsequently, 95.6% of classifications were observed to be successful. See Section 6.4 for discussion of the output characteristics such as the activated output signals, the error minimization and the memory R/W information. See Table 6.2 and 6.3 for the output parameters observed in the in-memory circuit inventory during simulation.

6.3.4 Finance technology

Finance applications cover a wide range of markets explaining different trading scenarios. The in-memory compute accelerator was demonstrated using the Wholesale Customers Dataset created by the Institute of Lisbon, Portugal [39], which comprises 440 instances and seven different business parameters, representing the various products traded by the wholesale distributor.

The in-memory compute architecture designed using the standard network having 32 input layer neurons, 40 hidden layer neurons and 10 output layer neurons was used to perform the inference and training with the Wholesale Customers Dataset. The seven parameters of the customer data were fed to the in-memory compute architecture, while the additional 25 neurons at the input layer and 8 neurons at the output layer of the crossbar element are the drop-out neurons which were temporarily turned off. The parameters were fed as inputs to the MCU-1, where the computation begins with forward propagation through the neuro-synaptic cores having RRAM-based crossbars and swish circuits; at MCU-2 backward propagation is computed through the neuro-synaptic cores having RRAM-based crossbars and swish circuits. The error circuit, the derivatives at hidden, output layer and memory update units were utilized for the training of the customer data, which was performed on 70% of the samples, with validation and testing each having 15% of samples from a total of 440 samples. The training concluded after 15 iterations where the error observed was minimum and the 32×40×10 neural network had successfully learnt the trained customers data. Subsequently, 86.6% of classifications were observed to be successful. See Section 6.4 for discussion of the output characteristics such as the activated output signals, the error minimization and the memory R/W information. See Table 6.2 and 6.3 for the output parameters observed in the in-memory circuit inventory during simulation.

6.3.5 Gaming technology (GT)

Gaming is a popular application which is also the most in demand. The Skill Craft Master Table dataset was used to demonstrate the in-memory compute accelerator. Skill Craft is a telemetry-based game designed to improve complex learning skills among children. The dataset was proposed by the Simon Fraser University, Canada [40, 41]. The video-game dataset consists of 3395 instances in total and 19 parameters identifying various gaming actions such as use of hot keys and their respective actions seen on the screen, screen movements and their fixations, possible success and failures on a hit, the time in which one gaming task is achieved, etc. The Skill Craft game is categorized into eight different leagues: Bronze, Silver, Gold, Platinum, Diamond, Master, Grand Master and Professional.

The in-memory compute architecture built using the standard network having 32 input layer neurons, 40 hidden layer neurons and 10 output layer neurons was used to perform the inference and training with the Skill Craft Master Table dataset. The 19 parameters of the gaming data were fed to the in-memory compute architecture, whereas the additional 13 neurons at the input layer and 2 neurons at the output layer of the crossbar element are the drop-out neurons which were temporarily turned off. The 19 parameters of the Skill Craft data were fed as inputs to the MCU-1, where the computation begins with forward propagation through the neuro-synaptic cores having RRAM-based crossbars and swish circuits; at MCU-2 the backward propagation is computed through the neuro-synaptic cores having RRAM-based crossbars and swish circuits. The error circuit, the derivatives at hidden, output layer and memory update units are utilized for the training of the video-gaming application. The training was performed on 70% of the samples, with validation and testing each having 15% of samples from a total of 3395 samples. The training concluded after 166 iterations where the error observed was minimum and the 32×40×10 neural network had successfully learnt the trained Skill Craft data. Subsequently, 62.2% of the classifications in the video-gaming application were successful. See Section 6.4 for discussion of the output characteristics such as the activated output signals, the error minimization and the memory R/W information. See Table 6.2 and 6.3 for the output parameters observed in the in-memory circuit inventory during simulation.

6.4 RESULTS AND DISCUSSION

The in-memory compute accelerator using RRAM devices was designed, implemented and verified using the Cadence® platform on ST Microelectronics 28 nm FD-SOI technology PDK and Matlab®. The following sub-sections discuss the results of the architecture.

6.4.1 Parametric analysis of ideal resistive memory and swish activation function

The I–V characteristics of the cell having a transistor and a resistive memory device were performed by running a DC analysis. The input voltage for the RRAM device was chosen as 1V and the parametric analysis was performed for a sweep between –2V and 2V. Figure 6.5 presents the observed characteristics. The current obtained for the voltage range is in μA based on the input voltage applied which was designed as per Equation (6.1). The I–V characteristics highlight the two key operating conditions of the RRAM device. First, the SET process which is the transition from high resistance (HRS) to low resistance state (LRS) of the device. This is the condition where the RRAM device is in the OFF state (0). Second, the RESET process, which

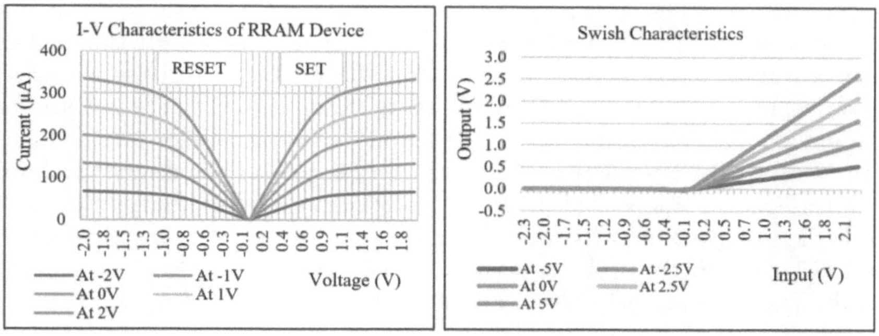

(a) (b)

Figure 6.5 Parametric analysis of ideal resistive memory and swish activation function: (a) parametric analysis of RRAM device; (b) parametric analysis of swish function.

is the transition from low resistance (LRS) to high resistance state (HRS) of the device. This is the condition where the RRAM device is in the ON state (1). Therefore, the on–off switching of the RRAM device happens for a short period of time. The device switching (Ron, Roff) is tabulated in Table 6.2. The RRAM device is non-volatile and has largely been responsible for improving the design performance of the in-memory compute accelerator.

The output characteristics of the swish activation circuit were identified by running a DC analysis. The input voltage for the RRAM-based swish circuit was chosen between −5V and 5V and parametric analysis was performed for a sweep between −2.3V and 2.3V. Figure 6.5 presents the observed characteristics and the voltage obtained at different conditions. The output characteristics obtained approximate closely to their theoretical counterparts, combining the benefits of ReLU and sigmoid functions, and the curve sees one-sided boundedness at zero, which is a swish-specific property. This function has been the principal driving factor invoking the firing potentials within the in-memory compute accelerator.

6.4.2 Inference analysis for the in-memory compute accelerator

The forward-propagation operation was computed for the in-memory compute accelerator on five applications: climate technology, social sciences, medical sciences, finance technology and gaming technology.

The inference results of the transient analysis performed for the climate technology application captured two outputs, O1 and O2, obtained in volts for a duration of 4000ns (Figure 6.6). The O1 and O2 contain the action potential (spike) signals modeled on 32×40×10 neural network. The inference results of the transient analysis performed for the social sciences application captured three outputs i.e., O1, O2 and O3, obtained in volts for a duration of 4000 ns (Figure 6.6). The O1, O2 and O3 are the spike signals modeled on 32×40×10 neural network. Transient analysis for the medical sciences application was performed and the inference results comprised two outputs, i.e., O1 and O2, obtained in volts for a duration of 4000ns (Figure 6.6). The O1 and O2 contain the spike signals modeled on 32×40×10 neural network. Transient analysis was performed for the finance technology application where the inference results comprised two outputs, i.e., O1 and O2, obtained in volts for a duration of 4000 ns (Figure 6.6). The O1 and O2 contain the spike signals modeled on the 32×40×10 neural network. The gaming technology application was evaluated by running the transient analysis for a duration of 4000 ns and the observed inference results consisted of eight outputs, i.e., O1, O2, O3, O4, O5, O6, O7 and O8, which are the spike signals modeled on the 32×40×10 neural network (Figure 6.6).

6.4.3 Training analysis for the in-memory compute accelerator

The backward-propagation operation was computed for the in-memory compute accelerator on five applications: climate technology, social sciences, medical sciences, finance technology and gaming technology.

Error evaluation was performed for the 32×40×10 neural network for all five applications and their characteristics are plotted in Figure 6.6. The error voltage is observed to decrease consecutively with time, and this is achieved by running the transient analysis for 200ps. The read /write (RW) update for the memory compute unit (MCU-1 and MCU-2) is captured for all the five applications over a span of 2000 ns (Figure 6.6). The memory updates in the in-memory compute architecture vary for the entire epoch window.

The accuracy achieved for the climate technology application is 91.5% for 34 iterations, accuracy for the social sciences application is 63.9% for 38 iterations, accuracy for the medical sciences application is 95.6% for 19 iterations, accuracy for the finance technology application is 86.6% for 15 iterations and the accuracy for the gaming technology is 62.2% for 166 iterations executed on a memory update frequency (i.e., interval) of 1 epoch. The overall performance of the 32×40×10 network is good as disclosed through the circuit inventory and their characteristics. Also, the neural net has learned the applications that were trained and recognized it correctly with a learning rate of 0.01. The performance results of the inference and training of the in-memory compute accelerator on the 32×40×10 network for all five applications is tabulated in Tables 6.2 and 6.3. Consequently, the neural network 32×40×10 can further be extended to more deeper nets having N neurons and N_h hidden layers by creating more deeper nets N_i x N_h x N_o as desired by an application. Likewise, the neural compute unit (Figure 6.4) can be elaborated to N number of cores depending upon the application-specific design.

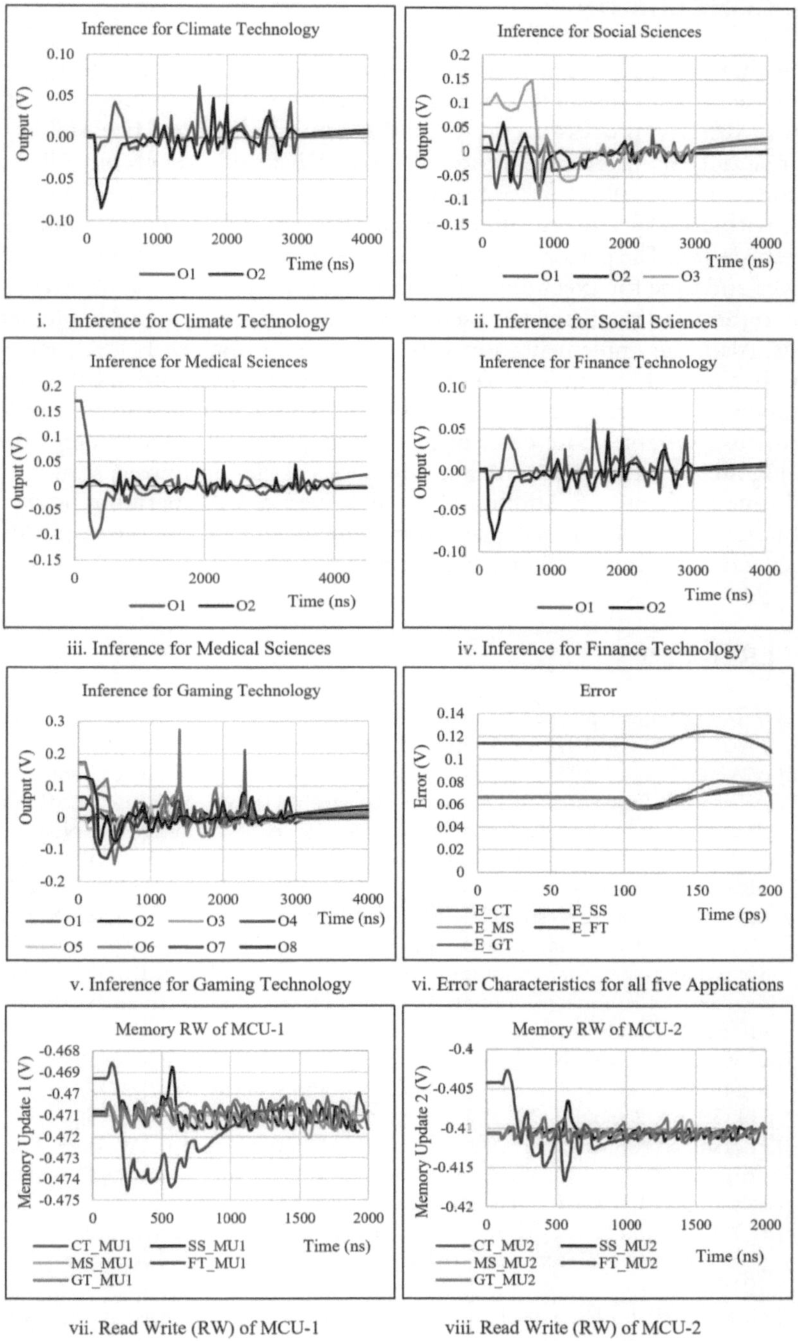

Figure 6.6 Inference and training analysis of the in-memory compute accelerator.

6.5 CONCLUSION

This chapter has presented a novel in-memory compute architecture that serves as a neuromorphic accelerator or co-processor capable of working with a RISC-V processor core or any commercial central processing unit over standard interfacing protocols like Advanced Microcontroller Bus Architecture (AMBA), Advanced eXtensible Interface (AXI), AXI – LITE, etc., with appropriate real-time constraints being met as part of embedded system design. The performance of the architecture in terms of computational power and time for executing a function has seen a significant improvement due to the resistive memories used for the neuron and synapse implementations. Also, the implementation was performed using the FD-SOI technology that helped maintain lower operating limits and improved the overall accuracy of the architecture. The standard accelerator can also be utilized as a neuro-synaptic core element in the design of neuro-based chips in the analog/mixed-signal domain that can serve as an optimized solution for real-time applications. Additionally, the power, speed, operating limits and accuracy achieved can serve as improvement factors for the application-specific designs using resistive memories.

REFERENCES

[1] S. B. Furber, F. Galluppi, S. Temple and L. A. Plana, "The SpiNNaker Project," *Proc. IEEE*, vol. 102, no. 5, pp. 652–665, 2014.

[2] MANCHESTER1824, "SpiNNaker Project – The SpiNNaker Chip," APT Advanced Processor Technologies Research Group, May, 2011. [Online]. Available: http://apt.cs.manchester.ac.uk/projects/SpiNNaker/SpiNNchip/

[3] C. Denk, F. L. Blandino, F. Galluppi, L. A. Plana, S. Furber, and J. Conradt, "Real-time interface board for closed-loop robotic tasks on the SpiNNaker neural computing system," *23rd Int. Conf. on Artificial Neural Networks (ICANN 2013)*, Sofia, Bulgaria, pp. 467–474, 2013.

[4] Spinnakermanchester, "SpOmnibot – TUM SpiNNaker Robot," External devices and robotics. [Online]. Available: https://spinnakermanchester.github.io/docs/spomnibot/ [Accessed April 13, 2023].

[5] F. Staudigl, F. Merchant and R. Leupers, "A survey of neuromorphic computing-in-memory: Architectures, simulators, and security," *IEEE Design & Test*, vol. 39, no. 2, pp. 90–99, 2022.

[6] H. Tsai, S. Ambrogio, P. Narayanan, R. M. Shelby, and G. W. Burr, "Recent progress in analog memory-based accelerators for deep learning," *Journal of Physics D: Applied Physics*, vol. 51, no. 28, pp. 1–27, 2018.

[7] P.-Y. Chen and S. Yu, "Technological benchmark of analog synaptic devices for neuroinspired architectures," *IEEE Design & Test*, vol. 36, no. 3, pp. 31–38, 2019.

[8] G. Pedretti and D. Ielmini, "In-memory computing with resistive memory circuits: Status and outlook," *Electronics*, vol. 10, no. 9, pp. 1–22, 2021.

[9] H. Wu et al., "Device and circuit optimization of RRAM for neuromorphic computing," *IEEE Int. Electron Devices Meeting (IEDM)*, San Francisco, CA, USA, pp. 11.5.1–11.5.4, 2017.
[10] T. P. Xiao, C. H. Bennett, B. Feinberg, S. Agarwal, and M. J. Marinella, "Analog architectures for neural network acceleration based on non-volatile memory," *Applied Physics Reviews*, vol. 7, no. 3, pp. 1–35, 2020.
[11] International Roadmap for Devices and Systems (2020) Applications Benchmarking. IEEE. https://irds.ieee.org/editions/2020/application-benchmarking
[12] I. Goodfellow, Y. Bengio, and A. Courville, *Deep Learning*. MIT Press, 2016.
[13] S. Hamdioui et al., "Applications of computation-in-memory architectures based on memristive devices," *2019 Design, Automation & Test in Europe Conference & Exhibition (DATE)*, Florence, Italy, pp. 486–491, March 2019.
[14] A. Sebastian, M. Le Gallo, R. K. Aljameh, and E. Eleftheriou, "Memory devices and applications for in-memory computing," *Nature Nanotechnology*, vol. 15, no. 7, pp. 529–544, 2020.
[15] S. Yu, X. Sun, X. Peng and S. Huang, "Compute-in-memory with emerging nonvolatile-memories: challenges and prospects," *IEEE Custom Integrated Circuits Conference (CICC)*, Boston, MA, pp. 1–4, 2020.
[16] V. Saxena, X. Wu, I. Srivastava, and K. Zhu, "Towards neuromorphic learning machines using emerging memory devices with brain-like energy efficiency," *Journal of Low Power Electronics and Applications*, vol. 8, no. 4, pp. 1–24, 2018.
[17] H.Y. Chang et al., "AI hardware acceleration with analog memory: Microarchitectures for low energy at high speed," *IBM Journal of Research and Development*, vol. 63, no. 6, pp. 8:1–8:14, 2019.
[18] G. Indiveri and S.-C. Liu, "Memory and information processing in neuromorphic systems," *Proc. IEEE*, vol. 103, no. 8, pp. 1379–1397, 2015.
[19] J. M. Correll et al., "A fully integrated reprogrammable CMOS-RRAM compute-in-memory coprocessor for neuromorphic applications," *IEEE Journal on Exploratory Solid-State Computational Devices and Circuits*, vol. 6, no. 1, pp. 36–44, 2020.
[20] N. Qiao and G. Indiveri, "Scaling mixed-signal neuromorphic processors to 28 nm FD-SOI technologies," *IEEE Biomedical Circuits and Systems Conference (BioCAS)*, Shanghai, China, pp. 552–555, October 2016.
[21] J.-W. Su et al., "15.2 A 28nm 64Kb Inference-training two-way transpose multibit 6T SRAM compute-in-memory macro for AI edge chips," *IEEE International Solid-State Circuits Conference (ISSCC)*, San Francisco, CA, pp. 240–242, February 2020.
[22] P.-Y. Chen and S. Yu, "Compact modeling of RRAM devices and its applications in 1T1R and 1S1R array design," *IEEE Transactions on Electron Devices*, vol. 62, no. 12, pp. 4022–4028, 2015.
[23] P. Ramachandran, B. Zoph, and Q. V. Le, "Swish: a self-gated activation function," *arXiv preprint arXiv:1710. 05941*, vol. 7, no. 1, pp. 1–5, 2017.
[24] A. Fatima and A. Pethe, "NVM device-based deep inference architecture using self-gated activation functions (Swish)," In *Proc. Machine Vision and Augmented Intelligence—Theory and Applications (MAI)*, pp. 33–44, November 2021.
[25] A. Fatima and A. Pethe, "Periodic analysis of resistive random access memory (RRAM)-based swish activation function," *SN Computer Science*, vol. 3, no. 3, 2022.

[26] C. Mead, and M. Ismail, *Analog VLSI implementation of neural systems*, vol. 80. Springer Science & Business Media, 1989.
[27] A. Fatima and A. Pethe, "Implementation of RRAM based swish activation function and its derivative on 28nm FD-SOI," *International Electrical Engineering Congress (iEECON)*, Khon Kaen, Thailand, pp. 1–4, 2022.
[28] A. Fatima and A. Pethe, "Inference architecture using RRAM devices", *58th Design Automation Conference (Poster)*, USA, 2021.
[29] A. Fatima and A. Pethe, "An RRAM based neuromorphic accelerator for speech based emotion recognition," in *Neuromorphic Computing Systems for Industry 4.0*, IGI Global, pp. 63–93, 2023.
[30] O. Krestinskaya, K. N. Salama and A. P. James, "Learning in memristive neural network architectures using analog backpropagation circuits," in *IEEE Transactions on Circuits and Systems I: Regular Papers*, vol. 66, no. 2, pp. 719–732, 2019.
[31] STMicroelectronics, "28nm FD-SOI Technology Catalog," 2016. [Online]. Available: http://surl.li/gijux
[32] STMicroelectronics, "Efficiency at all levels," [Online]. Available: https://www.st.com/content/st_com/en/about/innovationtechnology/FD-SOI.html
[33] D. D. Lucas et al., "Failure analysis of parameter-induced simulation crashes in climate models," *Geoscientific Model Development*, vol. 6, no. 4, pp. 1157–1171, 2013.
[34] D. Dua and C. Graff, "UCI machine learning repository Irvine," California: University of California, School of Information and Computer Science, 2019. [Online]. Available: http://archive.ics.uci.edu/ml
[35] J. Minguillón, J. Meneses, E. Aibar, N. Ferran-Ferrer, and S. Fàbregues, "Exploring the gender gap in the Spanish Wikipedia: differences in engagement and editing practices," *PLOS one*, vol. 16, no. 2, pp. 1–21, 2021.
[36] J. Minguillón, J. Meneses, E. Aibar, N. Ferran-Ferrer, and S. Fàbregues, "UCI machine learning repository: gender gap in Spanish WP data Set." [Online]. Available: https://archive.ics.uci.edu/ml/datasets/Gender+Gap+in+Spanish+WP
[37] W. N. Street, W. H. Wolberg, and O. L. Mangasarian, "Nuclear feature extraction for breast tumor diagnosis," *Biomedical Image Processing and Biomedical Visualization*, vol. 1905, pp. 861–870, 1993.
[38] Dua, D. and Graff, C, "UCI machine learning repository: breast cancer Wisconsin (diagnostic) data set," 2019. [Online]. Available: https://archive.ics.uci.edu/ml/datasets/breast+cancer+wisconsin+(diagnostic)
[39] G. M. S. Margarida Cardoso, "UCI Machine Learning Repository: Wholesale customers Data Set," 2014. [Online]. Available: https://archive.ics.uci.edu/ml/datasets/ wholesale+customers
[40] J. J. Thompson, M. R. Blair, L. Chen, and A. J. Henrey, "Video game telemetry as a critical tool in the study of complex skill learning," *PLOS one*, vol. 8, no. 9, pp. 1–12, 2013.
[41] J. J. Thompson, M. R. Blair, L. Chen, and A. J. Henrey, "UCI machine learning repository: skillcraft1 master table dataset data set," 2013. [Online]. Available: http://archive.ics.uci.edu/ml/datasets/skillcraft1+master+table+dataset

Chapter 7

Artificial intelligence-driven radio channel capacity in 5G and 6G wireless communication systems in the presence of vegetation

Prospect and challenges

Sachin Kumar
Head of Technology Product Strategy AI, Samsung Research, Amaravati, Korea

T. Senthil Siva Subramanian
Sharda Group of Institutions Mathura, Mathura, India

Kapil Sharma
Delhi Technological University (DTU), Delhi, India

G.V. Ramesh Babu
Sri Venkateswara University, Tirupati, India

7.1 INTRODUCTION

Radio wave-based communication systems have recently attracted worldwide attention from augmented reality (AR), virtual reality (VR) and metaverse-based users of 5G and 6G communications systems [1–3]. Research on provision of high-speed big-data bandwidth for users across the world is ongoing [4, 5]. The radio wave frequency spectrum ranging from 3–300GHz can support multi-gigabit data rates based on variable rates for various high data rate multiple AR, VR, and metaverse applications worldwide [6, 7]. The Federal Communications Commission (FCC) has recently announced 28GHz, 37GHz, 39GHz and 73GHz radio carrier frequency bands [8–11].

Radio wave propagation losses occur due to various channel obstructions and impairments from the atmosphere and from urban, and vegetation sources [2, 3]. In the field of wireless communication, the presence of vegetation can significantly impact the propagation of radio waves, leading to a phenomenon known as channel attenuation. This can be a major issue for 5G networks, where high frequency and low latency communication is critical for applications such as autonomous vehicles and virtual reality. ML algorithms and data analysis can be used to train a model that accurately

DOI: 10.1201/9781032628059-7

predicts the degree of channel attenuation caused by different types of vegetation and their densities. This helps network operators to identify the areas where 5G communication might be impacted and take appropriate measures to mitigate these effects. This solution is based on a vast amount of data collected from various sources such as field experiments and simulations. By combining this data with AI algorithms, this technique has been able to create a highly accurate and robust model that can be used to predict the impact of vegetation on 5G communication. This breakthrough has the potential to revolutionize the way 5G networks are deployed and managed, making it possible to deliver high-speed and reliable communication even in challenging environments. It is an excellent example of how AI and ML are transforming the wireless communication industry and shaping the future of 5G.

Artificial intelligence (AI) has revolutionized various fields including the communications industry. The integration of AI has enhanced the overall performance and efficiency of systems. One of the key areas where AI has been applied is in the management of radio channel capacity. The primary objective of AI-driven radio channel capacity management is to improve the utilization of available radio resources, increase network efficiency and ultimately enhance the user experience. AI algorithms are also used to optimize spectrum use, ensuring that the available resources are used in the most efficient way possible.

Artificial intelligence (AI)-driven radio (AI-DR) technology provides system capability in dynamic form, in which communication system parameters can be changed dynamically using software [12–17]. The flexibility of AI-DR systems includes varied transmission power, modulation, coding techniques, and communications interfaces, and AI-DR-based software functions, libraries and modules can be designed and updated under AI-DR technology [18]. This makes the system not only flexible but also very simple to design and use, with adaptive system parameters for optimum and efficient system performance [17–20]. AI-DR technology enables software-based control and system parameter changes. AI-DR-based 5G radio communication system control software can be designed to reconfigure transmission signal power. Channel losses occur due to radio propagation, and in areas of dense vegetation, it is important to predict Shannon channel capacity (SCC). The Shannon channel capacity (SCC) theorem defines the maximum channel wireless capacity for continuous additive white Gaussian noise channel (AWGN) [16, 21–24]. This chapter shows that the SCC can be increased or decreased in the presence of vegetation according to the demands and needs of the environment using AI-DR [16, 22]. An AI-DR-based technique is used to control transmission signal power at the transmitter based on the vegetation attenuation.

Embedded system and system on a chip (SOC) play a crucial role in the prediction of blockage channel capacity attenuation in wireless communication propagation. In the case of embedded systems, prediction is based on the characteristics of the wireless communication system, including the frequency band, modulation technique, and power output. The embedded system can measure the signal quality in real time, including the signal strength, signal-to-noise ratio (SNR), and bit error rate (BER). Based on these parameters, the embedded system can predict blockage channel capacity attenuation by taking into account the various obstacles that may be present in the propagation environment. For SOC, the prediction of blockage channel capacity attenuation is more advanced as it combines multiple functionalities on a single chip. SOCs can incorporate multiple sensors, communication interfaces, and processing capabilities. These features allow the SOC to perform real-time analysis of the signal quality and predict the blockage channel capacity attenuation based on the information gathered from the environment. The SOC can also take into account the impact of the physical environment on the signal quality and predict channel capacity attenuation accordingly. This prediction is critical for ensuring reliable and efficient communication.

To mitigate the impact of vegetation blockage, AI-driven SOC technologies can be used to predict blockage levels based on the surrounding vegetation and adjust the transmission power accordingly. Another key benefit of AI-driven SOC technologies is their ability to accurately predict the channel capacity in real time. This is achieved through the use of advanced machine learning algorithms, which can analyze the channel state information and predict the channel capacity based on historical data and current conditions. This information can then be used to optimize the transmission power and improve the overall network performance. In networks, signal quality is a critical factor that affects the overall performance of the system. AI-driven SOC technologies can be used to monitor the signal quality in real time and adjust the transmission parameters to ensure that the signal quality remains high. This can be achieved through the use of advanced signal processing techniques, such as beamforming, MIMO, and dynamic modulation and coding, which can improve the signal quality and ensure that the data is transmitted reliably. AI-driven SOC technologies have the potential to revolutionize the wireless communication industry by providing advanced solutions for vegetation blockage attenuation, channel capacity prediction, and signal quality improvement in 6G and 5G networks. With these technologies, wireless communication systems can become more reliable, efficient, and user friendly, enabling the development of new applications and services that will improve our lives and communities.

The prediction of Shannon channel capacity in wireless communication propagation in the presence of vegetation tree blockage can be done using machine learning techniques like ML algorithms. This involves the use of data collected from previous measurements of the signal quality impact on signal frequency power amplitude of wireless signals based on tree depth, size, leaf size, and depth. The data collected are fed into a machine learning model, which is trained to identify patterns and relationships between the different variables that affect the signal quality. The model then uses these patterns and relationships to predict the Shannon channel capacity for different scenarios based on the tree depth, size, leaf size, and depth.

The machine learning algorithms used in this process can be supervised or unsupervised, depending on the type of data available and the desired outcome. In supervised algorithms, the model is trained on a labeled dataset, where the data has already been categorized into different classes. In unsupervised algorithms, the model is trained on an unlabeled dataset, where the data has not been categorized into classes. Technical reasons that support the use of machine learning algorithms in this process include the ability to handle large amounts of data, the ability to identify patterns and relationships in complex data, and the ability to make predictions based on the data.

This chapter consists of three major sections. Theoretical considerations on attenuation due to vegetation for 5G radio wave communication systems are presented in Section 7.2, which also discusses the F-ITU-R vegetation attenuation model. Section 7.3 provides a brief overview of AI-DR vegetation attenuation control in radio wave communication systems. Section 7.4 discuses Shannon channel capacity (SCC) in the presence of vegetation attenuation for radio wave 5G communication systems. Conclusions are drawn in Section 7.5.

7.2 RADIO WAVE ATTENUATION DUE TO VEGETATION

AI is transforming the way we think about radio channel capacity in 5G and 6G wireless communication systems. With the growth of AI, there is an increasing demand for enhanced radio channel capacity during vegetation attenuation, which is the reduction in the strength of radio signals as they pass through trees, bushes, and other vegetation. AI-driven solutions can improve the performance of 5G and 6G wireless communication systems during vegetation attenuation, which presents significant problems in wireless communication systems as it can cause signal degradation and data loss, resulting in a lower quality of service (QoS) and a decrease in the data rate [3, 4, 9–13]. AI-driven solutions are under development that can reduce the effects of vegetation on radio signals and improve QoS for end users.

The following factors are responsible for random variation in areas of vegetation [14–17]:

- Types of vegetation: tree or plant
- Size or shape of leaves, branches and tree trunks
- In leaf or out of leaf

Vegetation affects radio wave propagation and can result in [15]:

- Multiple scattering
- Diffraction
- Delay
- Absorption of radiation.

Together, these different radio wave propagation characteristics generate significant overall signal attenuation due to vegetation, the value of which can be predicted by the vegetation attenuating model [13–19].

There are various methods for calculating theoretical vegetation attenuation. These models predict vegetation attenuation based on such factors as frequency of radio signal and depth of vegetation. The ITU recommendation (ITU-R) was developed from measurements carried out for vegetation attenuation, models were proposed, and the fitted ITU-R (F-ITU-R) model for vegetation attenuation was derived. This model is valid for 28GHz, 37GHz, 39GHz and 73GHz frequencies and provides separate in-leaf and out-of-leaf attenuation calculations [13–16].

Table 7.1 shows, 5G and 6G radio wave communication system vegetation attenuation based on the F-ITU-R model (in leaf) for 28GHz, 37GHz, 39GHz and 73GHz frequencies and random forest-based predicted values. Table 7.2

Table 7.1 Vegetation attenuation (F-ITU-R model, in leaf) for 28GHz, 37GHz, 39GHz and 73GHz

d (meter)	*28GHz*	*37GHz*	*39GHz*	*73GHz*
1	21.56 dB	23.45 dB	24.99 dB	30.74 dB
2	25.87 dB	28.85 dB	28.41 dB	36.56 dB
3	27.45 dB	31.96 dB	31.74 dB	40.46 dB
4	29.74 dB	33.75 dB	34.89 dB	43.47 dB
5	31.11 dB	35.45 dB	36.41 dB	45.97 dB
6	33.85 dB	36.41 dB	37.45 dB	48.11 dB
7	34.85 dB	38.12 dB	39.99 dB	50.74 dB
8	35.96 dB	39.66 dB	40.84 dB	51.70 dB
9	36.25 dB	40.25 dB	41.41 dB	53.25 dB
10	37.85 dB	41.85 dB	42.81 dB	54.67 dB

Table 7.2 Vegetation attenuation (F-ITU-R model, out of leaf) for 28GHz, 37GHz, 39GHz and 73GHz

d (meter)	*28GHz*	*37GHz*	*39GHz*	*73GHz*
1	2.85 dB	2.45 dB	2.85 dB	2.77 dB
2	3.41 dB	3.74 dB	3.7 dB	4.18 dB
3	4.96 dB	4.79 dB	4.99 dB	5.31 dB
4	5.57 dB	5.95 dB	5.41 dB	6.29 dB
5	6.86 dB	6.41 dB	6.85 dB	7.17 dB
6	6.84 dB	7.85 dB	7.74 dB	7.99 dB
7	7.25 dB	7.25 dB	7.15 dB	8.75 dB
8	7.75 dB	8.38 dB	8.15 dB	9.47 dB
9	8.74 dB	8.38 dB	9.15 dB	10.15 dB
10	9.95 dB	9.86 dB	9.89 dB	10.80 dB

shows 5G, 6G radio wave communication system vegetation attenuation based on the F-ITU-R model (out of leaf) for 28GHz, 37GHz, 39GHz and 73GHz frequencies and random forest-based predicted values. It is clear from the two tables that the in-leaf vegetation attenuation is greater than the out-of-leaf vegetation attenuation. Vegetation attenuation also continues to increase in line with increasing depth of vegetation as well as frequency increases.

Figures 7.1 and 7.2 show the 5G and 6G radio wave communication system vegetation attenuation in 2D along with the 2-axis frequency (GHz)

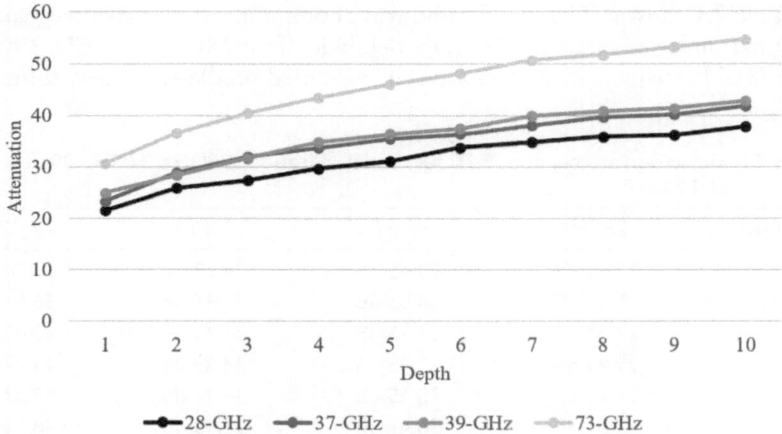

Figure 7.1 Attenuation due to vegetation (F-ITU-R model, in leaf).

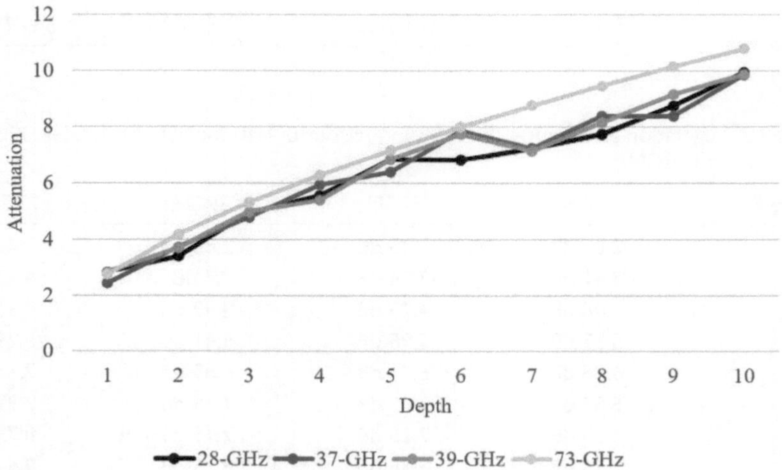

Figure 7.2 Attenuation due to vegetation (F-ITU-R model, out of leaf).

and vegetation depth (d in meter) for the random forest-based predicted values (with the graph varying for the depth of vegetation). Initially the slope of the vegetation attenuation is high because when the signal starts passing through vegetation, the initial attenuation rate is higher due to rapid reduction of the coherent component of the radio. The attenuation curve slope then reduces because as vegetation depth increases, the incoherent (diffuse) components suffer lower attenuation, resulting in low absorption losses in a vegetation medium. The attenuation continues to increase as vegetation depth increases from 1 to 10 meters and as frequency increases from 20GHz to higher GHz [13–19]. The F-ITU-R model with ground reflection consideration shows better radio vegetation attenuation prediction accuracy as compared to other models [16–19].

7.3 EMBEDDED ARCHITECTURE: AI-DR VEGETATION ATTENUATION PREDICTION SYSTEM

Machine learning (ML) can undertake complex data processing with various input data components that can be processed through mobile systems on a chip (SOC) [23, 24]. Embedded hardware architecture can be customized for machine learning algorithms, especially random forest, to improve computational efficiency for various 5G and 6G wireless applications. At system level, ML-based 5G and 6G applications, together with software-defined radio (SDR), can provide more product commercialization opportunities.

AI-driven radio channel capacity helps to enhance the overall performance in 5G and 6G wireless communication systems and reduce vegetation attenuation [23, 24]. Vegetation attenuation is a major issue in wireless communication as it reduces both strength and quality of the transmitted signal. It is caused by trees, bushes, and other vegetation blocking radio signals. This problem is particularly pronounced in rural areas where the density of vegetation is high. The SOC integrated with AI algorithms enables real-time channel monitoring and optimization. The AI algorithms analyze the channel conditions and make decisions on the transmission parameters such as power, frequency, and modulation scheme. The real-time optimization of the transmission parameters results in an increased radio channel capacity and improved communication performance. The real-time monitoring and optimization of the transmission parameters by AI algorithms ensures that the communication performance remains high, even in areas with high vegetation density [23, 24].

Attenuation of channel capacity in wireless communication propagation caused by leaf blockage can be predicted using machine learning algorithms and embedded systems. In this scenario, the impact on the wireless signal's performance in terms of signal frequency and power amplitude is analyzed based on plant depth and size, and leaf size. The first step in this process is to gather data on the impact of vegetation blockage on the wireless signal

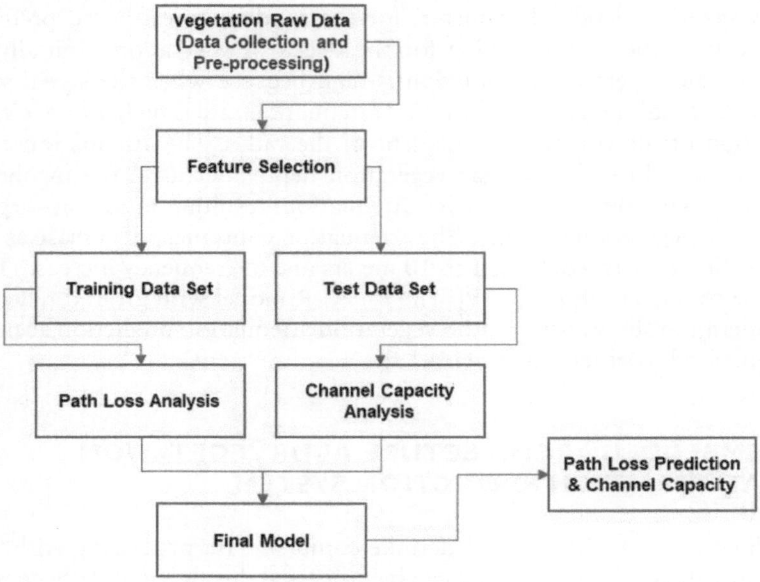

Figure 7.3 Block diagram of the proposed model.

quality. Various tools and techniques can be used to collect these data, such as signal strength meters, spectrum analyzers, and electromagnetic field simulators. The data are then used to train machine learning models, such as decision trees, random forests, and neural networks, to predict the impact of vegetation blockage on signal quality. The machine learning models can be embedded into an SOC or an embedded system integrated into the wireless communication infrastructure to provide real-time analysis and prediction of the impact of vegetation blockage on signal quality. The system can also be configured to continuously collect data and update the machine learning models to ensure that they remain accurate and up to date. Figure 7.3 shows vegetation attenuation and channel capacity per unit bandwidth prediction as a function of vegetation depth and signal-to-noise (SNR) power ratio.

A block diagram for a path loss prediction system based on machine learning techniques might include the following components:

- *Data collection and data preprocessing*. In this stage, the system collects data on various radio frequency (RF) propagation parameters, such as frequency, distance, and environmental characteristics. This data is used to train the machine learning algorithms. This stage involves cleaning, transforming, and organizing the collected data into a form suitable for use by the machine learning algorithms. This may involve removing missing or noisy data, normalizing the data, and splitting it into training and test datasets. In this stage, the system selects the appropriate machine learning algorithm to use for the path loss prediction task. Several algorithms are available, including linear regression, decision trees, support vector machines, and artificial neural networks.

- *Model training*. The selected machine learning algorithm is trained on the preprocessed data using the training dataset. The goal is to learn the relationships between the RF propagation parameters and the path loss, so that the algorithm can predict the path loss for new, unseen data. The trained model is validated using the test dataset. The performance of the model can be evaluated using metrics such as accuracy, precision, and recall.
- *Path loss and channel capacity prediction*. The trained and validated model is used to predict the path loss for a new set of RF propagation parameters. The inputs to the model are frequency, distance, and environmental characteristics, and the output is the predicted path loss and chanel capacity. The predicted path loss can be visualized with related values of path losses prediction and channel capacity.

The main objective of this strategy is to enhance data transmission rates, reduce latency, improve reliability, and enhance the overall performance of the communication system. The AI algorithms are designed to analyze the radio channel conditions in real time and make intelligent decisions regarding the selection of the radio channel, modulation technique, and coding scheme. The AI algorithms also adapt to the changing radio channel conditions, ensuring that the communication system is always operating at its maximum capacity. This results in improved data transmission rates, lower latency, increased reliability, and enhanced overall performance. In addition to the technical benefits, the implementation of AI-based techniques in the prediction of 5G wireless channel attenuation can bring about significant social and economic benefits. Improved network reliability will lead to increased access to high-speed internet for users, particularly in rural areas where vegetation obstruction is more prevalent. This will have a positive impact on the local economy as businesses can leverage high-speed internet for their operations and consumers can access a wider range of digital services.

In addition to these benefits, the AI-driven radio channel capacity strategy enables the communication system to be more flexible and scalable making it easily adaptable to different network configurations, changing traffic patterns, and requirements of different users. By incorporating AI algorithms and techniques into SOC, the communication system can be optimized to provide enhanced performance, reliability, and scalability. This will result in better and more efficient wireless communication systems in the future. The significance of the SDR-based vegetation attenuation control module in radio wave communication systems lies in the software control [13–15, 17–21]. The system has various sub-modules in software programs and reconfigurable hardware to allow power radiation change as per demand and need [21].

7.4 SCC PREDICTION

AI can be used to predict channel capacity, which is the maximum discharge that a channel can carry without experiencing significant damage. The presence of vegetation affects the flow of water and can result in changes to the channel capacity. Several AI techniques can be used for channel capacity prediction in vegetation channels, including machine learning algorithms such as random forest, decision trees, and support vector machines. These algorithms can be trained on data such as channel geometry, vegetation density, and flow velocity to make predictions about the channel capacity. In order to obtain accurate results, it is important to gather a sufficient amount of data and to clean and preprocess the data appropriately. Channel capacity prediction in the presence of vegetation blockage attenuation is a complex process that involves multiple factors such as tree depth, leaf size, and depth. In order to predict channel capacity, we need to use the equation and relation between these parameters. The Shannon channel capacity in wireless communication propagation in the presence of vegetation blockage is determined by the signal quality impact on the signal frequency power amplitude of the wireless signal in 6G and 5G. Advanced technologies and techniques can help improve the channel capacity in such scenarios. The equation for the channel capacity (C) can be given as:

$$C = B * \log_2\left(1 + (S/N)\right) \tag{7.1}$$

where B is the bandwidth of the channel, S is the signal power, and N is the noise power. The relation between the tree depth, leaf size, and depth can be given as:

$$\text{Attenuation} = A * e^{-kd} \tag{7.2}$$

where A is the initial signal strength, k is the propagation constant, and d is the distance between the transmitter and receiver. By incorporating the above relation into the equation for the channel capacity, we can predict the channel capacity in the presence of vegetation blockage attenuation as:

$$C = B * \log_2\left(1 + \frac{S}{N + A * e^{-kd}}\right) \tag{7.3}$$

In the presence of vegetation, the signal power is reduced due to the attenuation caused by the trees. The amount of attenuation depends on various factors such as tree depth, size of leaves and depth of the tree. These factors have a direct impact on the signal quality and frequency power amplitude, which in turn affects the channel capacity. The use of advanced antenna systems and beamforming techniques can help mitigate the impact of the vegetation blockage on the signal quality and frequency power amplitude, improving the channel capacity.

The use of machine learning techniques such as random forest and decision tree algorithms can provide a powerful tool for predicting the impact of vegetation tree plant leaf blockage on wireless communication propagation. By analyzing factors such as plant depth, size of leaves, and other environmental variables, these algorithms can accurately predict the effect of the blockage on the signal frequency, power amplitude, and overall signal quality. One of the key components of AI-driven channel capacity prediction is the use of mobile SOC. SOC is a compact, integrated circuit that combines multiple components, including processors, memory, and communication interfaces, into a single device. This technology makes it possible to integrate AI algorithms into mobile devices, enabling real-time channel capacity predictions in networks [14, 22].

In this section we predict Shannon channel capacity for 5G and 6G radio wave communication systems taking one of the major wireless channel attenuation sources, i.e., vegetation attenuation, into account [14]. According to Shannon's channel capacity theorem, channel capacity C is calculated in b/s/Hz for a continuous AWGN channel [22]. Vegetation channel capacity per unit bandwidth is predicted based on the F-ITU-R vegetation attenuation model for in leaf and out of leaf for the frequencies 28GHz, 37GHz, 39GHz and 73GHz [16]. AI algorithms can be used to analyze large amounts of data from the communication channel, including signal strength, interference, and noise levels. This information is then used to predict the maximum data rate that can be transmitted over the channel, which is known as the channel capacity. The use of mobile SOC, combined with AI algorithms, makes it possible to integrate this functionality into mobile devices, enabling real-time predictions in these networks.

Table 7.3 shows the vegetation channel capacity per unit bandwidth (F-ITU-R model, in leaf) for 28-GHz, 37-GHz, 39-GHz and 73-GHz. Table 7.4 shows the vegetation channel capacity per unit bandwidth (F-ITU-R model,

Table 7.3 Vegetation channel capacity/bandwidth (F-ITU-R, in leaf) for 28GHz, 37GHz, 39GHz & 73GHz

d (meter)	*28GHz*	*37GHz*	*39GHz*	*73GHz*
1	5.52 (b/s)	5.85 (b/s)	5.82 (b/s)	5.88 (b/s)
2	5.85 (b/s)	5.96 (b/s)	5.15 (b/s)	5.26 (b/s)
3	5.45 (b/s)	5.15 (b/s)	5.85 (b/s)	5.84 (b/s)
4	5.78 (b/s)	5.75 (b/s)	5.14 (b/s)	5.56 (b/s)
5	5.85 (b/s)	4.85 (b/s)	4.96 (b/s)	4.91 (b/s)
6	5.96 (b/s)	4.15 (b/s)	4.74 (b/s)	4.87 (b/s)
7	4.74 (b/s)	4.74 (b/s)	4.96 (b/s)	4.84 (b/s)
8	4.89 (b/s)	4.96 (b/s)	4.15 (b/s)	4.47 (b/s)
9	4.11 (b/s)	4.25 (b/s)	4.74 (b/s)	4.87 (b/s)
10	4.89 (b/s)	4.85 (b/s)	4.85 (b/s)	4.97 (b/s)

Table 7.4 Vegetation channel capacity / bandwidth (F-ITU-R, out of leaf) for 28GHz, 37GHz, 39GHz & 73GHz

d (meter)	*28GHz*	*37GHz*	*39GHz*	*73GHz*
1	5.78 (b/s)	5.89 (b/s)	5.99 (b/s)	6.19 (b/s)
2	5.87 (b/s)	5.96 (b/s)	5.96 (b/s)	5.94 (b/s)
3	5.74 (b/s)	5.84 (b/s)	5.94 (b/s)	5.91 (b/s)
4	5.69 (b/s)	5.74 (b/s)	5.98 (b/s)	5.92 (b/s)
5	5.55 (b/s)	5.85 (b/s)	5.90 (b/s)	5.94 (b/s)
6	5.89 (b/s)	5.91 (b/s)	5.99 (b/s)	5.91 (b/s)
7	5.87 (b/s)	5.86 (b/s)	5.95 (b/s)	5.91 (b/s)
8	5.46 (b/s)	5.75 (b/s)	5.94 (b/s)	5.98 (b/s)
9	5.15 (b/s)	5.58 (b/s)	5.83 (b/s)	5.88 (b/s)
10	5.43 (b/s)	5.82 (b/s)	5.94 (b/s)	5.98 (b/s)

out of leaf) for 28GHz, 37GHz, 39GHz and 73GHz. The table data is shown for 65dB signal-to-noise (SNR) power ratio. It is clear from the tables that the channel capacity per unit bandwidth (b/s) decreases as the depth of the vegetation increases [16].

Vegetation contributes to reducing the signal strength, due to which overall signal attenuation takes place. Signal suffers due to fading and losses. Vegetation attenuation prediction is based on the F-ITU-R in-leaf and out-of-leaf vegetation attenuation models for the 28GHz, 37GHz, 39GHz and 73GHz frequencies [16].

From Figures 7.4 and 7.5 it is clear that the channel capacity decreases as the vegetation depth increases [16]. It can also be observed that for a constant vegetation depth, the channel capacity increases with the signal-to-noise ratio (SNR) [22]. AI can also be utilized to perform real-time monitoring of the radio

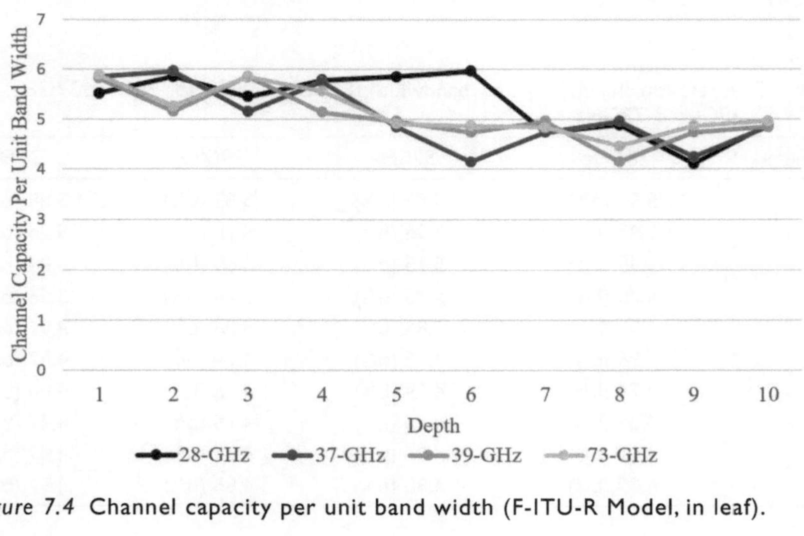

Figure 7.4 Channel capacity per unit band width (F-ITU-R Model, in leaf).

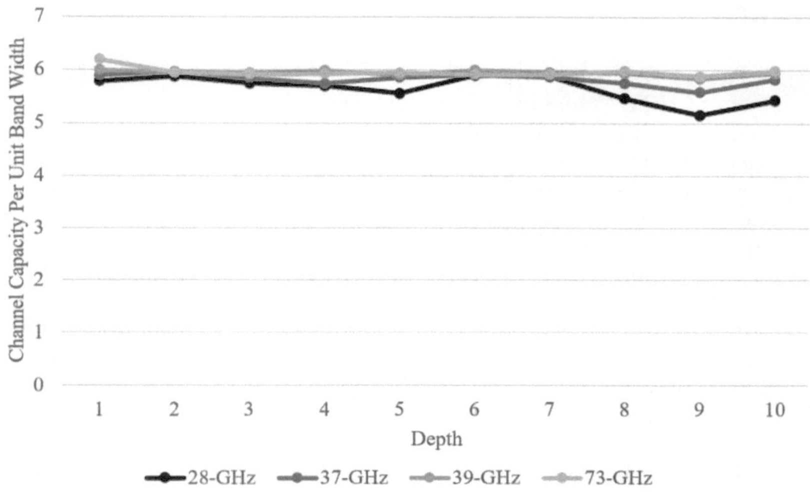

Figure 7.5 Channel capacity per unit band width (F-ITU-R Model, out of leaf).

channel conditions and to make dynamic adjustments to the radio channel capacity in response to changes in the vegetation environment. The systems can greatly improve the radio channel capacity during vegetation attenuation, ensuring reliable and high-quality wireless communication services [14, 16, 22].

7.5 CONCLUSIONS

This chapter presents vegetation attenuation prediction calculation and Shannon channel capacity prediction calculation in the presence of vegetation areas for 5G and 6G radio wave communication systems. The vegetation attenuation and Shannon channel capacity is predicted for the FCC-recommended channels. AI-DR-based vegetation attenuation prediction calculations based on the F-ITU-R model are useful to estimate link margins quickly without actual launching the system in a vegetation area. With these technologies, the prediction of channel capacity and vegetation blockage attenuation can be improved, resulting in a better user experience and improved network performance. The implementation of such predcitions in wireless propagation will provide a more robust and efficient system for users in areas with dense vegetation coverage.

ACKNOWLEDGMENTS

The authors are very grateful to the Samsung Research Institute Noida, India and Delhi Technological University (DTU), previously known as Delhi College of Engineering (DCE), New Delhi, India.

REFERENCES

[1] Z. Pi, and F. Khan, "An introduction to millimeter-wave mobile broadband systems," *IEEE Communications Magazine*, vol. 49, no. 6, pp. 101–107, 2011.
[2] J. G. Andrews et al., "What will 5G be?," *IEEE Journal on Selected Areas in Communications*, vol. 32, no. 6, pp. 1065–1082, 2014.
[3] C. Jeong, J. Park, and H. Yu, "Random access in millimeter-wave beamforming cellular networks: issues and approaches," *IEEE Communications Magazine*, vol. 53, no. 1, pp. 180–185, 2015.
[4] T. S. Rappaport et al., "Millimeter wave mobile communications for 5G cellular: It will work!," *IEEE Access*, vol. 1, pp. 335–349, 2013.
[5] W. Roh et al., "Millimeter-wave beamforming as an enabling technology for 5G cellular communications: Theoretical feasibility and prototype results," *IEEE Communications Magazine*, vol. 52, no. 2, pp. 106–113, 2014.
[6] Samsung, "Samsung-5G-Vision.pdf," August. 2015. [Online]. Available: http://surl.li/gikxl [Accessed April 12, 2023].
[7] L. Wei, R. Q. Hu, Y. Qian, and G. Wu, "Key elements to enable millimeter wave communications for 5G wireless systems," *IEEE Wireless Communications*, vol. 21, no. 6, pp. 136–143, 2014.
[8] Millimeter Wave Propagation, "Spectrum Management Implications. FCC, Office of Eng. And Tech," *Bulletin*, no. 70, 1997.
[9] Federal Communications Commission, "Report and order and further notice of proposed rulemaking," *Matter of Revision of the Commission's Rules to Ensure Compatibility with Enhanced*, vol. 911, pp. 94–102, 2016.
[10] D. L. Means, and K. W. Chan, "Evaluating Compliance with FCC Guidelines for Human Exposure to Radiofrequency Electromagnetic Fields," Supplement C Edition 01-01 to OET Edition 97-01, Bulletin 65, June 2001. [online]. Available: https://transition.fcc.gov/Bureaus/Engineering_Technology/Documents/bulletins/oet65/oet65c.pdf [Accessed April 12, 2023].
[11] S. K. Agrawal, and K. Sharma, "5G millimeter wave (mmWave) Communications," *3rd International Conference on Computing for Sustainable Global Development (INDIACom)*, New Delhi, India, pp. 3630–3634, 2016.
[12] S. K. Agrawal and K. Sharma, "5G Millimeter Wave Communication System with Software Defined Radio (SDR)," *International Conference on Recent Trends in Engineering & Science* (ICRTES-16), pp. 263–270, 2016.
[13] S. K. Agrawal and P. Garg, "Calculation of channel capacity considering the effect of different seasons for higher altitude platform system," *Wireless Personal Communications*, vol. 52, pp. 719–733, 2009.
[14] Y. S. Meng, Y. H. Lee and B. C. Ng, "Empirical near ground path loss modeling in a forest at VHF and UHF bands," *IEEE Transactions on Antennas and Propagation*, vol. 57, no. 5, pp. 1461–1468, 2009.
[15] S. K. Agrawal and P. Garg, "Calculation of the channel capacity in the presence of vegetation and urban-site environment for higher altitude platform communication System, to Wireless Personal Communications," *IET Microwaves, Antenna & Propagation*, vol. 3, no. 4, pp. 703–713, 2009.
[16] S. K. Agrawal and K. Sharma, "5G millimeter wave communication system with software defined radio (SDR)," *International Journal for Innovative Research in Science & Technology*, vol. 2, no. 9, pp. 263–270, 2016.

[17] K.-C. Chen, R. Prasad, and H. V. Poor, "Software radio," *IEEE Personal Communications*, vol. 6, no. 4, pp. 12–12, 1999.
[18] L. Mitola, "Technical challenges in the globalization of software radio," *IEEE Communications Magazine*, vol. 37, no. 2, pp. 84–89, 1999.
[19] Lie-Liang Yang and L. Hanzo, "Software-defined-radio-assisted adaptive broadband frequency hopping multicarrier DS-CDMA," *IEEE Communications Magazine*, vol. 40, no. 3, pp. 174–183, 2002.
[20] G. J. Minden et al., "KUAR: a flexible software-defined radio development platform," *2nd IEEE International Symposium on New Frontiers in Dynamic Spectrum Access Networks*, Dublin, Ireland, pp. 428–439, 2007.
[21] C. E. Shannon, "Communication in the presence of noise", *Proceedings of the IEEE*, vol. 37, no. 1, pp. 10–21, 1949.
[22] D. Giri, K. -L. Chiu, G. Di Guglielmo, P. Mantovani, and L. P. Carloni, "ESP4ML: Platform-based design of systems-on-chip for embedded machine learning," *2020 Design, Automation & Test in Europe Conference & Exhibition (DATE)*, Grenoble, France, pp. 1049–1054, 2020.
[23] Y. Zhu, M. Mattina, and P. N. Whatmough, "Mobile machine learning hardware at ARM: A Systems-on-Chip (SoC) perspective," in *Proc. First Conference on Systems and Machine Learning (SysML-18)*, Stanford, CA, pp. 1–3, 2018.
[24] N. Chawla, A. Singh, M. Kar and S. Mukhopadhyay, "Application inference using machine learning based side channel analysis," *International Joint Conference on Neural Networks (IJCNN)*, Budapest, Hungary, pp. 1–8, 2019.

Chapter 8

Smart cabin for office using embedded systems and sensors

Anirban Dasgupta, Abhranil Das and Parishmita Deka
IIT Guwahati, North Guwahati, India

Soham Das
IIT Kharagpur, Kharagpur, India

8.1 INTRODUCTION

The smart city [1], the new vision of many developing countries, uses information technology (IT) and communication to improve operational efficiency and increase the well-being, awareness, and safety of its citizens. This goal is achieved by collecting specific data through appropriate sensors, then pre-processing the data and performing data analytics in order to help manage assets, resources, and services efficiently. One aspect of smart cities is the development of smart offices.

8.1.1 Smart office

A smart office is a cabin room, in which sensors are placed for specific purposes. Office security, and the comfort and well-being of staff are the prime factors that need to be addressed to make an office cabin smart. Sensors and actuators communicate with each other in the smart office and make the overall group of subsystems work in cohesion. This functionality is popularly termed the internet of things (IoT), a system of interconnected subsystems capable of communicating with each other.

The term 'smart cabin' was initially envisioned for smart cars [3], aircraft [4], and ships [5]. One of the few prominent works to use the term 'smart cabin' for offices is the research article [6], whose authors developed a proof-of-concept system using Arduino and a number of sensors for temperature, sound, and flame. Their objective was to control the lighting and fans using sensor-based switches. This work has opened up the possibility of exploring additional sensors which can be incorporated into a smart office cabin with added functionality.

This chapter describes an attempt to develop a smart cabin for offices which includes the following three key functionalities:

- office security and attendance
- ambiance monitor
- well-being monitor

 DOI: 10.1201/9781032628059-8

Office security is of utmost importance, especially when confidential material is involved. The system should ensure that only authorized personnel can enter the office cabin and must be protected from impersonation. Hence, the first functionality is a three-stage authentication system which is more reliable than single-stage verification.

Since office staff spend a significant amount of time in their cabins, it is essential that the ambiance of the cabin room is maintained to provide sufficient comfort. Uncomfortable office environments have led to health issues for staff [7]. Ambiance monitoring systems record the real-time values of surrounding environmental factors through appropriate sensors. In this work, we capture the measurements from different sensors simultaneously using an ESP32.

The final functionality concerns the observation of the physical and mental health of the office staff. Physical health can be monitored by examining such parameters as blood pressure, heart rate, oxygen saturation, and body temperature. Blood pressure is a measure of the force exerted by the blood per unit area of the artery walls. A normal blood pressure level is around 120/80 mm of Hg [8]. Heart rate is the number of times per minute the heart beats, normal values being between 60 and 100 times per minute for adults [9]. Oxygen saturation is a measure of the percentage of hemoglobin bound to oxygen. Normal arterial blood oxygen saturation levels are 97–100 percent [10].

Positive mental health and well-being are desirable to enhance productivity. Psychological assessments [11] include the detection of fatigue and negative emotions in the office employee, which may lead to stress, anxiety, or depression and can result in a decline in performance and increased chance of human error [13]. Negative emotions can adversely impact work performance and hence need to be identified. Stress may be defined as a feeling of affective tension, manifested from any event or thought that presents frustration, anger, or nervousness [14], and can be a crucial contributor to declining work performance [15]. Fatigue and negative emotional state can be detected through physiological signals such as electroencephalograms (EEG) [16] with significant accuracy. However, in the office setting, the monitoring has to be conducted unobtrusively. The mental well-being assessment scheme proposed captures facial images in visual and near-infrared modalities and uses machine learning models to identify the occurrence of such events.

8.1.2 Motivation

As stated earlier, there are occasions when an employee in an office experiences health issues [17] requiring medical attention. If these issues are diagnosed late, they may lead to serious problems and perhaps even loss of life, for example, due to a stroke. Either the ambiance or the health of the employee may be responsible. It is therefore important to monitor both ambiance and health conditions. Security measures to prevent impersonation of employees to gain access to important documents requires authentication via biometrics.

8.1.3 Objectives

The objectives of this work are as follows:

- To develop an ambiance monitor using an embedded system that can take data from various sensors to estimate environmental variables such as temperature, relative humidity, dust level, CO_2 levels, etc.
- To develop a health monitor using an embedded system and respective sensors to identify mental health conditions such as negative emotion, and to monitor fatigue levels, as well as vital physiological parameters such as blood pressure, body temperature, oxygen saturation, heart rate, etc.
- To develop a biometric authentication system for security purposes.
- To transmit data from each smart system on a single platform which would display every parameter and their respective conditions.

8.2 CABIN SECURITY

Biometric authentication is essential to restrict unauthorized access to the smart cabin. The most popular biometric authentication systems use fingerprint mapping, facial recognition, speech recognition, and retinal scans. Retina scanning is comparatively expensive, but fingerprint scanners are relatively inexpensive and hence are widely used in biometrics. However, injuries or even sweat on the fingertips can interfere with fingerprint scans. Techniques to spoof and replicate fingerprints are also appearing. Speech recognition can fail due to factors like ambient noise, respiratory illness, etc. There can also be issues with facial identification due to variations in facial expressions, lighting, facial hair, etc.

Face verification compares a candidate face to another and verifies whether it is a match. It is a one-to-one mapping, two-class classification problem. Many uncertainty factors affect the verification performance, such as different expressions, poses, and lighting variations. The classical face verification approach uses features like local binary pattern (LBP) [18], 2D Gabor wavelets [19], etc., followed by dimensionality reduction using methods like principal component analysis (PCA) [20]. Finally, distance-based matching is performed, such as a k-nearest neighbour [21]. These approaches work well in conditions where the variations in facial expressions, pose, and light levels are constrained. Deep learning methods have to a great extent overcome the limitations of these methods. However, in this work, we have two constraints. The first is that being a security system, it has to work with near perfection. The second is that the system has to be implemented on an embedded platform; hence, it should work within the constraints of memory and processing speed.

Speaker verification involves examining the speech signal of the office staff to authenticate them as either genuine or an impostor. The speaker verification system works on the assumption that every speaker's voice is unique. The speaker's speech is compared with their template speech patterns that are already enrolled in the system [22]. These systems generally analyze the features of speech that differ between speakers. There are two types of speaker verification systems, based on the type of data used for enrolment and recognition: text dependent or text independent. In text-dependent recognition systems, the text which the test user will speak is kept the same for both enrolment and recognition phases. Text-independent systems have different texts for enrolment and recognition phases. The text-dependent scheme provides better accuracy as the feature matching will be more accurate with speech signals with similar text content. Accordingly, we have used the text-dependent scheme.

The fingerprint-based biometric is one of the most common techniques, as a fingerprint is a unique signature of the human body that distinguishes one individual from others. A fingerprint is also easily accessible, and sensing is inexpensive using a fingerprint sensor. Fingerprints are superior to face-based biometrics, as fingerprints are different even for identical twins [23]. Moreover, fingerprints are permanent in an individual's lifetime and relatively immune to aging and biological changes [24], when compared with face and speech.

Since none of these techniques can guarantee perfect recognition, an ensemble of the methods through a voting scheme can provide near-perfect results. Here, we use the information from fingerprints, facial images, and speech signals using the respective sensors, which is inexpensive and readily interfaceable with standard embedded communication protocols. Each problem is a two-class classification problem, where the classes are either the officer or not. The system successively inputs each modality in fingerprint, visual image, and speech (Figure 8.1). The user has to push a button to initiate the process. When the button is pushed, the cabin security system asks for the fingerprint. A green LED glows on successful capture of the fingerprint data. If the fingerprint is verified, the system says "fingerprint verified"; if not, it says "fingerprint verification failed". The system next asks the individual to look at the camera to capture their face. If the face is found, the green LED glows again. The face verification module executes, and accordingly, the decision is announced to the speakers. If neither stage verifies the system, the authentication process stops. Otherwise, the system asks for audio input to perform speech verification.

8.2.1 Fingerprint verification

A fingerprint sensor uses an optical system to scan a fingerprint; detection is done by reading the contours [25]. The fingerprint sensor used here is R307, connected to the Raspberry Pi using the RXD, TXD, 3.3 V, and GND on the GPIO pins. The operating voltage of this sensor module is 3.3 V. A single

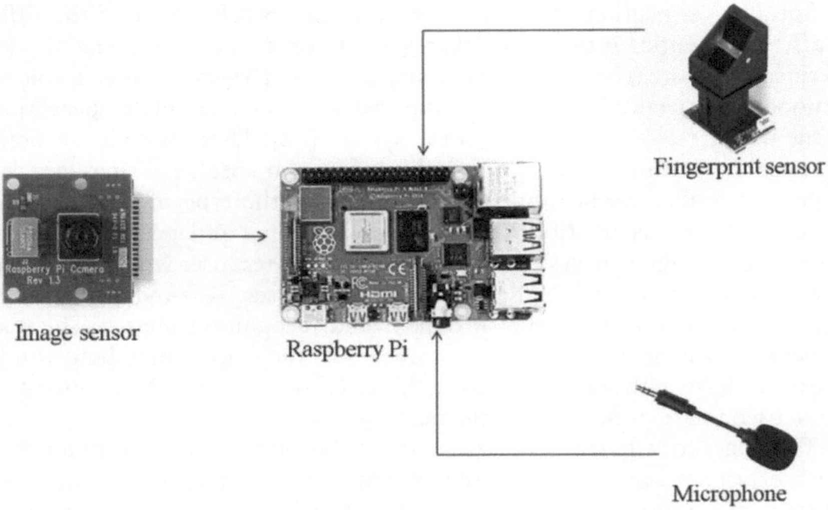

Figure 8.1 The subsystem for biometric authentication using image, speech, and fingerprint data from respective sensors connected to the Raspberry Pi.

fingerprint can be stored in different positions for clear, rapid detection. There is a provision to store up to 1000 different fingerprints.

Fingerprint capture results in 8 bytes of data which are converted into hexadecimal form and sent for fingerprint extraction. The next step is fingerprint matching, using the ridges and valley pattern. The most popular method for this is minutiae-based matching. The input is a fingerprint query along with an identity. The minutiae points of the query and template fingerprints are taken and represented in the form of vectors. Every element of this vector is a minutia point that can be described by different properties such as position, type, orientation, quality of the neighborhood region, etc. This method includes the following steps:

- Thinning of image.
- Minutiae detection.
- Minutiae matching using Euclidean distance.

8.2.2 Face verification

Face detection is based on ResNet [26]. If multiple faces are detected in the frame, the face region with the highest probability is selected. The first five detected faces are stored for face verification. FaceNet is used to verify the test face from the template face. This network directly learns to map from face images to a compact Euclidean space, where the Euclidean distances measure face dissimilarity, with the FaceNet embeddings forming the feature vectors. The verification process is carried out by comparing the

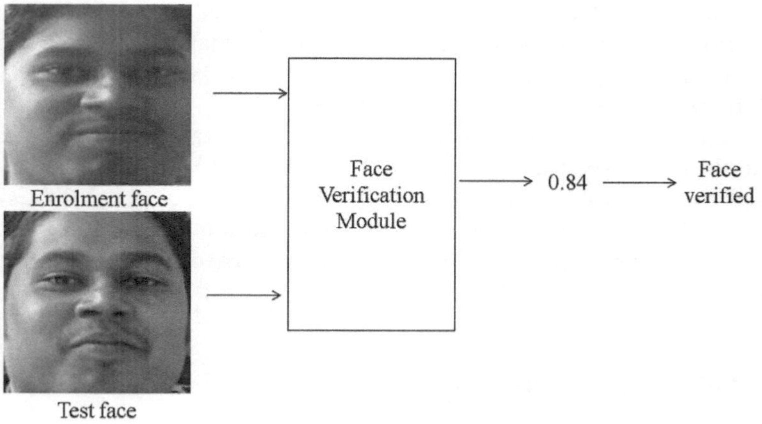

Figure 8.2 Sample face verification.

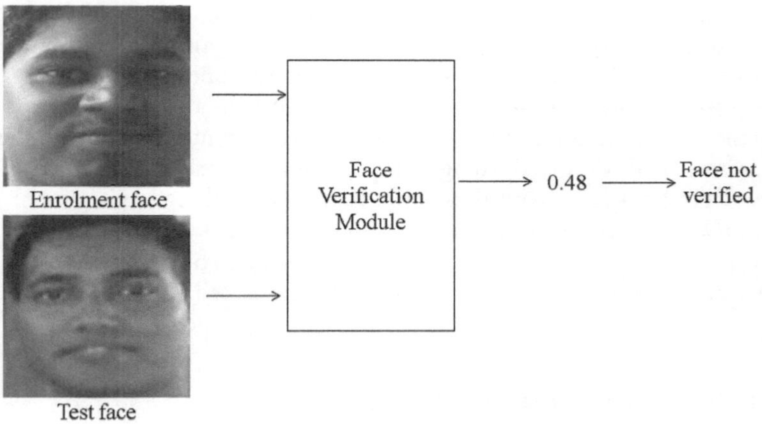

Figure 8.3 Sample case where face is not verified.

embedding of the test face image on this space with a threshold. The module outputs a probability of a match, with a value close to unity indicating a good match (Figure 8.2). A low match indicates that the test used may be an intruder (Figure 8.3). Empirically we have obtained a threshold of 0.75 to verify an individual.

8.2.3 Speech verification

The speaker verification framework consists of training, enrolment, and evaluation phases. In the training phase, the system is trained using the available data to learn the speaker-specific features from speech signals.

In the enrolment phase, the speaker utterances are fed to the trained system to obtain the speaker models, and finally, in evaluation, the test speaker utterance model is created and compared with the existing models to find similarities with already registered speakers.

In this final verification stage, the user is asked to state their full name and the audio is recorded and stored in the system RAM. Verification is performed by extracting the mel-frequency cepstral coefficients (MFCCs) and linear predictive coding (LPC). Vector quantization is performed on the coefficients, which are matched with pre-recorded data to verify the individual.

8.3 AMBIANCE MONITOR

The second subsystem is the ambiance monitor. An excellent interior ambiance is associated with highly flexible air quality and thermal comfort. The temperature, relative humidity, airflow, and other parameters in the indoor space significantly affect indoor air quality and thermal comfort. Air-conditioning systems provide a quality indoor ambiance. However, to minimize energy consumption, smart usage of air-conditioning systems may be required. Moreover, air-conditioning is not available in all offices. The ambiance monitor attempts to create an ambient indoor environment to boost staff productivity. Leakage of inflammable gases can lead to loss of life and property. To avoid these conditions, the indoor environment must be continuously monitored. In this system, the sensors used are for temperature and humidity, CO_2, total volatile organic compound (TVOC), air pressure, and dust level. These sensors are connected to an ESP32 module (Figure 8.4).

8.3.1 Temperature and relative humidity

Monitoring the relative humidity inside the office cabin is essential. Low humidity is a frequent cause of illnesses developed indoors, including dry and cracked skin, nosebleeds, chapped lips, and dry sinuses. On the other hand, high humidity may result in condensation on windows, wet stains on walls and ceilings, mold in bathrooms, musty odor, and a clammy feel to the air. Prolonged periods of high humidity can also cause rotting and structural damage to office furniture.

The temperature and relative humidity sensor used is DHT18. This sensor can sense temperature in the range of 0–50°C with a tolerance of 2°C and relative humidity in the range of 20–90% with a tolerance of ± 5%. The sensor provides a digital signal corresponding to the temperature and humidity values. The DHT11 is connected to the ESP32 board using GPIO pins. The connections are V_{CC} of 3.3 V with pin 1, ground with pin 4, and data with an internal pull-up resistor of 10 kΩ with pin 2.

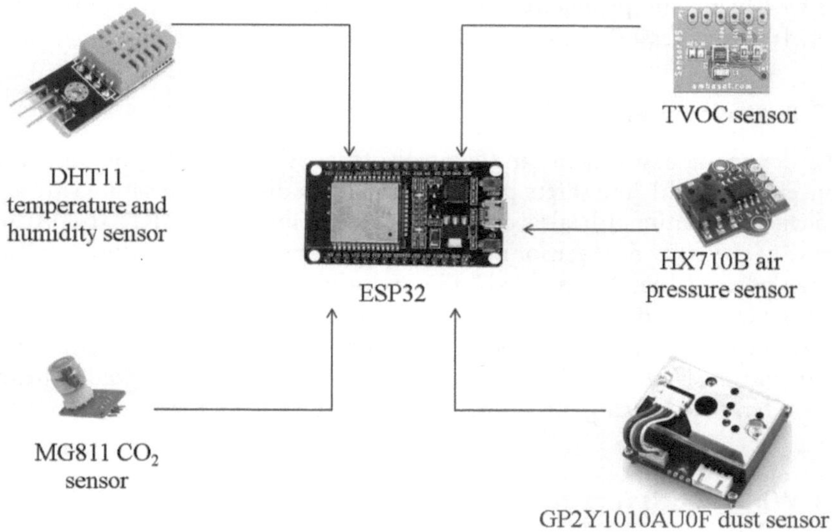

Figure 8.4 Subsystem for ambiance monitoring using ESP32.

8.3.2 Carbon dioxide

CO_2 monitoring is required to identify any areas of poor ventilation so that mechanical ventilation can be used to keep fresh air flowing, or doors and windows opened. Office cabins that are often closed for a prolonged duration result in poor ventilation. This effect leads to raised CO_2 levels. A level of more than 1000 ppm can be detrimental, causing headaches, impaired mental function, lethargy, etc. The CO_2 gas sensor used in this work is MG818. There is an onboard signal conditioning circuit to amplify the output signal and an onboard heating circuit to heat the sensor. The output voltage of the module falls as the ppm concentration of the CO_2 increases.

8.3.3 Total volatile organic compound (TVOC)

The TVOC is a significant indicator for determining hygiene and air quality in office cabins. Exposure to toxic volatile organic compounds may lead to severe health issues such as chronic respiratory disease and damage to the kidneys, liver, and central nervous system, which may even cause cancer. The sensor used for obtaining the TVOC is SGP30, which can detect a wide range of volatile organic compounds. The SGP30 sensor is connected to the ESP32 via the I2C protocol.

8.3.4 Air pressure

Measuring and maintaining air pressure in office cabins is essential to ensure the precision-controlled, comfortable office environment that is essential for the well-being and productivity of the office staff. In this work, we used the HX710B Air Pressure Sensor Module.

8.3.5 Dust level

The detection, assessment, and control of particulate matter, or dust, is a concern for health and safety professionals. Lung disease and asthma are significant occupational health problems. We employed the GP2Y1010AU0F optical air quality dust sensor to establish the concentration of dust particles in the office cabin. An infrared LED and a phototransistor are diagonally arranged so as to detect reflected light of dust in the air. The sensor can effectively detect very fine particles such as cigarette smoke and is commonly used in air purifier systems. This sensor provides an analog output and is hence connected to an analog pin of the ESP32. It requires 5V for its operation.

8.4 WELL-BEING MONITOR

The third subsystem of the smart cabin is the well-being monitor. Physical health is determined through contact-based sensors that intermittently ask the office staff to touch them to obtain the data. Mental health is assessed using a visual and an infrared camera connected to a Raspberry Pi board. These cameras capture facial images and look for signs of fatigue or negative emotion. The overall schematic of the well-being monitor is shown in Figure 8.5.

8.4.1 Physical health

Bio-sensors provide information about specific health parameters. The most common sensors include heart rate, pulse oximetry, and body temperature. The heart rate provides insight into the fitness level of the individual. An abnormal heart rate can indicate various cardiac disorders. Pulse oximetry measures the amount of oxygen saturation in the blood, indicating how well oxygen is transferred to different body parts. A reduction in value below normal can indicate a heart attack, chronic obstructive pulmonary disease (COPD), asthma, or pneumonia. Monitoring body temperature can help to detect illness or fever.

In this work, we used the MAX30102 heart rate and SPO_2 sensor, which directly provides heart rate and oxygen saturation values. The sensor is connected to the Raspberry Pi board using I2C protocol and operates on 3.3V drawn from the Pi board through its VIN pin. The SDA and SCL pins must be connected to obtain the serial data and serial clock, respectively. The remaining two pins which need connections are the INT and GND pins. The INT pin checks value availability, indicating that an employee has placed their finger on the sensor.

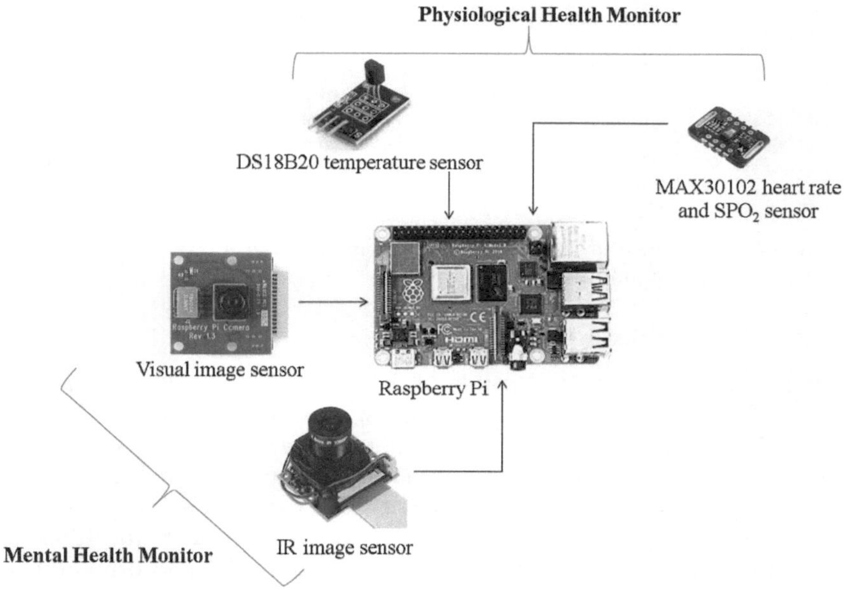

Figure 8.5 Well-being monitor using biomedical and image sensors connected to Raspberry Pi.

Body temperature is measured using the DS18B20 temperature sensor, which is connected to the Pi board via the 1-wire protocol. The sensor is of 12-bit resolution and hence can measure temperature with a resolution of 0.0625°C.

8.4.2 Mental health

Mental health is the aspect of well-being related to thoughts, feelings, and actions. Although mental health and well-being is a very broad topic, in this work, we are only looking at the identification of fatigue or negative emotion in the office employee.

8.4.2.1 Fatigue state

The most convenient and accurate way of establishing the state of fatigue is by observing the percentage of eyelid closure over time, a standard metric called PERCLOS [27], which has been widely used to identify fatigue levels in drivers [28]. The metric is computed by detecting the face and eye regions, followed by classifying the eye state as open or closed. The percentage of closed eyes over a period of one minute is the PERCLOS value. Office workers are judged to be fatigued if the PERCLOS is greater than 20%.

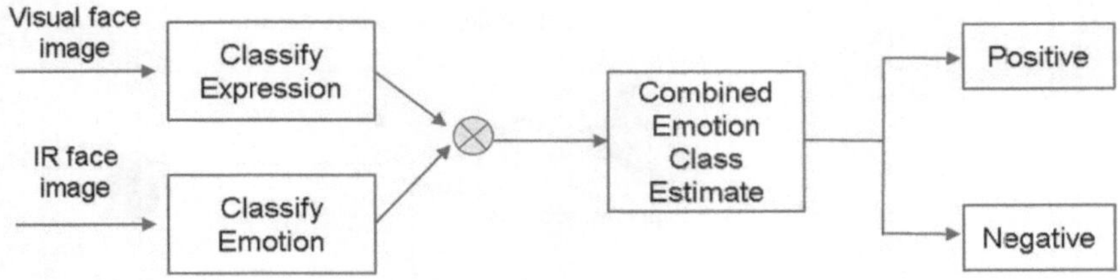

Figure 8.6 Block diagram for classifying the input facial image as a positive or negative emotion.

8.4.2.2 Emotional state

Classification of emotion depends on how many states one is interested in. For this specific situation, the simplest classification would be a two-class one: negative or non-negative emotional state (positive or neutral). This is achieved using data from visual images and infrared images. The visual images provide information about facial expression, while the infrared images show the facial heatmap distribution that is due to the difference in positive and negative emotions. Considering only visual information can be misleading as expressions can be faked. The block diagram is shown in Figure 8.6. Facial images of the office workers are captured in both visual and infrared modalities. The visual image provides information about the probability of the facial expression being positive or negative. Similarly, the infrared image indicates the probability of positive or negative emotion. The combined probabilities give a score from which the final classification is calculated. The expression and emotion classification models are based on CNNs, which are loaded into the Raspberry Pi as .h5 files.

8.5 DATA TRANSMISSION AND DISPLAY

There are three subsystems, the first of which is for verification purposes. This system is designed to act independently of the other two. The verification system communicates the access log to a server.

Similarly, the other two subsystems communicate the results of the analysis to the server, where the regular information is stored. In the event of an anomaly in the ambiance or well-being monitoring, the client unit generates an alarm. If successive alarms are generated, a critical notification is shown in the server.

8.6 RESULTS

The first subsystem is for biometric verification-based security. Here, we tested how the combined use of the three modalities of fingerprint, face and voice performed individually and jointly. Tests were conducted on 12 participants, for each of whom there were ten face images, voice samples and fingerprints, with different variations. Hence there was a total of 120 samples for each modality. The objective of the test was to compare how many individuals were correctly verified, designated true positives (TP), how many imposters were correctly denied as true negatives (TN), how many imposters were wrongly verified, designated false positives (FP), and how many original individuals were mistakenly not verified, termed false negatives (FN). The results are displayed in Figure 8.7. The percentage accuracies of the methods are provided in Table 8.1. We found that the fingerprint-based method performed best, while the combined verifications were perfect.

The ambiance monitor was tested by observing the variations in readings for three successive days in an office cabin. The data was sampled every 15 minutes (4 samples per hour) for 24 hours, resulting in 24×4 = 96 samples. The histograms for three consecutive days are shown in Figure 8.8. The mean reading has the highest frequency of occurrence or likelihood, while some outliers are observed due to external factors.

We observe a similar pattern for the relative humidity and CO_2 concentration in Figures 8.9 and 8.10 respectively. A larger variation was seen in the

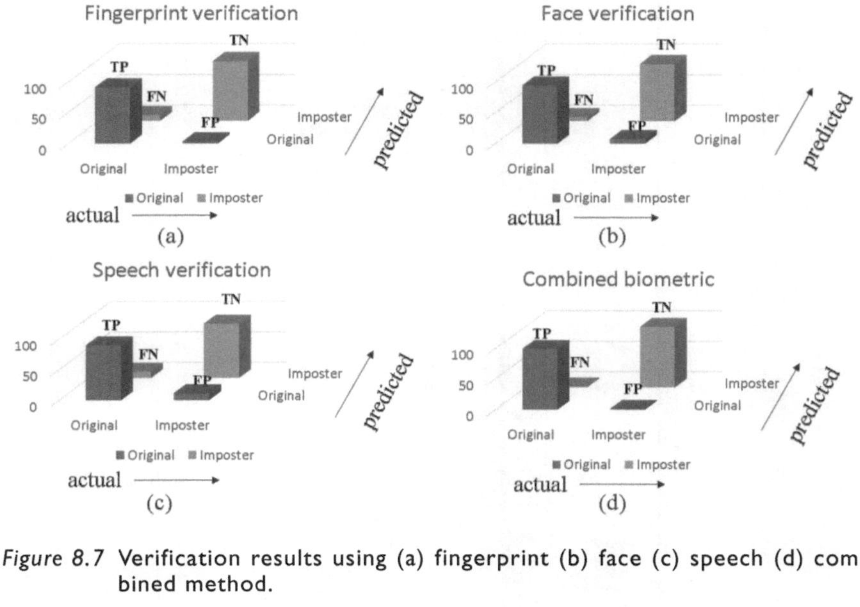

Figure 8.7 Verification results using (a) fingerprint (b) face (c) speech (d) combined method.

Table 8.1 % Accuracies of different methods

Method	*Fingerprint*	*Face*	*Speech*	*Combined*
% Accuracy	96.67	92.5	89.16	99.16

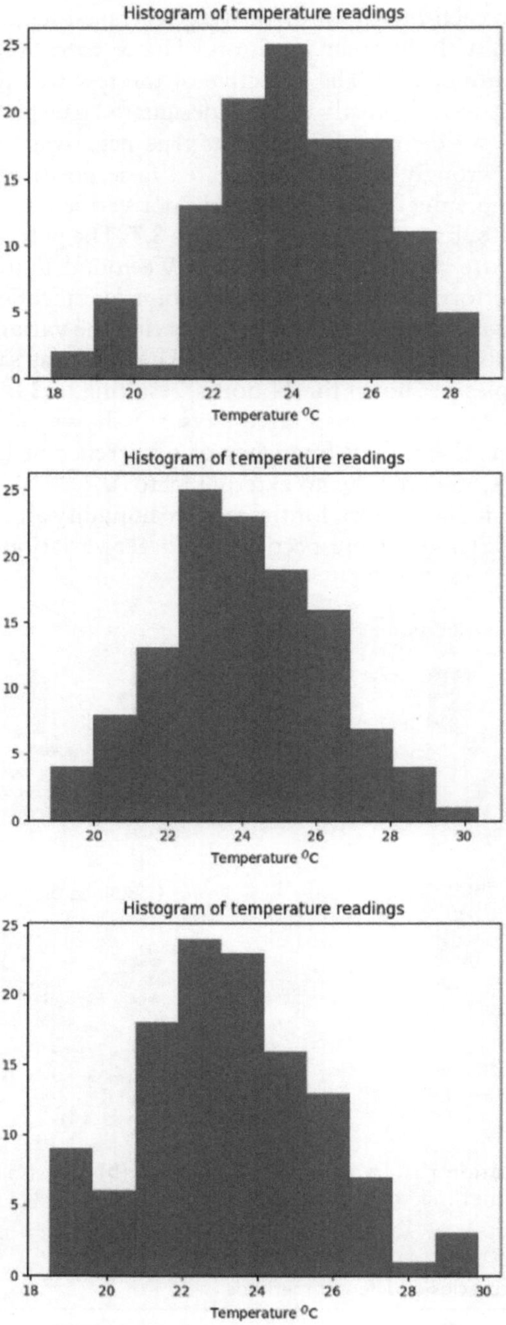

Figure 8.8 Temperature readings from sensor for three successive days.

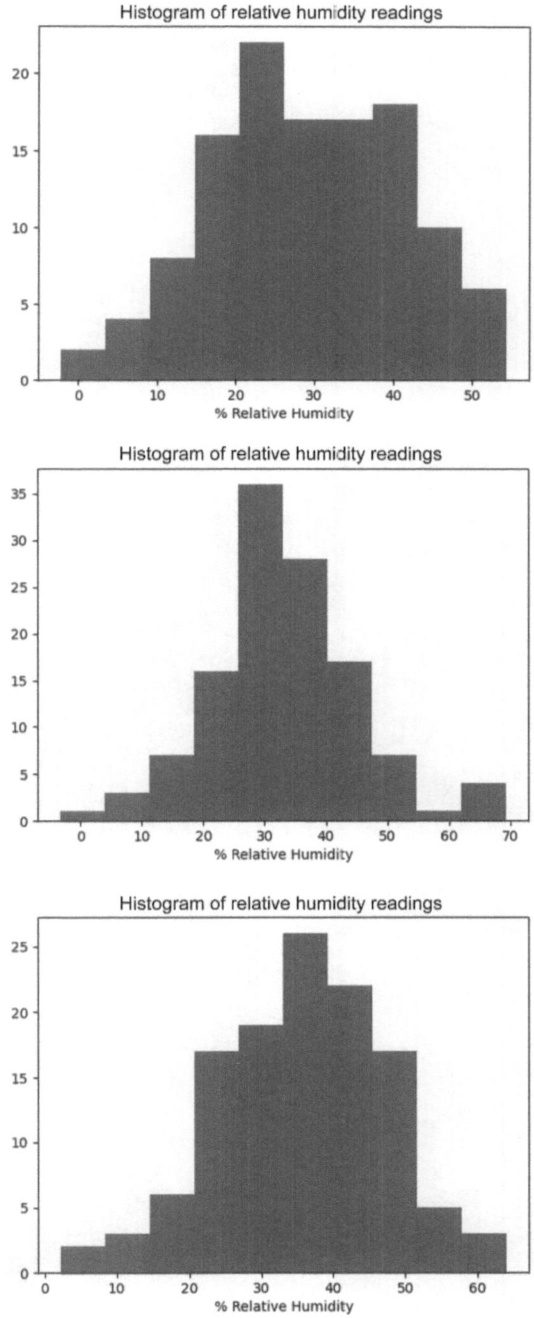

Figure 8.9 Histograms of relative humidity across three successive days.

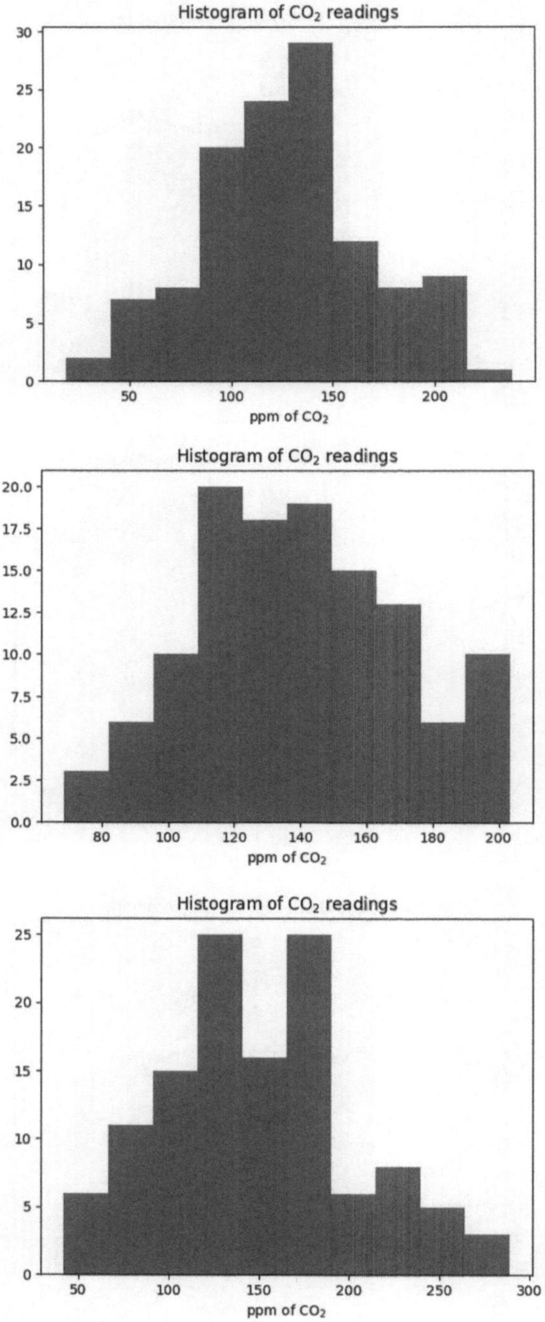

Figure 8.10 Histogram of CO_2 concentration in ppm for three consecutive days.

CO_2 readings on the third day, with the mean value also increasing. This may be because on the last day, the room was closed for a longer time, to test if the readings are affected by such a change.

Fatigue was observed through the PERCLOS measurements of 23 participants, with each reading taken at hourly intervals, referred to as stages, for a total of 12 stages. In each stage, they were given a pdf to read, while their facial images were recorded. The variation in PERCLOS is shown in Figure 8.11, as a box plot. The box plot suggests how each participant transits from an alert stage to a fatigued stage as the day progresses. The large variations in the last three stages indicate that different users experience different levels of fatigue after similar passage of intervals on similar tasks.

The classification performance for emotion is shown in Figure 8.12 (a-b). In this case, we have three classes of emotions: positive, negative, and neutral. The figure reveals that although there are certain false positives and false negatives in the estimation, the accuracies obtained from the visual and infrared modalities are 98.2% and 89.6% respectively. It is impossible to achieve near-perfect accuracy in emotion identification, as the estimation is indirect.

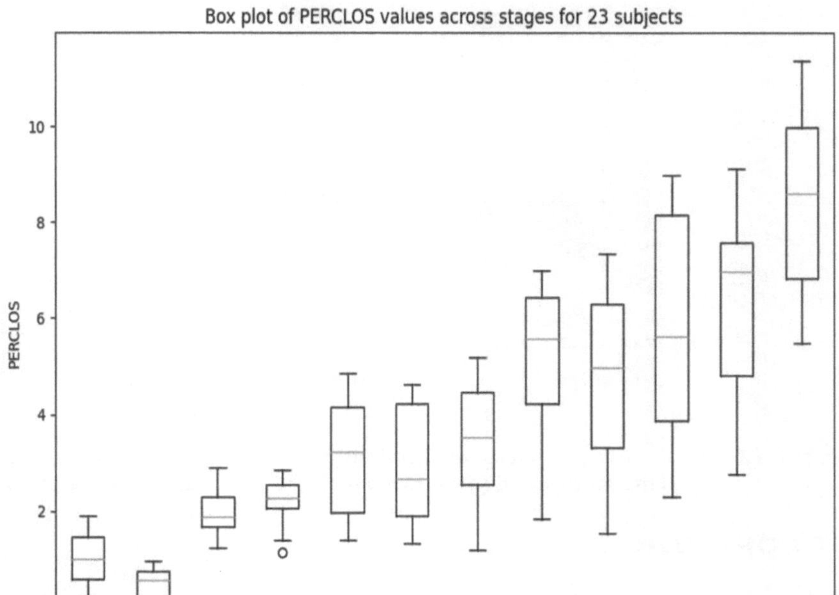

Figure 8.11 Variation of PERCLOS as an indicator of fatigue.

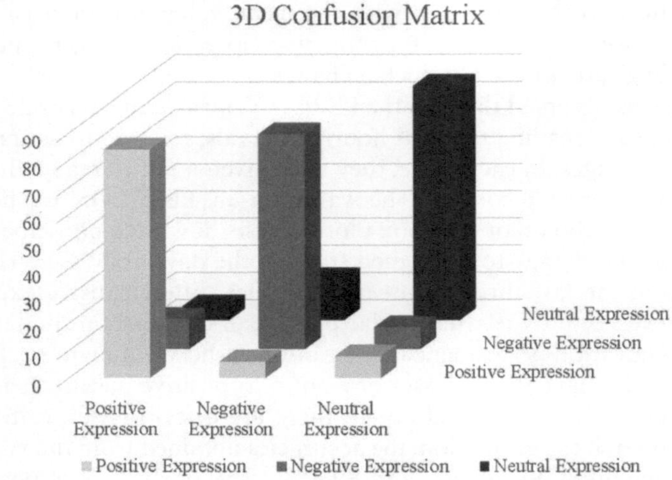

(a)

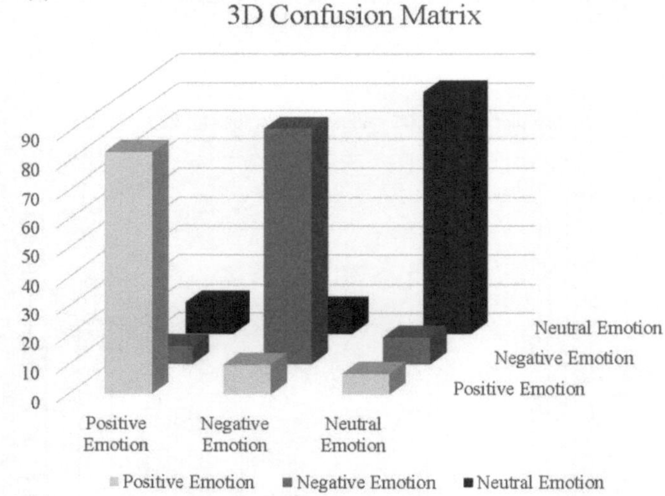

(b)

Figure 8.12 Classification of facial expressions and emotional states from, (a) Visual modalities, and (b) Infrared modalities, respectively, are shown.

8.7 CONCLUSION

This chapter discusses the development of three subsystems of the smart office. The first subsystem is a three-stage verification process involving fingerprints, face, and speech. The three-stage ensemble makes the verification process more robust than a single method.

The ambiance monitor provides valuable information about the office environment and reports any parameter which may cause discomfort to workers. The health monitor keeps a check on both mental and physical health through different modes of sensing. Physical health is monitored through a group of biomedical parameters, while mental health relies on the identification of fatigue and negative emotions.

ACKNOWLEDGMENTS

The authors wish to acknowledge the financial support of the Indian Institute of Technology Guwahati for its start-up grant. The y would also like to acknowledge Mr Jayesh Vishwakarma, Mr Prashant Pandey, Mr Sayan Alam, Miss M. Navya Shri, Mr Daksh Sharma, Mr Kautilya Pandey, Mr Navjot Singh, and Mr Pravin Kumar Sethi for their assistance with the implementation. We also thank the participants for the experiments and data collection.

REFERENCES

[1] A. Gaur, B. Scotney, G. Parr, and S. McClean, "Smart city architecture and its applications based on IoT," *Procedia Computer Science*, vol. 52, pp. 1089–1094, 2015.

[2] C. Le Gal, J. Martin, A. Lux, and J. L. Crowley, "Smart office: Design of an intelligent environment," *IEEE Intelligent Systems*, vol. 16, no. 4, pp. 60–66, 2008.

[3] M. Gupta, J. Benson, F. Patwa, and R. Sandhu, "Dynamic groups and attribute-based access control for next-generation smart cars," in *Proc. Ninth ACM Conference on Data and Application Security and Privacy*, Richardson Texas USA, pp. 61–72, 2019.

[4] J. Wang, J. Feng, Y. Gao, X. Jiang, and J. Tang, "A Intelligent Cabin Design Based on Aviation Internet of Things," in *Proc. 5th China Aeronautical Science and Technology Conference*, Springer Singapore, pp. 963–969, 2022.

[5] A. Mahroo, D. Spoladore, M. Nolich, R. Buqi, S. Carciotti, and M. Sacco, "Smart cabin: a semantic-based framework for indoor comfort customization inside a cruise cabin," in *Fourth Int. Congress on Information and Communication Technology (ICICT)*, London, Vol. 1, pp. 41–53, 2020.

[6] D. Rai, N. Kumar, D. Rasaily, P. Gurung, T. W. Tamang, C. Chettri, P. Kami, and G. Bhutia, "Arduino-gsm interfaced secure and smart cabin for smart office," in *2nd Int. Conf. on Intelligent Communication and Computational Techniques (ICCT)*, Jaipur, India, pp. 203–206, 2019.

[7] K. Vimalanathan, and T. Ramesh Babu, "The effect of indoor office environment on the work performance, health and well-being of office workers," *Journal of Environmental Health Science and Engineering*, vol. 12, pp. 1–8, 2014.

[8] R. S. Vasan et al., "Impact of high-normal blood pressure on the risk of cardiovascular disease," *New England Journal of Medicine*, vol. 345, no. 18, pp. 1291–1297, 2008.

[9] U.R. Acharya, O. W. Sing, L. Y. Ping, and T. Chua, "Heart rate analysis in normal subjects of various age groups," *Biomedical Engineering Online*, vol. 3, pp. 1–8, 2004.

[10] J.-A. Collins, A. Rudenski, J. Gibson, L. Howard, and R. O'Driscoll, "Relating oxygen partial pressure, saturation and content: the haemoglobin--oxygen dissociation curve," *Breathe*, vol. 11, no. 3, pp. 194–201, 2015.

[11] P. A. Areán, K. H. Ly, and G. Andersson, "Mobile technology for mental health assessment," *Dialogues in Clinical Neuroscience*, vol. 18, no. 2, pp. 163–169, 2022.

[12] M. A. S. Boksem, T. F. Meijman, and M. M. Lorist, "Mental fatigue, motivation and action monitoring," *Biological Psychology*, vol. 72, no. 2, pp. 123–132, 2006.

[13] J. Van Cutsem, S. Marcora, K. De Pauw, S. Bailey, R. Meeusen, and B. Roelands, "The effects of mental fatigue on physical performance: a systematic review," *Sports Medicine*, vol. 47, no. 8, pp. 1569–1588, 2017.

[14] M. Esler, "Mental stress and human cardiovascular disease," *Neuroscience & Biobehavioral Reviews*, vol. 74, pp. 269–276, 2017.

[15] Z. Ashtari, Y. Farhady, and M. R. Khodaee, "Relationship between job burnout and work performance in a sample of Iranian mental health staff," *African Journal of Psychiatry*, vol. 12, no. 1, pp. 71–74, 2009.

[16] A. F. Rabbi, K. Ivanca, A. V. Putnam, A. Musa, C. B. Thaden, and R. F. Rezai, "Human performance evaluation based on EEG signal analysis: a prospective review," in *Annual Int. Conf. of the IEEE Engineering in Medicine and Biology Society*, Minneapolis, MN, pp. 1879–1882, 2009.

[17] X. Zhang, P. Zheng, T. Peng, Q. He, C. K. M. Lee, and R. Tang, "Promoting employee health in smart office: A survey," *Advanced Engineering Informatics*, vol. 51, pp. 1–17, 2022.

[18] T. Ahonen, A. Hadid, and M. Pietikainen, "Face description with local binary patterns: Application to face recognition," *IEEE Transactions on Pattern Analysis and Machine Intelligence*, vol. 28, no. 12, pp. 2037–2041, 2006.

[19] A. Serrano, I. M. de Diego, C. Conde, and E. Cabello, "Recent advances in face biometrics with Gabor wavelets: A review," *Pattern Recognition Letters*, vol. 31, no. 5, pp. 372–381, 2010.

[20] R. Kaur, and E. Himanshi, "Face recognition using principal component analysis," *IEEE International Advance Computing Conference (IACC)*, Bangalore, India, pp. 585–589, 2015.

[21] I. L. Kambi Beli and C. Guo, "Enhancing face identification using local binary patterns and k-nearest neighbors," *Journal of Imaging*, vol. 3, no. 3, pp. 1–12, 2017.

[22] F. Bimbot et al., "A tutorial on text-independent speaker verification," *EURASIP Journal on Advances in Signal Processing*, vol. 2004, pp. 1–22, 2004.

[23] S. N. Srihari, H. Srinivasan, and G. Fang, "Discriminability of fingerprints of twins," *Journal of Forensic Identification*, vol. 58, no. 1, pp. 1–19, 2008.

[24] J. M. Pietrasik et al., "Nontraditional systems in aging research: an update," *Cellular and Molecular Life Sciences*, vol. 78, pp. 1275–1304, 2021.

[25] V. Govindaraju, Z. Shi, and J. Schneider, "Feature extraction using a chain-coded contour representation of fingerprint images," in *Proc. 4th International Conference, AVBPA*, Guildford, UK, pp. 268–275, 2003.

[26] R. Ranjan, V. M. Patel and R. Chellappa, "HyperFace: A deep multi-task learning framework for face detection, landmark localization, pose estimation, and gender recognition," *IEEE Transactions on Pattern Analysis and Machine Intelligence*, vol. 41, no. 1, pp. 121–135, 2019.
[27] D. Sommer and M. Golz, "Evaluation of PERCLOS based current fatigue monitoring technologies," *Annual Int. Conf. of the IEEE Engineering in Medicine and Biology*, Buenos Aires, Argentina, pp. 4456–4459, 2010.
[28] A. Dasgupta, A. George, S. L. Happy and A. Routray, "A vision-based system for monitoring the loss of attention in automotive drivers," *IEEE Transactions on Intelligent Transportation Systems*, vol. 14, no. 4, pp. 1825–1838, 2013.

Chapter 9

Wireless protocols for swarm of sensors

Sigfox, Lorawan, and Nb-IoT

Luiz Alberto Pasini Melek
CayennE-k Tecnologia, Curitiba, Brazil

9.1 INTRODUCTION

In recent years, the decrease in the cost of electronic components has allowed the development of several new products and applications. Following Moore's Law, technology has evolved to allow the use of integrated circuits and systems in ever smaller products that consume less energy and have numerous functionalities and characteristics. The vast majority of these applications need sensors for monitoring, whether for data acquisition for the application itself or for simple verification of the device operation. In this sense, pervasive wireless networking and ultralow power technologies applied to the majority of objects will result in billions, or even trillions, of connected objects. This is known as the sensory swarm [1]. In 2020, it was estimated that there would be around 9.7 billion devices using technologies such as the Internet of Things (IoT), the network of physical objects that contain embedded technology to communicate and sense or interact with their internal states or the external environment and which excludes portable computers, tablets and smartphones [2]. That number is expected to triple in 2030, to around 29 billion devices. Around 5 billion of these devices are expected to be in China.

Among the various important applications with worldwide coverage that really have the potential to reach thousands and even millions of people, agriculture ranks highly [3]. With an ever-increasing need for food production for billions of people, cultivation must be increased considerably. Likewise, losses in production, harvesting, transport and storage until final consumption must be minimized, as well as in checking the quality of the food. A number of sensors are used for these objectives, including soil humidity, temperature, luminosity, wind, atmospheric pressure and amount of rainfall. Annual growth of 20–30% in the agricultural area is expected due to the installation of IoT systems; the number of connected devices in this sector is expected to rise from 13 million in 2014 to 225 million in 2024 [4].

Another important application is in smart city traffic control in large urban centers and on highways, traffic offenses and the capture of stolen vehicles. Traffic congestion causes commuters to lose hours every day. Automatic number

DOI: 10.1201/9781032628059-9

plate recognition and counting of vehicles can be done by magnetic sensors (determining vehicle speed), imaging, ultrasound, lidar-type radar, time of flight (ToF) and weight (to limit the access of very large vehicles to certain areas of the city).

In medicine, sensors can be used for continuous monitoring of patients to ensure prompt response in an emergency. Thus, doctors or nurses do not need to be constantly in contact or present where the patients are being monitored. Some examples of sensors are: body temperature, blood pressure, oxygenation, heart rate and blood glucose level.

In the area of home automation, there is a real incentive to create smart homes capable of generating comfort, well-being and safety through automatic control of ambient luminosity, room temperature, opening and closing doors, windows and blinds, permitting entry and exit of people, and turning appliances on and off remotely. In the area of industrial automation, sensors are used to control manufacturing processes, allowing considerable gains in efficiency and reducing waste of time and materials. A real revolution is underway, with what is known as Industry 4.0.

Regarding utilities, telemetry has become increasingly important not only for the simplification of operations but for the number of people that can benefit from it. Measuring the consumption of water, electricity, gas and public lighting is the oldest idea in this regard. However, with greater coverage of communication systems and better infrastructure, these systems are able to provide information in real time and, in the case of emergencies or problems such as, for example, water or gas leaks and burned-out lightbulbs, operators can make decisions as soon as possible. The most widely used sensors are voltage, current, magnetic and water and gas flow types. A common feature of these applications is that the amount of information traveling over communication networks is not large and messages can be sent from devices to control centers a few times a day.

However, of all the applications mentioned, those that have best applied the concepts of wireless communication and low energy consumption electronic circuits are logistics and tracking systems, which are strongly linked. In the logistics area, monitoring loads is essential for efficient operations, and in many cases worldwide compatibility of communication systems is required. The number of objects being monitored can reach millions or even billions of units. Public transport companies may own hundreds or thousands of buses. Road transport companies taking agricultural produce from field to the food industries or ports for export own up to 10,000 trucks. It is reported that car security and recovery companies monitor up to 600,000 units each. Smaller numbers of higher-value vehicles, such as bicycles, tractors and harvesters, are also monitored. The minimum architecture of the devices is basically the same: a GNSS positioning receiver, device power and communication with a server. They may also include sensors for detecting vibration, shocks, temperature, humidity and pressure, among many others.

What all these applications have in common is that the sensors present in the devices communicate over a wireless network and record data in databases. Many devices rely on small batteries or harvest energy from the sun, heat, bio-sources or elsewhere, so they require an efficient method of communication. The infrastructure can be proprietary (defined, created and managed by the users themselves) or part of an open network (defined, created and managed by specific companies for this purpose). There are several examples of networks and, among the most used are Zigbee, Wi-Fi, Lorawan, Nb-IoT (narrow-band IoT), Bluetooth and Sigfox.

This chapter describes and compares the Sigfox, Lorawan, and Nb-IoT wireless protocols.

9.1.1 Basic communication infrastructure

The basic infrastructure of the wireless communication system or, more specifically a LPWAN (low-power wide area network), is shown in Figure 9.1. Depending on the application, different types of sensors can be installed in end devices, regardless of whether they are mobile or not. The information acquired from the sensors is processed and transmitted through an antenna to the gateways, fixed devices normally installed in towers, on top of buildings or places easily visible by the end devices. Gateways can be owned by the user or by third-party companies offering services, such as Sigfox and mobile phone companies (Nb-IoT). For correct communication, the end devices must be within the coverage area of the gateways, and both the end devices and the gateways must follow the specifications of the communication protocol, physical and logical, according to the chosen system (in the case of Sigfox, Lorawan or Nb-IoT). Gateways are responsible for

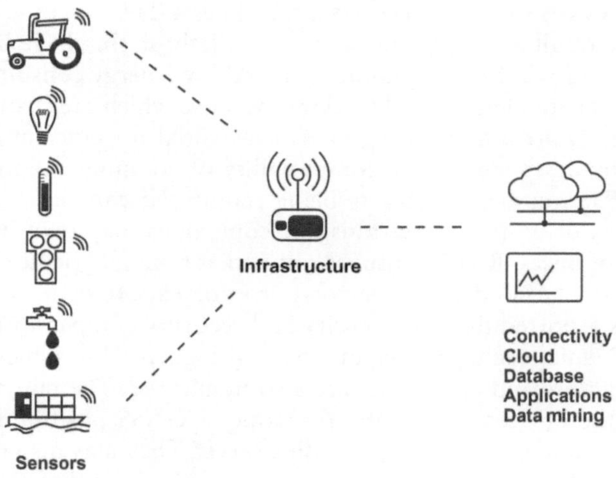

Figure 9.1 Basic sensor communication structure.

exchanging information with servers, usually via IP networks. The servers have databases, process incoming information and provide connectivity to end users through applications. The efficiency of the communication network must be high due to the limited resources of the end devices.

9.2 SIGFOX

9.2.1 Overview

Sigfox is a communication protocol developed in France in 2010 with the aim of offering connectivity to all objects from the physical world to the digital world [5]. The technology is present in more than 70 countries, has more than 10 million end devices activated, covering an area of more than 5.5 million square kilometers, and reaching more than 1.2 billion people. Around 83.4 million messages are exchanged daily [6, 7]. Sigfox provides all the connectivity between the gateways and the messaging server; users are only responsible for the development of end devices. This way, Sigfox protocol can only be used in places where the service is offered by the company or its partners.

However, the Sigfox protocol goes against what is expected from other protocols, in which data volume and speed only increase with each evolution: it is a low-speed protocol with very low data volume. It is therefore known as 0G technology [5] and consequently it can offer low energy consumption (the end devices can work with small batteries or even capture energy from the environment). Sigfox uses the UNB (ultra-narrow band) technique in the unlicensed sub-GHz ISM band with triple diversity (time, frequency and space). A range of 10 km is reported in urban centers and up to 40 km in rural areas; the link budget is 158 dB.

Sigfox is a very limited protocol in terms of size and number of messages. Only 12 bytes can be sent (payload) from the end device to the gateway (upload) per message and 8 bytes in the reverse direction (download). A maximum of 140 upload messages and four download messages are allowed daily. In this way, Sigfox communication is virtually unidirectional in the upload sense; firmware update of the end devices is not feasible and only configuration and status parameters are sent in the download direction. For the same reason, it does not provide data encryption.

Every Sigfox end device receives a unique ID (identification number) and PAC (porting authorization code) pair. The PAC proves the ownership of the end device. The ID is a 6–8-character hexadecimal number (for example, 28A7C3), while the PAC is a 16-hexadecimal number (for example, 27D83A947C027AB4). Both ID and PAC are used for unique identification of the end device in the network, resulting in the possibility of billions of connected end devices. For proper operation, ID and PAC must be registered and activated in the network.

One of the main advantages of Sigfox is that it is very cheap to implement and use. It is reported that Sigfox chipsets can cost less than USD3 per unit and message plans can cost up to USD1 per year per end device.

9.2.2 Regions of operation

The Sigfox operation is divided into seven world regions (see Figure 9.2), each allocated a region code (RC1–7). A summary of the region codes with their associated countries are:

- RC1:
  - Europe: European Union, Andorra, Liechtenstein, Norway, Serbia, Switzerland, Ukraine, United Kingdom.
  - Overseas France: French Guiana, French Polynesia, Guadeloupe, Martinique, Mayotte, New Caledonia, Reunion.
  - Middle East and Africa: Botswana, Ivory Coast, Kenya, Mauritius, Namibia, Nigeria, Oman, Senegal, South Africa, Swaziland, United Arab Emirates.
- RC2: Brazil, Canada, Mexico, Puerto Rico, USA.
- RC3: Japan.
- RC4:
  - Latin America: Argentina, Chile, Colombia, Costa Rica, Ecuador, El Salvador, Guatemala, Honduras, Nicaragua, Panama, Paraguay, Peru, Trinidad & Tobago, Uruguay.
  - Asia Pacific: Australia, Hong Kong, Indonesia, Malaysia, New Zealand, Singapore, Taiwan, Thailand.
- RC5: South Korea.
- RC6: India.
- RC7: Russia.

To circumvent the limitations imposed by the division into region codes, the Monarch system was developed. This system allows worldwide operation, with automatic reconfiguration of the transmitters in each region served. For that, hardware prepared for frequency and power adequation and special software must be used.

9.2.3 Technical characteristics

Every region code operates with specific frequency and power characteristics, determined by the regulatory agencies of each country. Thus, an end device operating in one region may not be used in another region. Table 9.1 shows the different regions of the world with their characteristics.

The maximum EIRP (effective isotropic radiated power) defined for each of the region codes is shown in Table 9.2.

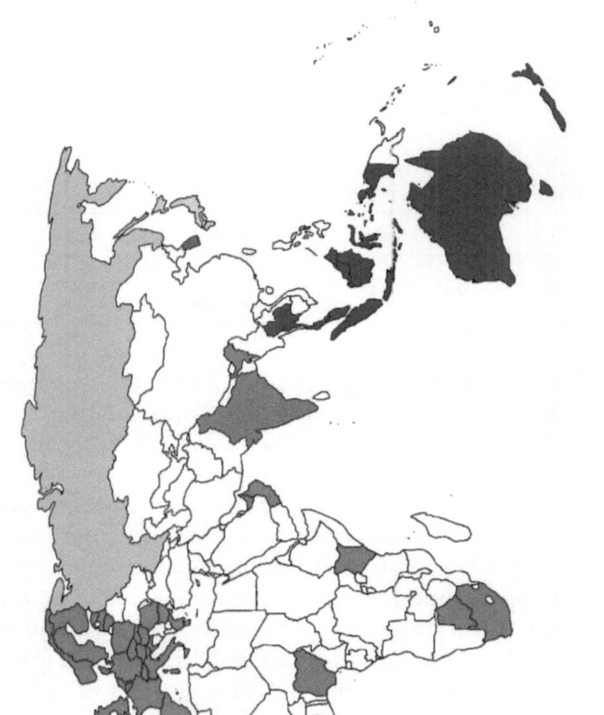

Figure 9.2 Sigfox coverage (2023) and region codes.

Table 9.1 World zones and their defined frequency characteristics

	Zone 1 *Europe, Middle East, Africa*	*Zone 2* *Americas, Asia, Pacific*	*Zone 3* *Japan, South Korea*
Frequency (MHz)	862–876	902-928	
Region Code	RC1, RC6, RC7	RC2, RC4	RC3, RC5

Table 9.2 Power allotted to different region codes

	RC1	*RC2*	*RC3*	*RC4*	*RC5*	*RC6*	*RC7*
Power (dBm EIRP)	16	24	16	24 22 (Singapore)	14	16	16

As stated before, Sigfox is a low-speed (0G) technology, which uses the transmission (uplink) and reception (downlink) rates shown in Table 9.3. Each region code defines the possible transmission (uplink) rates, as shown in Table 9.4.

Sigfox is characterized as a low-power communication protocol, in that for 99% of the time the transceivers are in idle mode, with current consumption around 6 nA. During transmission, which can take up to around 6 seconds, current consumption varies between 10 and 50 mA. On average, net current consumption is extremely low. Peak power is around 25 mW for uplink and 500 mW for downlink in countries that follow ETSI (European Telecommunications Standards Institute); and around 150 mW for uplink and 1000 mW for downlink in countries that follow FCC (Federal Communications Commission). Furthermore, message pairing (or synchronization)

Table 9.3 Uplink and downlink specifications

Communication	*Symbol rate name*	*Symbol rate (baud)*	*Cumulated error over full length of radio burst*
Uplink	$BR100_{UL}$	100	+/- 3%
	$BR600_{UL}$	600	+/- 3%
Downlink	$BR600_{DL}$	600	+/- 0.01%

Table 9.4 Symbol rate allotted to different region codes

	RC1	*RC2*	*RC3*	*RC4*	*RC5*	*RC6*	*RC7*
Allowed Symbol Rates	$BR100_{UL}$ $BR600_{UL}$	$BR600_{UL}$	$BR100_{UL}$ $BR600_{UL}$	$BR600_{UL}$	$BR100_{UL}$ $BR600_{UL}$	$BR100_{UL}$ $BR600_{UL}$	$BR100_{UL}$ $BR600_{UL}$

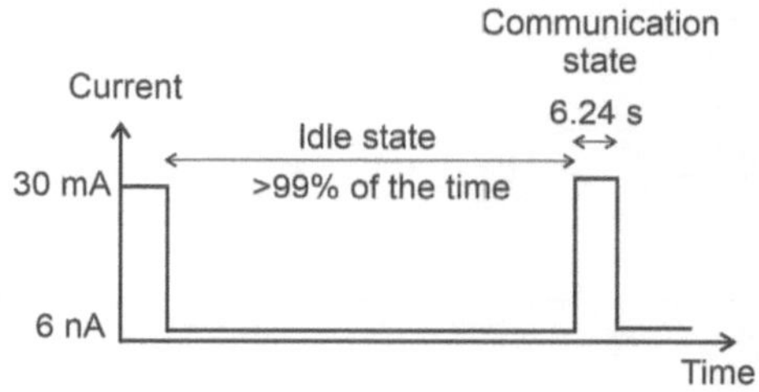

Figure 9.3 Current × time format during transmission.

between the end device and the gateway is not required, which eliminates transmission times. The diagram of current consumption over time during data transmission is shown in Figure 9.3.

Regarding modulation, Sigfox uses the DBPSK (differential binary phase shift keying) method for uplink and GFSK (Gaussian frequency shift keying) method for downlink.

9.2.4 Sigfox diversity

The Sigfox radio protocol is named 3D-UNB. UNB stands for ultra-narrowband while 3D stands for triple diversity, i.e., temporal, frequency and spatial. Sigfox uses a total frequency band of 192kHz, which is divided into channels of only 100Hz for ETSI countries and 600Hz for FCC countries. That is, in this range, 1,920 channels are available for communication for the 868 MHz ISM band for an ETSI country (see Figure 9.4). Any available channel can be used for communication. For temporal diversity, an end device does not need to use the same communication channel at different times, that is, at a given moment a channel (frequency) is used for the transmission of a data packet (frame), while at the next moment another channel (frequency) is used, as shown in Figure 9.5. Finally, the Sigfox protocol also offers spatial diversity; a message sent by an end device can be received by several gateways simultaneously, as shown in Figure 9.6. Each of the gateways forwards the message to the message server, which detects the redundancy, processes and stores only one message.

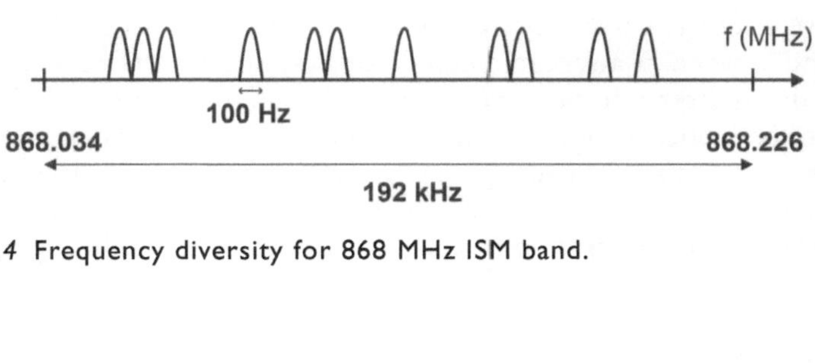

Figure 9.4 Frequency diversity for 868 MHz ISM band.

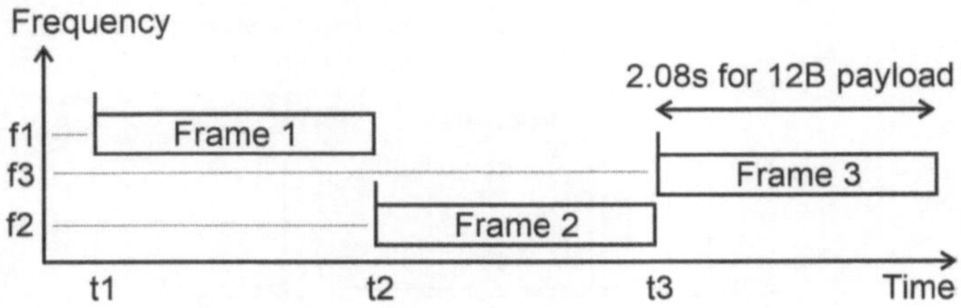

Figure 9.5 Temporal diversity example.

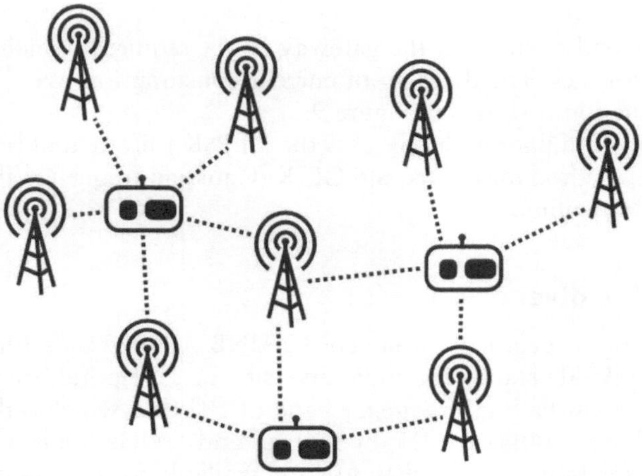

Figure 9.6 Spatial diversity.

9.2.5 Network architecture

The Sigfox network architecture is shown in Figure 9.7. The end devices connect to the gateways installed in the base stations which communicate with the various back-end servers through IP links. The functions of the back-end servers are: to manage the base stations, to monitor network traffic and the functioning of the base stations, and to process all the messages. Incoming messages are stored in their own database. After processing, the messages travel through the metadata modems (which carry the status of the network and of the end devices) to the front-end servers. The back-end servers and the local message storage system form the Sigfox cloud. Front-end servers, in a separate cloud, are responsible for managing users, user groups (users with the same access and configuration permissions) and groups of end devices (devices with the same characteristics), as well as the modems. Notice that a group of devices treat status information and payloads and dispatch information in exactly the same way. Access to the front-end servers and configuration of end devices can be done through the web

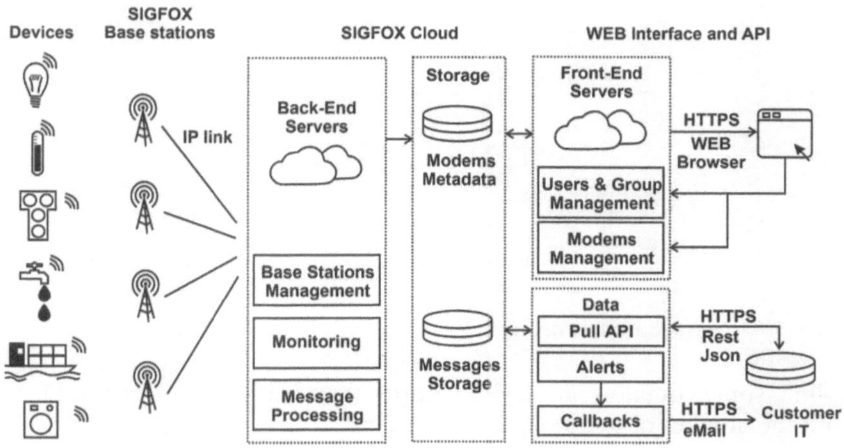

Figure 9.7 Sigfox network architecture.

browser, with secure https requests, as part of the web interface. To make the information coming from the end devices available, the processed data (payloads) that were stored can be accessed through the Sigfox API (application programming interface), by means of alerts and callbacks. The data, in fact, are dispatched through https requests, in Rest and Json formats, or even by email.

In order to provide valuable information to the end user, the end device (ID and PAC), the way payloads are processed, the alerts and the callbacks must be properly configured.

9.2.6 Message system

The Sigfox message system, or the protocol stack, is divided into uplink and downlink messages. Both messages consist of up to 232 bits [8]. In the OSI model, it consists of: application layer, frame layer, media access control (MAC), and physical layer (PHY), as shown in Figure 9.8. The downlink message fields are the following [9]: preamble (91 bits), frame synchronization (13 bits), error correction code (32 bits), data payload (up to 64 bits), message authentication code (16 bits), frame check sequence (8 bits). The uplink message fields are the following [9]: preamble (19 bits), frame synchronization (29 bits), device ID (32 bits), data payload (up to 96 bits), message authentication code (16–40 bits), frame check sequence (16 bits).

The payload of the messages, consisting of a maximum of 8 bytes for the downlink and 12 bytes for the uplink, is very limited. However, this is sufficient for many applications. For the downlink, the payload is only enough to send some configuration data or ways of operation of the device. For the

Downlink Frame Format					
91 bits	13 bits	32 bits	0 - 64 bits	16 bits	8 bits
Preamble	Frame Sync.	ECC	Payload	Authentication	FCS
PHY		MAC	Application	MAC	PHY

(a)

Uplink Frame Format					
19 bits	29 bits	32 bits	0 - 96 bits	16 - 40 bits	16 bits
Preamble	Frame Sync.	ID	Payload	Authentication	FCS
PHY header		MAC	Application	MAC	PHY

(b)

Figure 9.8 Message frame formats.

uplink, some common physical quantities that can be monitored are listed with the number of bytes required for them:

- GPS coordinate – 8 bytes (less if current and previous coordinate difference is used);
- Temperature – 10 bits (from –40 °C to +105 °C, with 0.1 °C precision);
- Speed – 1 byte (0–255 km/h);
- Object status – 1 bit (on or off);
- Keep alive – 0 bytes (working and activated).

The transmission and the reception of messages are based on callback services, which are divided into data, service, error, and event. A summary of the main messages is shown in Table 9.5.

9.2.7 Summary of characteristics

The main characteristics of the Sigfox protocol are shown in Table 9.6.

Table 9.5 Message descriptions and tasks

Callback services		*Description*	*Trigger*
Data	Uplink	Sends a message to the network	At the reception of the first frame
	Downlink	Receives a message from the network	
Service	Status	Sends the temperature and the battery voltage of the end device	
	Acknowledge	Acknowledge a downlink message	
	Advanced Data	Sends metadata	25 seconds' delay
Error		Sends error messages	Variable
Event	Device events	Sends alerts	Event

Table 9.6 Sigfox protocol specification

Sigfox	
FREQUENCY	ISM not licensed 862–876 MHz (ETSI) and 902–928 MHz (FCC)
MODULATION	DBPSK (uplink) and GFSK (downlink)
TRANSMISSION RATE	100 bps (ETSI) and 600 bps (FCC)
BANDWIDTH	192 kHz
CHANNEL WIDTH	100 Hz (ETSI) and 600 Hz (FCC)
TRANSMISSION POWER	25 mW uplink and 500mW downlink (ETSI) 150 mW uplink and 1000 mW downlink (FCC)
LINK BUDGET	158 dB
COVERAGE	> 10 km in urban areas > 40 km in rural areas
MAXIMUM MESSAGE LENGTH	12 bytes uplink 8 bytes downlink
QUANTITY OF MESSAGES	1–140 messages uplink (Platinum) 4 messages downlink
TRANSMISSION/RECEPTION	Almost unidirectional
MESSAGE DELIVERY	Via callback (http, email etc. event triggers)
COST	Extremely low Up to USD1 per year per end device
CONNECTION CAPACITY	Billions of sensors
TECHNOLOGY	Owned by Sigfox/UnaBiz

9.3 LORAWAN

9.3.1 Overview

Lora is an acronym for long range and is a physical radio communication technique based on spread spectrum. It was developed in France by a company called Cycleo, which was later acquired by Semtech. Lorawan is the software protocol stack and the system architecture [10]. Both terms are used interchangeably and together they define a protocol used to exchange data between sensors and gateways. Lora Alliance was created in 2015 in order to maintain compatibility and operation between equipment. In 2018 it was reported that Lora was present in more than 163 countries. For proper operation, the user is also responsible for the installation of the gateways and their communication with the servers in the cloud.

Lorawan was developed to allow long-range communication in both urban (up to 5km) and rural areas (up to 20km). It uses the 2.4 GHz band for worldwide interoperability and also ISM sub-GHz bands, which results

Table 9.7 Frequencies allotted to different continents in Lorawan protocol

Region	*Frequency*
Europe	863–873 MHz
South America	915–928 MHz
Asia	433–434 MHz and 915–928 MHz
North America	902–928 MHz
India	865–867 MHz
Worldwide	2400

in good indoor usage. Lorawan is a free network and once the end devices and the gateways comply with the standard set by the Lora Alliance and regional agencies, they can be used. The frequencies used in each country are shown in Table 9.7. Link budget is around 155–170 dB. The total transmission time (uplink or downlink) during a single day (duty cycle) is limited to 0.1–10% depending on the channel used and country legislation.

Since the user is responsible for the operation of both the end devices and the gateways (client is responsible for the infrastructure), Lorawan is not limited by the number of messages exchanged and is considered a bidirectional protocol. It also allows high throughput from 300 bps to 254 kbps. The reception of the message sent by the end device is confirmed by the gateway and is limited to 242 bytes. A secure mechanism is provided with cryptography and 128-bit protection keys.

Every Lorawan end device receives an unique serial number of 32 bits for identification in the network, resulting in the possibility of billions of connected end devices. However, in practice around 8,000 connections for each gateway are done.

Lorawan is still considered a relatively cheap solution even though it is necessary to purchase, install and maintain the gateways.

9.3.2 Characteristics

The output power during transmission is limited to 14 dBm in Europe and 30dBm in the Americas, potentially achieving longer ranges. The transmission speed, coverage, sensitivity and data length depends on the spreading factor (SF), numbered from 7 to 12. Spreading factors from 7 to 10 (SF7–SF10) are used in uplink messages, while spreading factors from 8 to 12 (SF8–SF12) are used for downlink messages. Regarding the bandwidth 125 kHz, 250 kHz, and 500 kHz can be used for uplink, and 500 kHz for downlink. Bitrate (BR) can be calculated from Equation (9.1), where SF is the spreading factor, BW is the channel bandwidth, and CR is the coding rate. Table 9.8 shows the bitrate, maximum range, and maximum payload, as a function of the bandwidth. The sensibility of the receiver depends on the spreading factor. Higher

Table 9.8 Different specifications and their allotted values in Lorawan

Code rate (CR)	*Spreading factor (SF)*	*Channel frequency*	*Uplink or downlink*	*Bitrate (bps)*	*Maximum payload size (bytes)*
1	SF10	125 kHz	Uplink	977	11
2	SF9	125 kHz	Uplink	1464	53
3	SF8	125 kHz	Uplink	2232	125
4	SF7	125 kHz	Uplink	3417	242
1	SF8	500 kHz	Uplink	12500	242
1	SF12	500 kHz	Downlink	1172	53
2	SF11	500 kHz	Downlink	1790	129
3	SF10	500 kHz	Downlink	2790	242
4	SF9	500 kHz	Downlink	4395	242
1	SF8	500 kHz	Downlink	12500	242
1	SF7	500 kHz	Downlink	21875	242

Table 9.9 For spreading factor various sensitivity values

Spreading factor	*Receiver sensibility (125 kHz bandwidth)*
SF7	−123 dBm
SF8	−126 dBm
SF9	−129 dBm
SF10	−132 dBm
SF11	−134 dBm
SF12	−137 dBm

SF means higher sensibility. Table 9.9 shows the receiver sensibility as a function of the SF, for a 125kHz bandwidth.

$$\mathrm{BR} = \mathrm{SF} \cdot \frac{\mathrm{BW}}{2^{\mathrm{SF}}} \cdot \frac{4}{4 + \mathrm{CR}}. \tag{9.1}$$

9.3.3 Uplink and downlink communication

As already stated, Lora uses the 868MHz ISM bands in Europe and India and the 915MHz ISM bands in the USA, South America, and Asia [11]. Within these bands, the channels shown in Figure 9.9 are created for communication. For example, in Australia eight channels of 125 kHz are used. In Europe, channels can be in any frequency between the ISM band limits, but they must include the default channels (868.1MHz, 868.3MHz, and 868.5MHz). The USA, South America, and Australia have 64 uplink

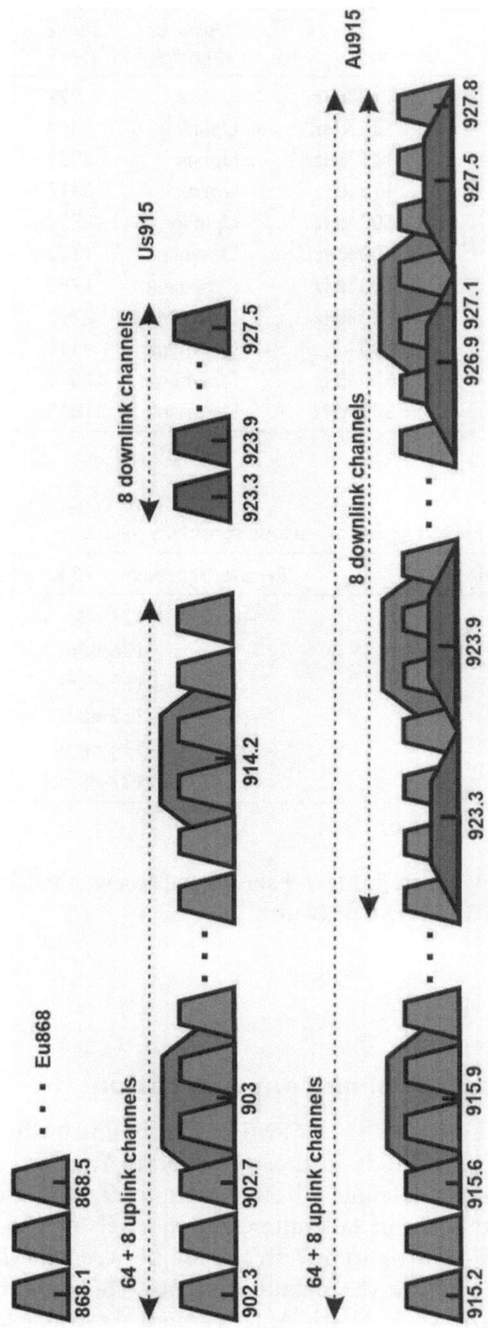

Figure 9.9 Uplink and downlink communication channels.

channels of 125kHz, centrally spaced by 200kHz. They also include eight uplink channels of 500kHz, centrally spaced by 1.6MHz. Both areas use eight channels of 500kHz for downlink communication, but the main difference is that the USA uses a separate band, while in Australia it overlaps the uplink band [12]. To sum up, the countries that use the 915MHz ISM band have 72 uplink channels and eight downlink channels.

9.3.4 Classes of operation

The operation of Lorawan end devices is divided into classes A, B, and C [13].

An end device operating in Class A can transmit information to the gateway at any time, and then waits for its return in two time slots, but not in both, as shown in Figure 9.10. End devices operating in Classes B and C must support Class A functionality.

In addition to receiving time slots (like Class A), Class B end devices open additional time slots at specified times, called ping periods (Figure 9.11). The end device also receives a signal (beacon) from the gateway, which lets the gateway know when the end device is listening. A Class B end device does not support the Class C functionality.

Class C end devices, in addition to having Class A time slots, will continuously listen for responses from the gateway, as shown in Figure 9.12. Class C devices do not support Class B functionality. Because they are constantly in reception mode, they are hardly powered by batteries, because power consumption is quite high.

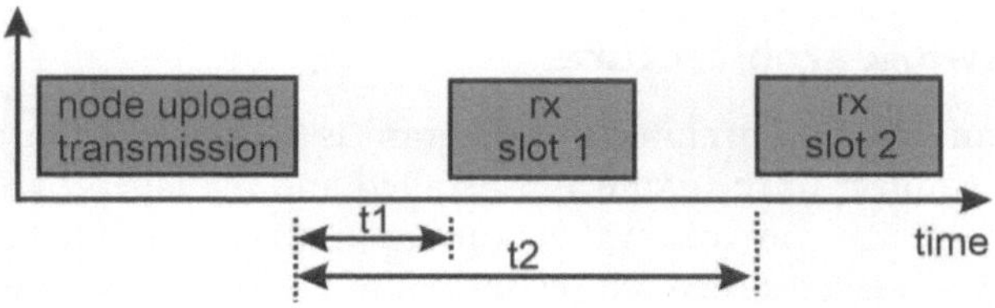

Figure 9.10 Class A operation. Adapted from [14].

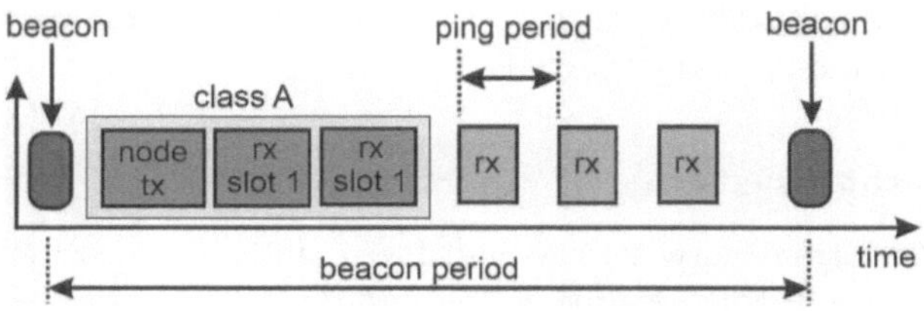

Figure 9.11 Class B operation. Adapted from [14].

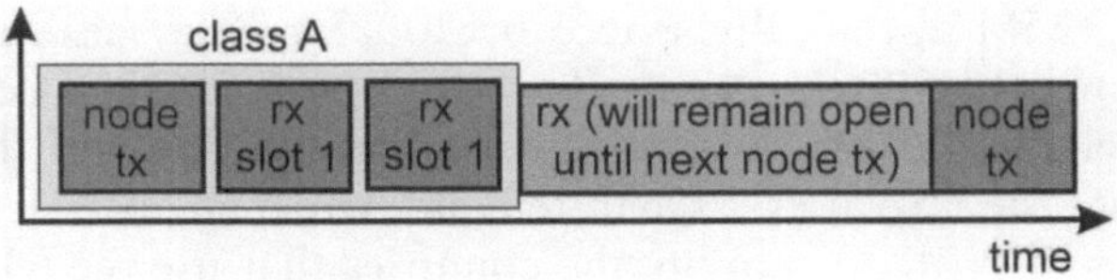

Figure 9.12 Class C operation. Adapted from [14].

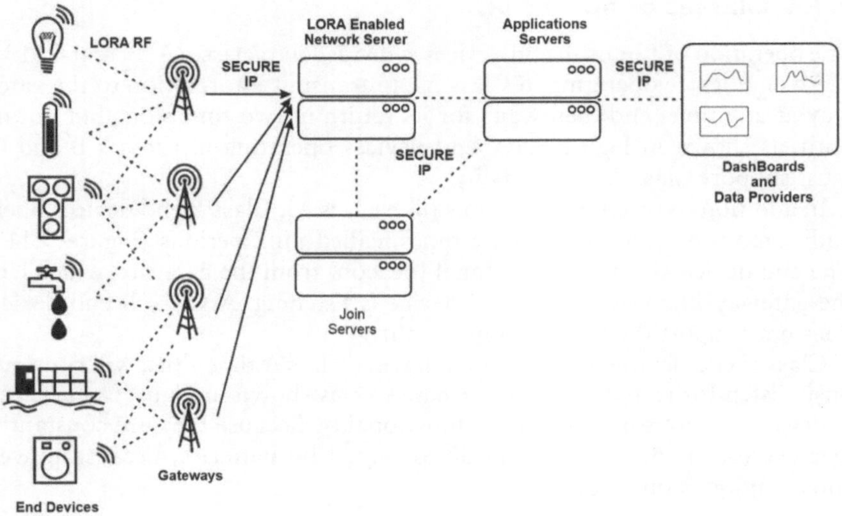

Figure 9.13 Lorawan network architecture.

9.3.5 Network architecture

The Lorawan network architecture diagram is shown in Figure 9.13. The end devices connect to the gateways installed and managed by the users by means of the Lora protocol. In their turn, gateways communicate with the various network servers through secure IP links. Notice that before any real communication can occur, the end device must connect to the join server, which manages all the end devices. Once the end device joins the network, all information received by the network servers are forwarded to the application servers, whose function is to provide valuable information to dashboards, web pages, and data portals.

9.3.6 Message system

The message format between the end device and the gateway is shown in Figure 9.14. It is formed by the 32-bit end device address (DevAddr), followed by a 16-bit frame counter (FCnt), the encrypted payload of up to 242 bytes, and the 32-bit message integrity check (MIC) [15, 16].

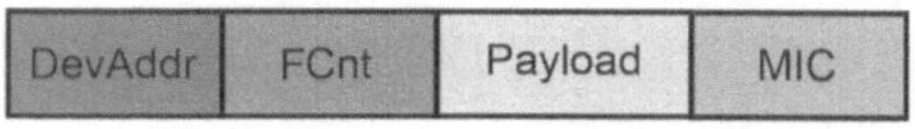

Figure 9.14 Message format.

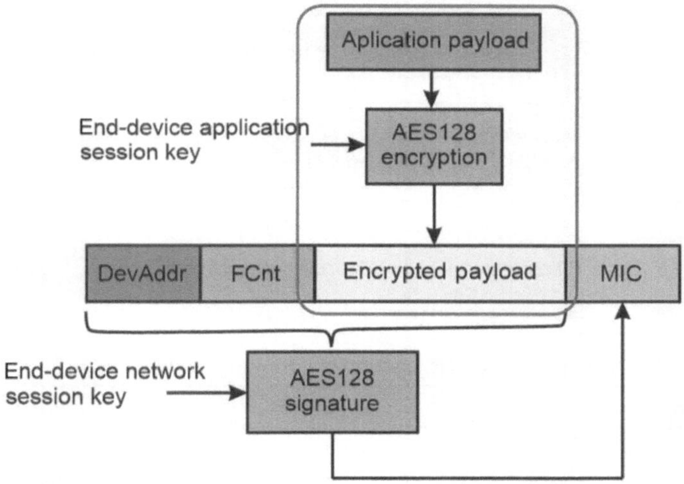

Figure 9.15 Data encryption. Adapted from [15].

In order to guarantee privacy, Lora offers two levels of security with AES-128 encryption: data security and data transmission security. For data security, as shown in Figure 9.15, the payload is encrypted with the AS-128 algorithm, with the application session key. For secure data transmission, the MIC is generated from the DevAddr, FCnt, and the encrypted payload by means of AES-128 signature with the end-device network session key (Figure 9.16). This key guarantees that the packet does not contain any errors (the server just checks for error without knowing the payload).

Messages between the end devices and the gateways are quite simple. First, the end device must send a 'join' message to the network. Once it is accepted, messages can be exchanged between the end devices and the message servers by means of the gateways. Message types are shown in Table 9.10.

The payload of the messages, consisting of a maximum of 242 bytes, and the total daily transmission time, limit the use of the Lora network. However, it is sufficient for many applications, for example [14]:

- Environmental monitoring such as air quality control
- Agriculture
- Energy, waste, and water management
- Street lighting
- Smart building (heating and air conditioning)

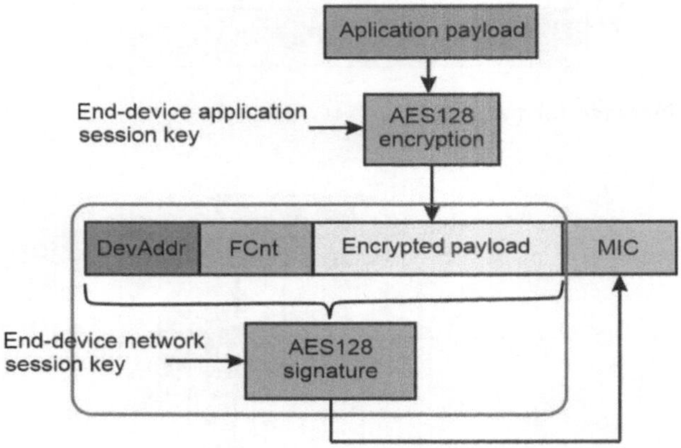

Figure 9.16 Data transmission encryption. Adapted from [15].

Table 9.10 Message types and their function in Lorawan protocol

Message type	*Description*
Uplink	Sends a message to the gateway
Downlink	Receives a message from the gateway
Join	Asks to join the network
Rejoin	Asks to rejoin the network if connection is lost

- Fire, flood, CO/CO_2, and air quality monitoring
- Fleet management, logistics, transportation and asset tracking
- Agriculture

9.3.7 Characteristics summary

The main characteristics of the Lorawan protocol are shown in Table 9.11.

9.4 NB-IOT

9.4.1 Overview

Nb-IoT stands for narrow-band internet of things and is a protocol based on mobile phone networks and infrastructure [14]. It is defined by the 3GPP consortium. It is aimed at indoor coverage and high density of connections (around 50,000 connections per cell). Good indoor coverage is achieved by intermediate or high energy consumption and sub-GHz frequency plan. However, it is not intended for long-range communication. The communication range is only around 1 km in urban areas and 10 km in rural areas.

Table 9.11 Summary of main feature in Lorawan protocol

LORA	
FREQUENCY	ISM not licensed 863–870 MHz (ETSI), 902–928 MHz (FCC) and 2.4 GHz
MODULATION	Chirp spread spectrum
TRANSMISSION RATE	300 bps–254 kbps
BANDWIDTH	125 kHz or 500 kHz for uplink 500 kHz for downlink
TRANSMISSION POWER	14 dBm in Europe 30 dBm in the Americas
LINK BUDGET	155–170 dB
COVERAGE	> 5 km in urban areas > 20 km in rural areas
MAXIMUM MESSAGE LENGTH	242 bytes
QUANTITY OF MESSAGES	free (but daily limit to total on-air time)
TRANSMISSION/RECEPTION	Bidirectional
MESSAGE DELIVERY	With gateway confirmation
COST	Cheap
CONNECTION CAPACITY	Billions of sensors, but limited by the gateways in practice
TECHNOLOGY	Standardized by the Lora Alliance

Nb-IoT is a more expensive solution because of the electronic components used, it uses licensed frequencies of the spectrum (700MHz, 800MHz, and 900MHz generally) and every end device requires a sim card. It is reported that Nb-IoT modules can cost up to USD13 per unit without the sim card. However, with the number of messages allowed per day, it is actually a less limited solution considering the quantity of messages allowed per day and is really a bidirectional communication protocol. The payload can transport up to 1,600 bytes, allowing firmware upgrades of the end devices.

Nb-IoT uses SC-FDMA (single carrier frequency division multiple access) for uplink messages and OFDM (orthogonal frequency division multiplexing) for downlink messages. The maximum transmission throughput is 66 kbit/s or 159 kbit/s depending on the infrastructure provided; channel bandwidth is 200kHz. The link budget is 164dB and transmission power is limited to 23dBm.

9.4.2 Characteristics

Nb-IoT has three types of operations when deployed in network infrastructures (Figure 9.17). The standalone operation reserves a specific center

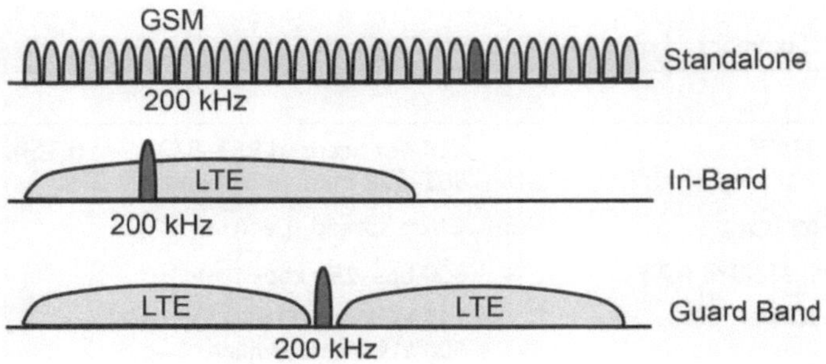

Figure 9.17 Nb-IoT operation.

frequency independent of other channels. In-band operation uses physical resources from the LTE (long-term evolution) networks provider. Finally, the Band-guard operation uses the guard band between channels of the LTE provider.

9.4.3 Message system

Connections of the end devices and gateways can be made with the attachment with PDN (packet data network). Four types of connections with PDN are available: a) IP over control plane; b) IP over user plane; c) non-IP over control plane; and d) non-IP over user plane. In order to optimize power consumption, non-IP, UDP, and TCP can be used for specific usage. UDP reduces the time of the connection, while TCP will keep the connection until it is closed.

9.5 CONCLUSION

Sections 9.2, 9.3, and 9.4 described the main characteristics of Sigfox, Lora, and Nb-IoT wireless protocols and their advantages and disadvantages, The main features for choosing between the protocols are also set out in Figure 9.18. For example, Sigfox is far better than the other two protocols in terms of range, coverage, cost efficiency and battery life. However, it is very limited in terms of payload length. Lorawan is cost-effective and has long battery life, with better deployment and longer payload than Sigfox. Finally, Nb-IoT has much better scalability, latency, payload length, and quality of service (QoS) than Sigfox and Lorawan; its main drawback is the cost of the electronics involved, the message plans and sim card requirement.

To summarize, Sigfox has better range, lower energy consumption, and uses its own infrastructure. Lorawan has a virtually unlimited number of messages, is bidirectional, and allows firmware updates of the end devices. On the other hand, Nb-IoT has a higher transmission throughput, longer

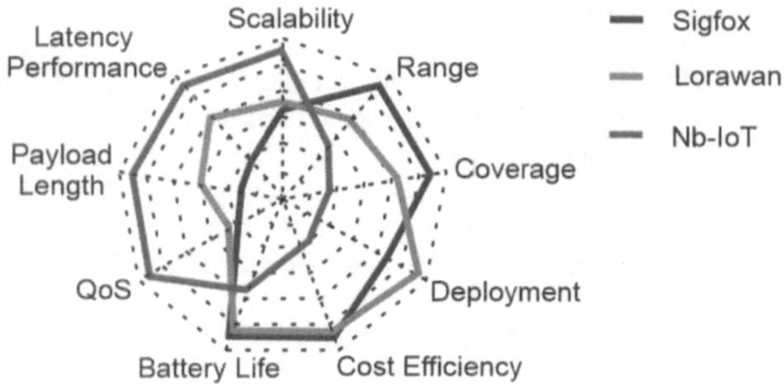

Figure 9.18 Sigfox, Lorawan, and Nb-IoT main characteristics comparison.

payload, allows the firmware updates for the end devices, and uses an infrastructure that already exists.

All three protocols have good characteristics for creating a swarm of sensors, depending on the application type and requirements. Sigfox shows better results in urban and rural areas in applications such as logistics, fleet management and tracking devices. Lorawan may have better results in rural areas in applications such as agriculture, private networks, and smart cities. Nb-IoT has better deployment in urban areas with applications such as smart cities, smart metering, utilities and wearables.

REFERENCES

[1] J. Rabaey, "Keynote Speaker," in *Proc. IEEE Int. Conf. on Microelectronic Systems Education (MSE)*, San Diego, CA, pp. 1–4, 2011.

[2] L. S. Vailshery, "Number of IoT connected devices worldwide 2019–2021, with forecasts to 2030," November 22, 2022. [Online]. Available: https://www.statista.com/statistics/1183457/iot-connected-devices-worldwide/ [Accessed April 12, 2023].

[3] Cesar Agostino, "The future of IoT in agriculture," *4iplatform.com*, November. 15, 2021. [Online]. Available: https://4iplatform.com/blog/the-future-of-iot-in-agriculture/ [Accessed April 12, 2023].

[4] Sigfox 0G Technology, "The story discover the origin of sigfox 0G technology," 2022. [Online]. Available: http://surl.li/gikri.[Accessed April 12, 2023].

[5] Wikipedia, "Sigfox," March 2023 [Online]. Available: https://en.wikipedia.org/wiki/Sigfox. [Accessed April 12, 2023].

[6] P. Di Gennaro, V. Daniele, D. Lofu, P. Tedeschi, and P. Baccadoro, "WaterS: A Sigfox-compliant prototype for water monitoring," *Internet Technology Letters*, vol. 2, no. 1, pp. 1–4, 2018.

[7] P. Boccadoro, V. Daniele, P. Di Gennaro, D. Lofù, and P. Tedeschi, "Water quality prediction on a sigfox-compliant IOT device: The road ahead of waters," *Ad Hoc Networks*, vol. 126, pp. 1–14, 2022.

[8] J.C. Zuniga, and B. Ponsard, "Sigfox System Description," in *Proc. LPWAN@ IETF97'*, pp. 1–9, 1997.

[9] P. Bertoleti, *Projetos com ESP32 e Lora*, Editora NCB, 2019. [E-book] Available: Google e-book.
[10] EMBARCADOS, "Conheça a tecnologia LoRa® e o protocolo LoRaWAN™ – Embarcados,". [Online]. Available: https://embarcados.com.br/conheca-tecnologia-lora-e-o-protocolo-lorawan/. [Accessed April 12, 2023].
[11] M. C. Bor, U. Roedig, T. Voigt, and J. M. Alonso, "Do LoRa low-power wide-area networks scale?," In *Proc. 19th ACM International Conference on Modeling, Analysis and Simulation of Wireless and Mobile Systems*, pp. 59–67, 2016.
[12] Mobilefish.com, "LoRaWAN". [Online]. Available: https://www.mobilefish.com/developer/lorawan/lorawan_quickguide_tutorial.html. [Accessed April 12, 2023].
[13] B. Eric, "LoRa Documentation," December 3, 2019. [Online]. Available: https://buildmedia.readthedocs.org/media/pdf/lora/latest/lora.pdf. [Accessed April 12, 2023].
[14] Groupe Speciale Mobile Association, "NB-IoT Deployment Guide to Basic Feature set Requirements," *GSMA*, 2019. [Online]. Available: https://www.gsma.com/iot/wp-content/uploads/2019/07/201906-GSMA-NB-IoT-Deployment-Guide-v3.pdf [Accessed April 12, 2023].
[15] M. B. de Almeida, "LoRaWAN: O que você precisa saber como desenvolvedor de dispositivos," YouTube, November 29, 2020. [Video File]. Available: https://youtu.be/j0YqZ2VXiCI. [Accessed April 12, 2023].
[16] LoRa Alliance, "RP2-1.0.1 LoRaWAN® Regional Parameters," 2020. [Online]. Available: https://lora-alliance.org/resource_hub/rp2-101-lorawan-regional-parameters-2/. [Accessed April 12, 2023].

Chapter 10

Design and test of thermal energy harvester for self-powered autonomous electronic load

Arun Kumar Sinha
VIT-AP University, Near Vijayawada, India

10.1 INTRODUCTION

Harvesting electrical energy from renewable sources i.e., thermal, solar, electromagnetics, kinetic, wind energy, acoustic noise, etc., is known as energy harvesting. Figure 10.1 illustrates different ambient sources along with their respective energy harvesting mechanisms. Over the past few years, energy harvesting has gained more attention because of the wide availability of the sources, the short time needed to replenish them, and the ability to self-start the system. A vast number of applications can be found in various domains [1], such as low-power devices, health monitoring systems, wearable devices, communication networks, standalone embedded systems, wireless sensor networks, industrial process machine monitoring, battlefield surveillance, etc. Nowadays, smart e-cities, smart homes, and smart electronic devices can be seen everywhere which requires more electrical energy than the previous generations. Using renewable sources to meet the demand is the best way to overcome the limitations of non-renewable sources.

For continuous monitoring of the environment, multiple sensors have been used in the current smart world. Batteries were previously used to power these sensors, but since they are expensive, require maintenance and are polluting, they are not an effective way to power wireless sensor networks. An energy harvester is an efficient way of powering the sensor nodes that overcomes the limitations of using batteries. Research on integrating energy harvesting mechanisms with embedded systems, the internet of things (IoT), the medical domain, the agricultural field, etc. is ongoing. Harvesting thermal energy to generate electrical energy is known as thermal energy harvesting (TEH). There are multiple ways to harvest thermal energy. The human body, pavements, industrial equipment, and diesel engines are the major sources of thermal energy. The electrical energy generated can be used to run a standalone embedded system or to design a self-powered wireless

DOI: 10.1201/9781032628059-10

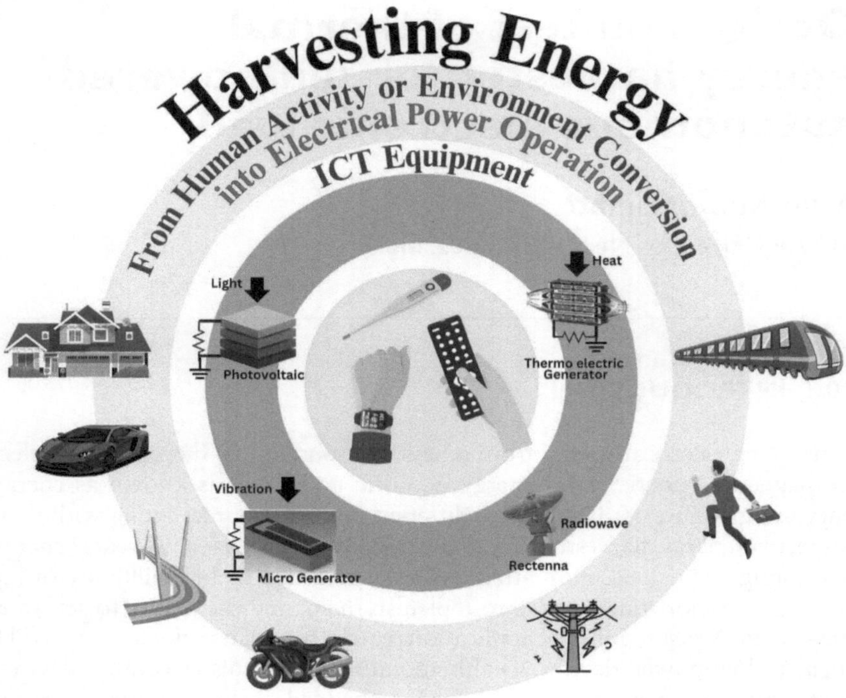

Figure 10.1 An overview of different methods of harvesting energy [2].

sensor network. Two different lines of research are seeking to obtain maximum electrical energy from renewable sources:

1. A TEG can be designed in a way that allows it to withstand temperature changes and generate maximum power by changing the material composition.
2. Enhancing low-level energy by developing new architecture.

Many authors have proposed systems to harvest energy from different thermal sources, targeting efficiency, start-up voltage etc. The human body is one of the major sources of thermal energy, and authors have proposed various architectures to harvest thermal energy it. They mainly focus on developing a wearable and flexible device that can harvest the maximum amount of thermal energy. A soft multi-modal thermoelectric skin (TES) was developed in [3] to harvest energy from underwater and to provide thermoregulation. The skin, placed on the human body, generates electricity by considering the thermal difference between ocean water and the human body. In order to regulate the body temperature of divers, to provide thermoelectric cooling/heating the generated energy will be used. The stretchability of the developed device was good, when the temperature difference is 5°K, TES source has an output voltage of 0.971V with a maximum power density of 3.42 $mVcm^{-2}$. A flexible thermoelectric generator (f-TEG) suitable for wearable applications was developed and examined by Yujin in [4]. In order to fabricate f-TEG the authors used n-type $Bi_2Te_{2.7}Se_{0.3}$ (BTS) of 124 $\mu Wm^{-1}K^{-2}$, and p-type $Bi_{0.5}Sb_{1.5}Te_3$ (BST) of 133 $\mu Wm^{-1}K^{-2}$ thermoelectric films and harvested an output power of 12.6nW, output voltage of 18.13mV and 2.74μA current signal at a temperature difference of 25°K. The main drawback of TEH is that maximum energy can only be harvested by maintaining a proper temperature difference. A system that can run the RF wireless networks or IoT sensing system by harvesting the daily ambient temperatures was reported in [5]. The proposed structure contains TEG, phase change material (PCM), and a heat sink. The PCM plays a dual role: when the heatsink functions as a heat source it acts as a cold source, and when the heatsink functions as a cold source it acts as a heat source, so that it is possible to harvest thermal energy throughout the day. For three days an average maximum output power of 340μW was obtained.

Solar radiation in summer leads to surface temperatures of up to 60°C on the roads. Electrical energy harvested from geothermal energy can be utilized to turn on LEDs, traffic lights, etc. By extracting the temperature difference between road surface and subgrade soil, two prototypes, design A and design B, were developed and tested [6]. A heat conduction plate, cold-end cooling module, and thermoelectric conversion were included in the prototype. To keep the temperature of the heatsink down, it was filled with PCM. At a temperature of 55°C, the prototypes of design B and design A produced a maximum power of 27.35 mW and 24.95 mW, respectively. The thermal energy generated can supply power to low-power roadside applications such as warning signals. In an industrial environment and in automobiles, running engines will release a certain amount of energy in the form of heat which is always wasted. Using appropriate mechanisms, this heat can be harvested into electrical energy in a process known as waste heat recovery, and can be used to power the sensor nodes to generate warning signals in industrial environments. Authors have proposed energy scavenging from engines, which releases a significant amount of energy into the environment in the form of gases and coolants. A heat exchanger was proposed in [7] to exhaust the waste heat released by diesel engines. In order to store heat from the engines, ribbed plates containing PCM were used to design heat exchangers. At a rib height of 7.5mm and rib gap of 90mm, better results were obtained. It takes less than one minute to warm the intake air from 273°K–283°K, to 302.2°K and 300.7°K, respectively, by using PCM-stored energy in cold conditions.

Harvesting energy from thermal has been a challenging task that has motivated researchers to develop various type of converter circuit working with low temperature difference to source a driving potential to powering various electronic loads, such as self-powered sensor nodes, small embedded modules monitoring various parameters of a system in remote, etc. [8]. This chapter will discuss the various aspects of circuit design involved in developing a thermal harvesting system, including an oscillator design for a self-starting module. Earlier work reported a fully electric energy harvesting circuit starting from a low voltage and requiring off-the-shelf components like inductors, capacitors, a mechanical switch etc., and the rest of the harvesting circuit is integrated on the chip. Using techniques like zero current switching, an efficient circuit was designed that extracted more than 50% of the maximum available power. To start the circuit from low voltage to first power the peripheral circuits, different start-up methods have been used. In [9], external voltage of 650mV is used to kick-start the circuit, in [10] a mechanical switch is used, whereas in [11] an oscillator with external inductor is used. The harvesting system can thus be categorized into: 1) externally starting [9, 10]; and 2) self-starting [11]. This chapter focus on the design and development of a self-starting TEH system with the following specifications:

- V_{TEG} = 50–200mV.
- R_{TEG} = 5 Ohm.
- V_{OUT} = 1V (Regulated).
- Efficiency > 50%.
- Transient settling time < 5 ms, at 50 mV.

The overall architecture of our circuit is shown in Figure 10.2 [12]. The circuit consists of two stages: auxiliary and main. The auxiliary stage is driven by a low-voltage starter (LVS), which starts up at around 30mV. The LVS consists of a two-stage enhanced swing ring oscillator (ESRO). The AC output of this oscillator is converted into DC voltage ($V_{O,DCP}$) through the Dickson charge pump (DCP), which powers the first current starved ring oscillator (CSRO-1). The CSRO-1 drives the n-channel metal oxide semiconductor (NMOS) switch (NM0) to boost the voltage at V_{DDi} using an external inductor (L_{AUX}) and diode-connected low-voltage p-channel metal oxide semiconductor (PMOS) device (LPD). The diode is made from low threshold voltage (V_t) transistor, this reduces the forward voltage drop compared to standard V_t devices. Once V_{DDi} reaches a particular voltage, the peripheral circuits start to work. The second current starved ring oscillator (i.e., CSRO-2) along with the control logic block generates complementary clocks to drive two NMOS switches,NM 1 and 2. A zero current switching (ZCS) network generates complementary pulses to open and close PMOS switches PM 1 and 2. When V_{TEG} voltage increases, this can increase the voltage at V_{DDi} beyond 1V. To avoid this, a RO_{CNTRL} pulse disables the CSRO-1 whenever such situation arises, hence regulating the voltage at V_{DDi} node. Reference voltages at V_{DDi} and V_{OUT} nodes regulate the power by using comparators in the control logic block. It is reported in [9], that discontinuous conduction mode (DCM) is an efficient way to power the load. Therefore, our work will use DCM to operate the inductor.

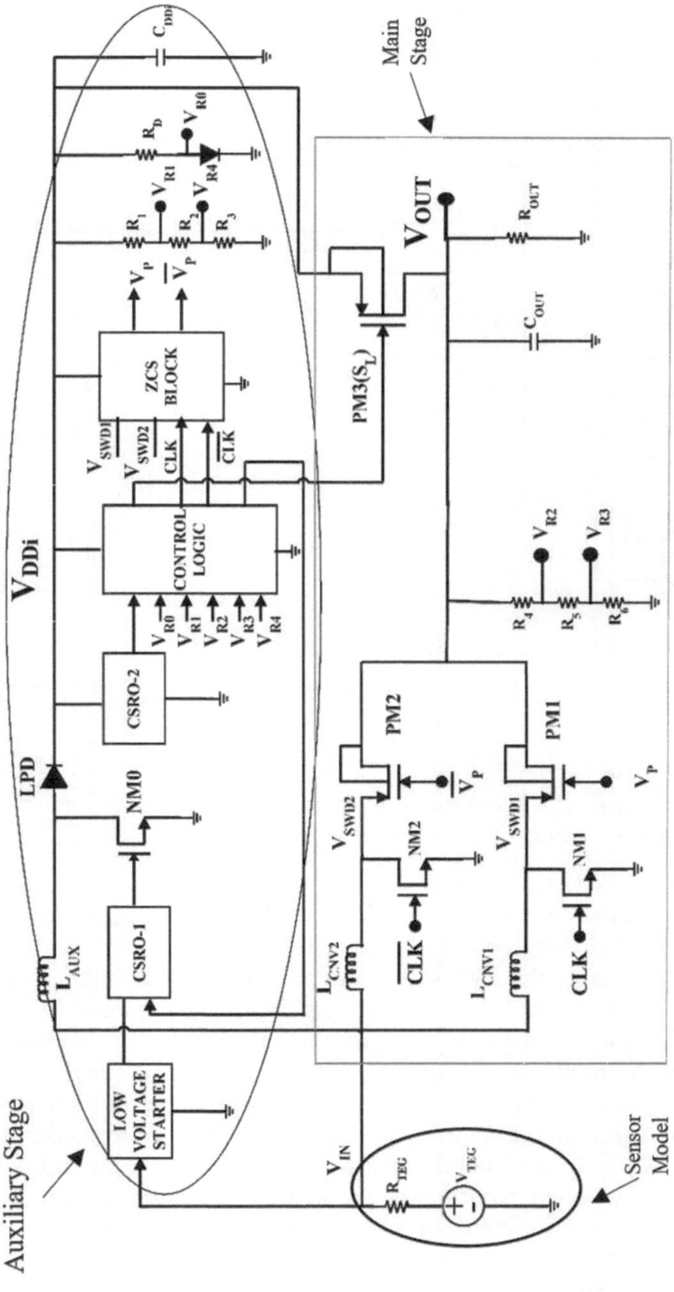

Figure 10.2 Block diagram of the energy harvesting system, with off-chip inductor for circuit start-up. The inductors L_{AUX}, L_{CNV1}, L_{CNV2} and capacitor C_{OUT} are off-the-shelf components [12].

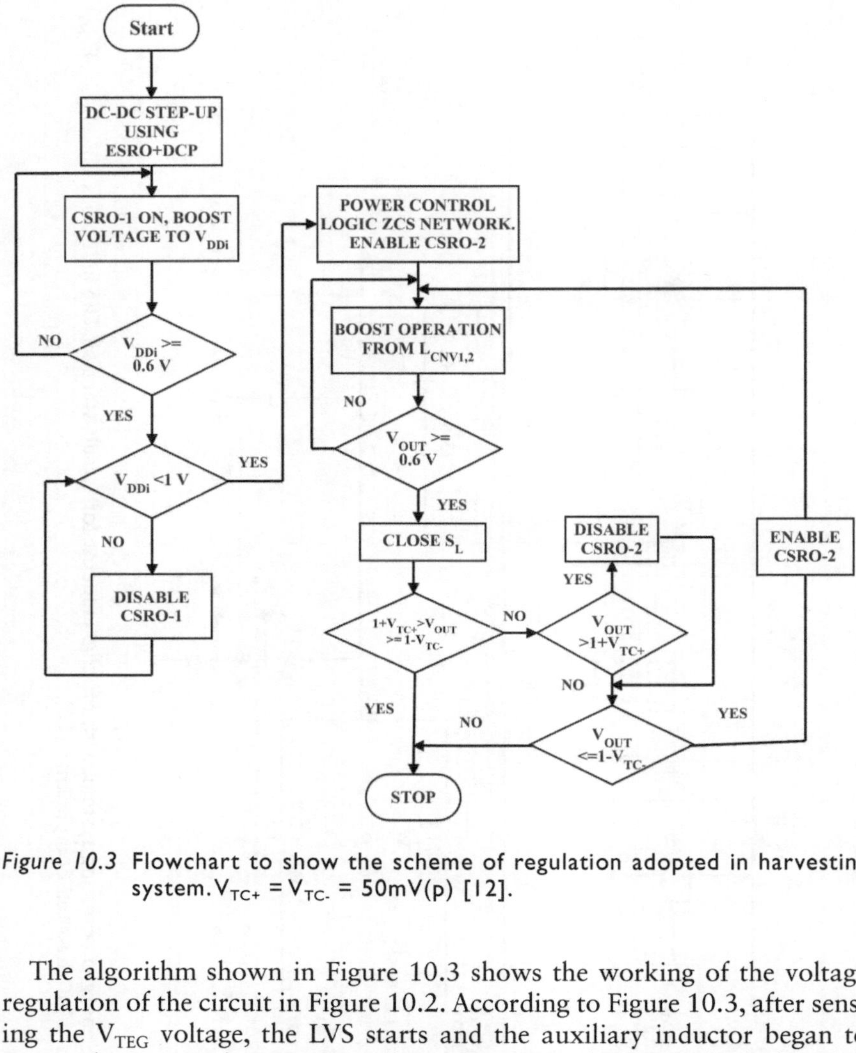

Figure 10.3 Flowchart to show the scheme of regulation adopted in harvesting system. $V_{TC+} = V_{TC-}$ = 50mV(p) [12].

The algorithm shown in Figure 10.3 shows the working of the voltage regulation of the circuit in Figure 10.2. According to Figure 10.3, after sensing the V_{TEG} voltage, the LVS starts and the auxiliary inductor began to power the V_{DDi} node, increasing the node potential. This is checked with a condition that the voltage in V_{DDi} is greater than 0.6 V. Circuit simulations show that with limited power consumption the peripheral circuit can work at 0.6V. The voltage at V_{DDi} node is checked for another condition, that this voltage is not more than 1V, otherwise CSRO-1 is disabled by a comparator through RO_{CNTRL}. If both conditions are satisfied, the peripheral circuits, including CSRO-2 begin to drive the inductors L_{CNV1} and L_{CNV2}. And when the voltage at node V_{OUT} rises above 0.6 V, the switch S_L is closed to set up the self-regeneration loop. Under the condition that the charge dumped on the capacitor is sufficient to power the load and the peripherals, with the passage of time the inductor discharges more current to the capacitor.

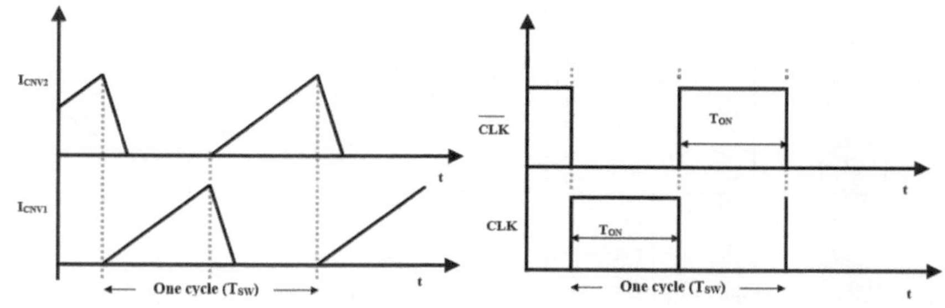

Figure 10.4 Waveform depicting storage and release of energy by the inductors during two half cycles of clock (T_{SW}) [13].

The self-regeneration loop can elevate the voltage to the set 1V. The self-regeneration loop will fail to elevate the voltage, if losses are more than the current discharged by the inductor. Once the voltage reaches 1V, another condition is checked so that V_{OUT} do not exceed the threshold voltage V_{TC+} of a comparator, otherwise CSRO-2 is disabled and is enabled when V_{OUT} goes down to V_{TC-} of the comparator. Thus, the regulation scheme presented in Figure 10.3 can provide a regulated voltage of 1 V at the load. In the present design, the inductors are energized separately for two cycles of the waveform shown in Figure 10.4. Using this technique, it is possible to extract more than 50% of the maximum available power from the TEG by regulating V_{IN} close to $V_{TEG}/2$.

To extract most of the available power, the architecture utilizes two inductors, L_{CNV1} and L_{CNV2}, to work with the 50% duty cycle pulse generated by the CSRO-2. As shown in Figure 10.4, CLK and $\overline{CLK}$ are generated after passing through a series of gates present in the control logic block. Therefore, for one half pulse, when inductor L_{CNV1} is storing current I_{CNV1}, the other inductor L_{CNV2} is discharging current to the capacitor and to the load. The opposite situation happens in the next half clock cycle. Thus, by utilizing each half cycle, we can ensure that the maximum value of the average current is available to the load. This will also ensure that the two inductors are extracting energy from the low-voltage source in an efficient way.

10.1.1 Calculation of the maximum power conversion efficiency

In this sub-section, mathematical equations are used to estimate how power extracted by an ideal convertor (Figure 10.5a), can extract without any input capacitor (C_{IN}) at V_{IN} node. The available power is given by

$$P_{AV} = \frac{V_{TEG}^2}{4 \cdot R_{TEG}}. \tag{10.1}$$

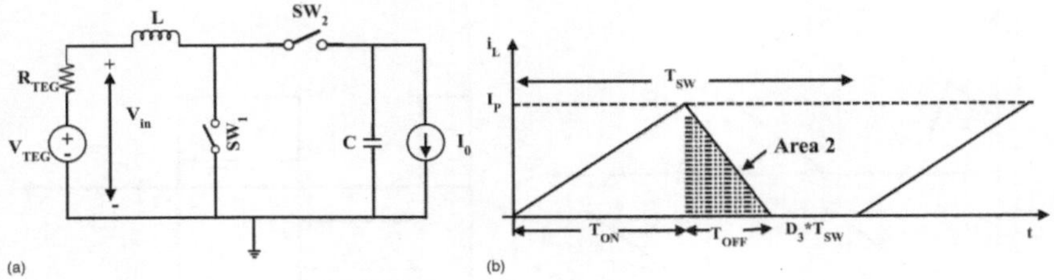

Figure 10.5 (a) Assuming lossless L, S_{W1} and S_{W2}, (b) Inductor current waveform in DCM [14, 15].

The Area2 in Figure 10.5b, will give the average value of current available at the output.

$$I_O \cdot T_{SW} = \text{Area2} = \frac{I_P \cdot D2 \cdot T_{SW}}{2}. \tag{10.2}$$

Solving (10.2),

$$I_O = \frac{D2 \cdot I_P}{2}. \tag{10.3}$$

The power delivered to the load (lossless converter) is,

$$P_O = \frac{1}{T_{SW}} \int_0^{T_{SW}} p(t)\ dt = \frac{1}{T_{SW}} \int_0^{T_{SW}} \left(V_{TEG} - R_{TEG} \cdot i_{TEG}\right) \cdot i_{TEG}\ dt. \tag{10.4}$$

$$\begin{aligned} T_{SW} \times P_O = &\int_0^{T_{ON}} \left(V_{TEG} - R_{TEG} \frac{I_P \cdot t}{T_{ON}}\right) \times \frac{I_P \cdot t}{T_{ON}} dt \\ &+ \int_{T_{ON}}^{T_{ON}+T_{OFF}} \left[V_{TEG} - R_{TEG} \frac{I_P \cdot (T_{ON} + T_{OFF} - t)}{T_{OFF}}\right] \times \frac{I_P \cdot (T_{ON} + T_{OFF} - t)}{T_{OFF}} dt. \end{aligned} \tag{10.5}$$

$$\begin{aligned} = &\left.\frac{V_{TEG} I_P t^2}{2 \cdot T_{ON}}\right|_0^{T_{ON}} - \left.\frac{R_{TEG} I_P^2 t^3}{3 \cdot T_{ON}^2}\right|_0^{T_{ON}} - \left.\frac{V_{TEG} I_P \left(T_{ON} + T_{OFF} - t\right)}{2 \cdot T_{OFF}}\right|_{T_{ON}}^{T_{ON}+T_{OFF}} \\ &+ \left.\frac{R_{TEG} I_P^2 \left(T_{ON} + T_{OFF} - t\right)^3}{3 \cdot T_{OFF}^2}\right|_{T_{ON}}^{T_{ON}+T_{OFF}}. \end{aligned} \tag{10.6}$$

$$T_{SW} \times P_O = \frac{V_{TEG} I_P T_{ON}}{2} - \frac{R_{TEG} I_P^2 T_{ON}}{3} + \frac{V_{TEG} I_P T_{OFF}}{2} - \frac{R_{TEG} I_P^2 T_{OFF}}{3} \tag{10.7}$$

$$T_{SW} \times P_O = \left[\frac{V_{TEG}}{2} - \frac{R_{TEG} I_P}{3}\right](T_{ON} + T_{OFF}) I_P \tag{10.8}$$

For the power to be max,

$$\frac{dP_O}{dI_P} = 0 \rightarrow \frac{V_{TEG}}{2} - \frac{2 \cdot R_{TEG} I_P}{3} = 0 \tag{10.9}$$

$$I_{P\max} = \frac{3}{4} \frac{V_{TEG}}{R_{TEG}} \tag{10.10}$$

$$I_{O\max} = \frac{D_2 I_{P\max}}{2} = D_2 \frac{3}{8} \frac{V_{TEG}}{R_{TEG}} \tag{10.11}$$

From (10.8), the maximum output power,

$$P_{O\max} = (D1 + D2) \cdot \frac{3}{4} \frac{V_{TEG}^2}{4 \cdot R_{TEG}} \tag{10.12}$$

$$\frac{P_{O\max}}{P_{AV}} = (D1 + D2) \frac{3}{4} \tag{10.13}$$

In the case of a symmetrical clock used for switching in a high-gain situation, D1+D2 ≈ D1 (= 0.5), therefore (10.13) can be numerically expressed as

$$\frac{P_{O\max}}{P_{AV}} = \frac{3}{8}. \tag{10.14}$$

Therefore, it can be concluded that an ideal converter can extract 37.5% of the available power, when the inductor is charged for one half cycle. If both half cycles are used, the maximum output power is 75% of the input available power. In the following sections, formulas are derived for the blocks present in the energy harvesting system (Figure 10.2). The derivation of the formula and the detailed circuit diagram will help calculation by hand of the numerical value of the circuit parameters.

10.2 LOW-VOLTAGE STARTER

The LVS consists of a cross-coupled, two-stage ESRO [16], and a DCP for AC-DC conversion [17]. Detailed formulations for designing this block will be presented in this section.

10.2.1 Enhanced swing ring oscillator

The inductor connected between the drain of the MN1/MN2 transistor and the V_{IN} in Figure 10.6 can be represented by a parallel resistance R_P and inductor L_P:

$$L_P \approx L_1 \tag{10.15}$$

$$R_P \cong Q_1^2 \cdot R_1 \tag{10.16}$$

Equations (10.15), and (10.16) hold when the quality factor of inductors > 3, for some range of frequencies. The admittance between the drain node and ground is given by

$$Y(j\omega) = \left(g_0 + \frac{1}{R_P}\right) - j\left(\frac{1}{\omega L_P}\right) = G_P - jX_P \tag{10.17}$$

In (10.17), g_0 is the channel conductance.

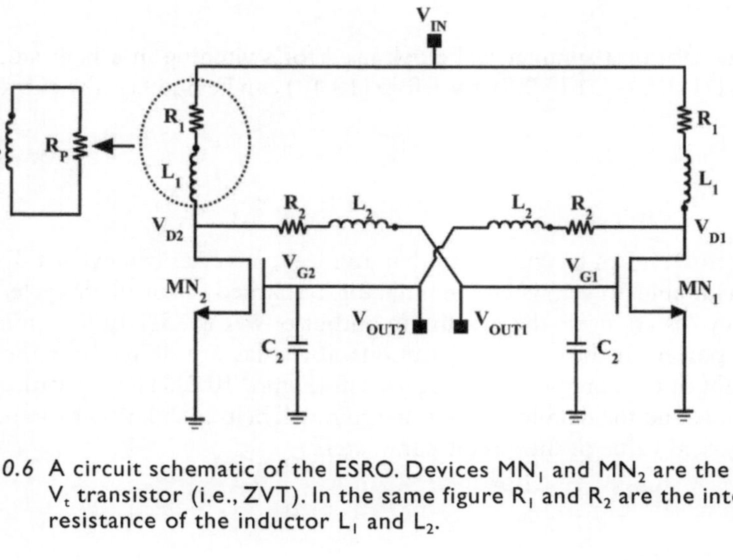

Figure 10.6 A circuit schematic of the ESRO. Devices MN_1 and MN_2 are the zero V_t transistor (i.e., ZVT). In the same figure R_1 and R_2 are the internal resistance of the inductor L_1 and L_2.

The circuit diagram in Figure 10.6 can be drawn in the small signal domain shown by Figure 10.7. At node A in Figure 10.7 we can apply KCL, and solve to get V_{D2} which is given by

$$\frac{V_{D2}}{V_{OUT1}} = \frac{1 - g_m(R_2 + sL_2)\cdot(V_{OUT2}/V_{OUT1})}{1 + Y(s)(R_2 + sL_2)} \tag{10.18}$$

Similarly, at node B

$$\frac{V_{D2}}{V_{OUT1}} = 1 + sC_2(R_2 + sL_2) \tag{10.19}$$

The transfer function after solving (10.18) and (10.19) is

$$\begin{aligned}\frac{V_{OUT1}}{V_{OUT2}} &= \frac{g_m(R_2 + sL_2)}{1 - (1 + Y(s)\cdot(R_2 + sL_2))\cdot(1 + sC_2\cdot(R_2 + sL_2))} \\ &= -\frac{g_m}{sC_2 + Y(s)(1 + sC_2\cdot(R_2 + sL_2))}\end{aligned} \tag{10.20}$$

Equation (10.20), can be written in a simplified form as,

$$\begin{aligned}\frac{V_{OUT1}}{V_{OUT2}} &= -\frac{g_m}{sC_2 + (G_P - jX_P)\cdot(A_0 + jB_0)} \\ &= -\frac{g_m}{G_PA_0 + X_PB_0 + j(\omega C_2 + B_0G_P - X_PA_0)}\end{aligned} \tag{10.21}$$

In (10.21), $A_0 = 1 - L_2C_2\omega^2$; $B_0 = \omega C_2R_2$; $G_P = g_0 + (1/R_P)$; $X_P = 1/\omega L_P$. In Equation (10.21), the imaginary part of the equation can be equated to zero, to get a phase difference of 0^0 or 360^0. Solving equation $\omega C_2 + B_0G_P - X_PA_0 = 0$ will give a peak resonance frequency (f_0)

$$f_0 = \frac{1}{2\pi\sqrt{L_2C_2 + L_PC_2(1 + R_2G_P)}} \tag{10.22}$$

where $f = 1/2\pi\omega$, f is in Hz and ω is in rad/s. In (10.22) we can assume that $R_2G_P << 1$ to give $1 + R_2G_P \approx 1$ also let $K = L_2/L_P$, then (10.22) can be expressed as,

$$f_0 = \frac{1}{2\pi\sqrt{C_2L_P(K + 1)}} \tag{10.23}$$

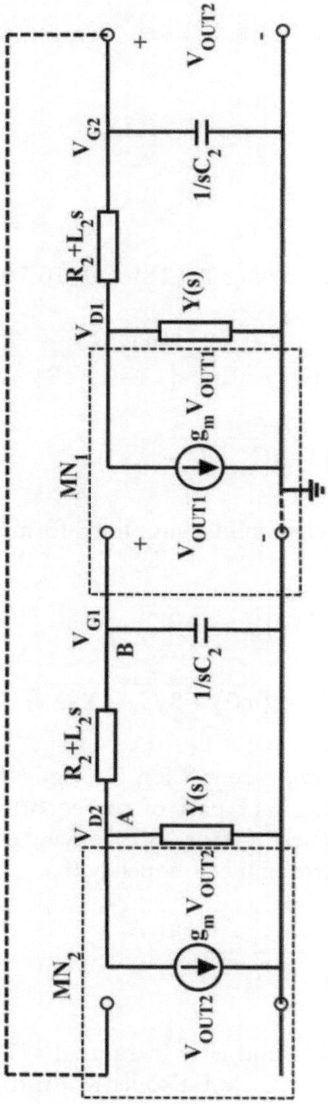

Figure 10.7 Small signal representation of the circuit shown in Figure 10.6.

In (10.23), $C_2 = C_{gate} = (2/3)\times W\times L\times n_f\times C_{ox}$, where $C_{ox} = \varepsilon_{ox}\varepsilon_d/T_{ox}$, neglecting overlap capaci-tance, the constant $\varepsilon_{ox} = 8.854\times10^{-12}$ F/m and $\varepsilon_d = 3.9$ for SiO_2, T_{ox} is thickness in meters for a layer of SiO_2, and n_f is the number of fingers to increase the transistor width. Let f_{max} be the maximum frequency when K = 0, then

$$f_{max} = \frac{1}{2\pi\sqrt{C_2 L_P}} \tag{10.24}$$

The K can also be expressed in term of resonant frequency as,

$$K = \left(\frac{f_{max}}{f_0}\right)^2 - 1 \tag{10.25}$$

For sustained oscillation, g_m (gate transconductance) should be greater than the real part in transfer function (10.21) [18]

$$g_m > G_P A_0 + X_P B_0 \tag{10.26}$$

Equation (10.26) can be expressed as

$$g_m > G_P\left(1 - L_2 C_2 \omega^2\right) + \frac{C_2 R_2}{L_P} \tag{10.27}$$

The resonance angular frequency term in (10.23) can be substituted with (10.27) to be written as

$$g_m > \frac{G_P}{K+1} + \left(\frac{K R_2 C_2}{L_2}\right) \tag{10.28}$$

The $g_m = (g_{ms} - g_0)/n$ [18], Equation (10.28) can be written in term of g_{ms}/g_0 as

$$\frac{g_{ms}}{g_0} > 1 + n\left[\left(1 + \frac{1}{g_0 R_P}\right)\frac{1}{K+1} + \frac{C_2 R_2}{g_0 L_P}\right] \tag{10.29}$$

where $R_P = \omega_0 \cdot Q_1 \cdot L_P$ and n is the slope factor which can be equal to one for a simple solution. The value of V_{DD} can be found from general equation

$$V_{DD} = U_T \ln\left(\frac{g_{ms}}{g_0}\right) + \frac{U_T^2\left(g_{ms} - g_0\right)}{2I_S} \tag{10.30}$$

In (10.30), U_T is the thermodynamic voltage at room temperature and I_S, is the specific current. To achieve a low start-up value, all the parameters in (10.30) need to be determined.

10.2.2 Dickson charge pump

Figure 10.8 shows the diagram of a Dickson-based voltage multiplier [17] with a zero-threshold voltage transistor (ZVT) acting as a diode (D_1 to D_{N+1}), by connection gate to drain terminals.

The I-V equation of a diode can be given by

$$V_D = nU_T \ln\left(\frac{I_D}{I_S} + 1\right) \tag{10.31}$$

Let I_P be the peak current of complementary voltages V_ϕ and $\overline{V_\phi}$. It is related to load current (i.e., I_L) and saturation current (i.e., I_S) as, $I_P = 2 \cdot I_L + I_S$. From (10.31) the resistance (i.e., R_D) across diode can be expressed by

$$\frac{1}{R_D} = \frac{dI_D}{dV_D} = \frac{2I_S}{nU_T}\left(1 + \frac{I_L}{I_S}\right) \tag{10.32}$$

The Dickson equivalent circuit of Figure 10.8 is shown in Figure 10.9a.

In Figure 10.9a, for N number of stages, the $V_{I,DCP}$ is given by

$$V_{I,DCP} = V_{IN} + N \cdot \left[\frac{C \cdot V_\phi}{C + C_P}\right] - (N+1) \cdot n \cdot U_T \ln\left[2\left(1 + \frac{I_L}{I_S}\right)\right] \tag{10.33}$$

and R_T is given by

$$R_T = N\left[\frac{1}{(C + C_P)f_0} + \frac{n \cdot U_T}{2I_S(1 + (I_L / I_S))}\right] \tag{10.34}$$

If $V_{O,DCP}$ is the output voltage, it can be expressed by

$$V_{O,DCP} = \frac{R_{RO}}{R_T + R_{RO}} \cdot V_{I,DCP} \tag{10.35}$$

where R_{RO} is the load resistance; at present it is CSRO-1. The equivalent capacitance C_{DE} of the Dickson stage after considering $C >> 2C_P$ will be

$$C_{DE} \cong \frac{N}{2} \times \frac{C}{2} \tag{10.36}$$

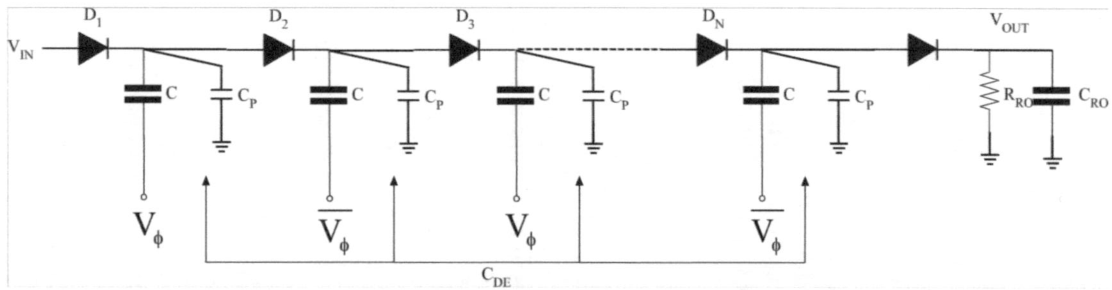

Figure 10.8 DC input voltage $V_{IN} = 0$ in our design. C_{DE} is the Dickson equivalent capacitance. C and C_p are the coupling and parasitic capacitance respectively. The V_ϕ and $\overline{V_\phi}$ are the peak voltage of the alternating signal and are complementary to each other. The C_{RO} and R_{RO} is the input capacitance and the resistance of the CSRO-1.

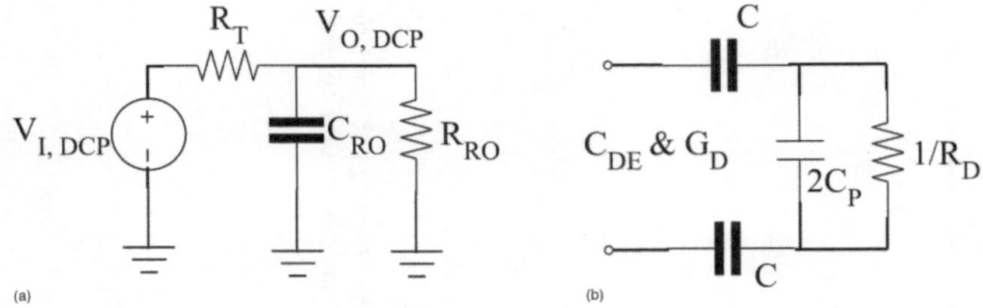

Figure 10.9 (a) Equivalent circuit of Figure 10.8, (b) Input capacitance and conductance of one stage in DCP. In the same figure G_D is the equivalent input conductance.

The ESRO connection with the Dickson charge pump (DCP) may require for precise calculation the C_{DE} and equivalent conductance G_D ($\approx 1/NR_D$) in hand calculation, as shown in Fig. 10.9(b). Therefore, in (10.21), the C_2 can be replaced by C_{net} where $C_{net} = C_{DE} + C_{gate}$ and G_D can be included as

$$\frac{V_{OUT1}}{V_{OUT2}} = -\frac{g_m}{(G_D + sC_{net}) + (G_P - jX_P)\cdot(A_0 + jB_0) + G_D\cdot(G_P - jX_P)\cdot(R_2 + sL_2)} \tag{10.37}$$

On solving (10.37), the transfer function will have some new terms, but the basic structure of the equation will be similar to (10.21)

$$\frac{V_{OUT1}}{V_{OUT2}} = -\frac{g_m}{\begin{array}{c}G_PA_0 + X_PB_0 + G_D(1 + R_2G_P + \omega L_2X_P) + \\ j\left[\omega C_{net} + B_0G_P - X_PA_0 + G_D(\omega G_PL_2 - R_2X_P)\right]\end{array}} \tag{10.38}$$

Equating imaginary part to zero to find the oscillation frequency,

$$\omega_0^2 = \frac{1 + G_DR_2}{C_{net}L_P(K+1) + G_DG_PKL_P^2} \tag{10.39}$$

If, $G_DR_2 \approx 0$, and, $G_DG_PK L_P^2 << C_{net}L_P(K+1)$, (10.39) can be reduced in similar form to (10.26). Similarly, from the real part of (10.39), the following condition can be derived:

$$g_m > G_PA_0 + X_PB_0 + G_D(1 + R_2G_P + \omega L_2X_P) \tag{10.40}$$

On solving (10.40) with (10.39),

$$g_m > \frac{G_P}{K+1} + \frac{KR_2(C_{net} + C_{DE})}{L_2} + G_D(K+1) \tag{10.41}$$

Therefore (10.41) is similar to (10.29), with one additional term, $G_D(K+1)$. The g_{ms}/g_0 ratio will be

$$\frac{g_{ms}}{g_0} > 1 + n\left[\left(1 + \frac{1}{g_0 R_P}\right)\frac{1}{K+1} + \frac{C_{net}R_2}{g_0 L_P} + \frac{G_D(K+1)}{g_0}\right] \tag{10.42}$$

With the increase in the value of capacitance, the oscillation frequency decreases because the quality factor of the inductor and R_P value decreases; this requires the value of g_0 to increase to satisfy the relation given in (10.42).

10.2.3 AC model of ESRO

The circuit in Figure 10.10 represents an equivalent model of the ESRO; the equivalent transformer has N number of turns given by

$$N = \frac{L_1}{L_1 + L_2} = \frac{1}{K+1} \tag{10.43}$$

The transformer model will be valid only when quality factor $Q_T = \omega_0 C_{net}/(N^2 G)$, and $Q_E = (L_1 + L_2)/(\omega_0 L_1 L_2 G)$ is greater than 10 [19]. As given in [19] that $L = L_1 + L_2$ with primary open. Therefore $G_L = 1/(r_p + R_p)$, where r_p is the parallel resistance of inductor L_2 for $Q_2 > 3$. The Q_T of this tuned circuit is high (>10). Therefore, the oscillator can be represented by a large signal model (shown in Figure 10.10) to determine the $V_{OUT1}(t)$. The self-sustained oscillation with peak frequency f_0 given by (10.23) stabilizes at the value of $x = V_{OUT1}/U_T$ for which the following equality holds:

$$\frac{\left[G_{ms}(x)/N\right] \cdot N^2}{N^2 \cdot G + G_L} = 1 \tag{10.44}$$

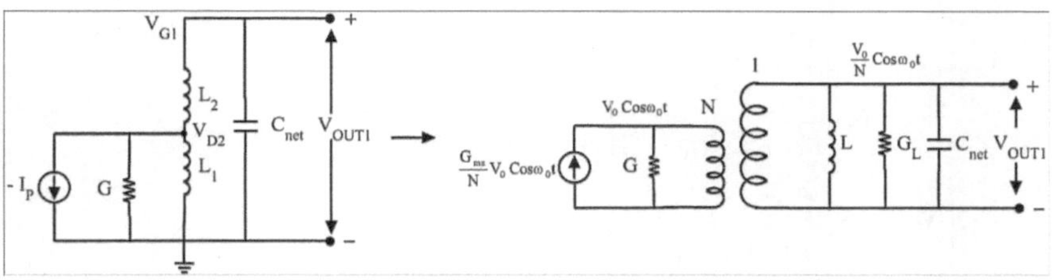

Figure 10.10 AC equivalent model of ERSO set-up. The dependent current source is opposite in phase to the gate voltage V_{OUT2}, the transformer model of an ESRO can be derived [19].

On solving,

$$\frac{G_m(x)}{g_{mQ}} = \frac{G_L + N^2 \cdot G}{g_{mQ} \cdot N} \tag{10.45}$$

The ratio given in (10.45) can be calculated, which corresponds to the fundamental component of the drain current as

$$\frac{G_m(x)}{g_m} = \frac{2 \cdot I_1(x)}{x \cdot I_0(x)} \tag{10.46}$$

From the tabular method, the corresponding value of x can be evaluated and hence the peak value of the output voltage swing can be determined [19].

10.2.4 Important formulas for the LVS block

The condition for start-up after ignoring G_D in (10.42) can be given by

$$\frac{g_{ms}}{g_0} - \left(1 + n\left[\left(1 + \frac{1}{g_0 R_P}\right)\frac{1}{K+1} + \frac{C_{net} R_2}{g_0 L_P}\right]\right) > 0 \tag{10.47}$$

In (10.47), the C_{net} is the sum of gate capacitance of ZVT, and the total capacitance presented from the DCP. Typical value of C_{net} is in pF, when $C_{net} << g_0 L_p$ irrespective of the R_2 value, then (10.47) can be simply expressed as

$$\frac{g_{ms}}{g_0} - \left(1 + n\left[\left(1 + \frac{1}{g_0 R_P}\right)\frac{1}{K+1}\right]\right) > 0 \tag{10.48}$$

Eq. (10.48), gives an idea of selecting L_1 and K with minimum supply voltage, $L_2 = K \cdot L_1$. The extended equation for the DCP output using (10.45) can be written as

$$V_{O,DCP} = \frac{R_{RO}}{N\left[\frac{1}{(C + C_P) f_0} + \frac{n \cdot U_T}{2 I_S (1 + (I_L / I_S))}\right] + R_{RO}} \cdot \left(V_{IN} + N \cdot \left[\frac{C \cdot V_\phi}{C + C_P}\right] - (N+1) \cdot n \cdot U_T \ln\left[2\left(1 + \frac{I_L}{I_S}\right)\right]\right) \tag{10.49}$$

The optimized values of different parameters selected for the LVS block are given in Table 10.1

Table 10.1 Parameters and their values used in LVS block

Parameter	*W/L ($MN_{1,2}$) (μm/μm)*	*N*	*C (pF)*	*C_P (pF)*	*n*	*U_T*	*I_S (μA)*	*I_L @ 0.5 V*	*R_{RO} (MΩ)*	*C_{RO} (pF)*
Value	(6×40)×4/0.42	12	6.4	0.5	1.2	26 mV	3.2	66 nA	8	5

The charge pump specifications such as rise time (T_{Rampup}) and the peak-peak ripple voltage (dV), can be estimated by (10.50) and (10.51) with hand calculation.

$$I_L = C_{RO}\frac{V_{O,\ DCP}}{T_{Rampup}} \tag{10.50}$$

$$I_L = C_{RO}\frac{dV}{dT} \tag{10.51}$$

In (10.51), dT is the inverse of ESRO oscillator frequency f_0 [20].

10.3 POWER EQUATIONS FOR CONVERTOR OPERATING IN DCM

In this section we present equations for the boost operation with inductor [14, 15]. This section also derives the equations to extract maximum power from a low-voltage source.

10.3.1 Boost convertor

Figure 10.11 shows the circuit diagram of a boost convertor. The peak current (I_{Peak}) for the ON event can be given as

$$I_{Peak} = I_{Valley} + S_{ON} \cdot T_{ON} \tag{10.52}$$

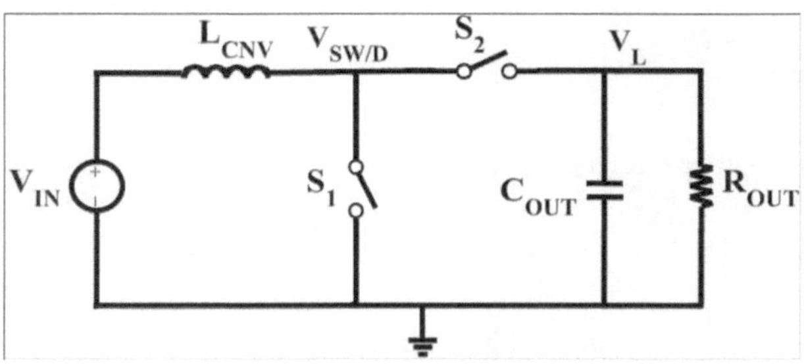

Figure 10.11 Boost convertor operated with NMOS switch S_1 and PMOS switch S_2. Presently both switches are ideal [15].

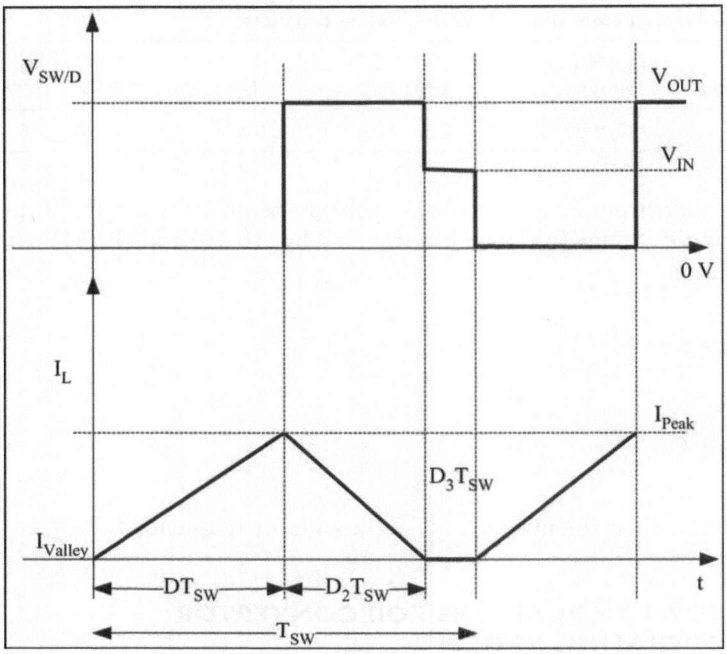

Figure 10.12 **Discontinuous conduction mode waveform for boost operation.**

In (10.52), S_{ON} is given by V_{IN}/L_{CNV}. The equation for the OFF event can be written as,

$$I_{Valley} = I_{Peak} + S_{OFF} \cdot T_{OFF} \tag{10.53}$$

In (10.53), S_{OFF} is given by: $-(V_{OUT} - V_{IN})/L_{CNV}$. The I_{Peak} is a maximum peak value of the current and I_{Valley} is the minimum value of the inductor current. The boost ratio $M = V_{OUT}/V_{IN}$ can be written as

$$M = \frac{T_{ON}}{T_{OFF}} + 1 \tag{10.54}$$

The average voltage across inductor can be written as

$$\langle V_L \rangle_{T_{SW}} = \langle V_{IN} \rangle_{T_{SW}} - \langle V_{SW/D} \rangle_{T_{SW}} = 0 \tag{10.55}$$

which is zero at steady state. The voltage drop at $V_{SW/D}$ can be written as

$$\langle V_{SW/D} \rangle_{T_{SW}} = D_2 V_{OUT} + D_3 V_{IN} \tag{10.56}$$

The $D_3 = 1 - D - D_2$, from (10.55) $\langle V_{IN} \rangle_{TSW} = \langle V_{SW/D} \rangle_{TSW}$, therefore solving (10.56) will give,

$$\frac{V_{OUT}}{V_{IN}} = \frac{D}{D_2} + 1 \tag{10.57}$$

Assuming 100% efficiency, $V_{IN} \langle I_{IN} \rangle = V_{OUT} \langle I_{OUT} \rangle$. As $I_{OUT} = V_{OUT}/R_{OUT}$ and $\langle I_{IN} \rangle = \langle I_{CNV} \rangle$ (average input inductor current), therefore

$$\langle I_{CNV} \rangle = M \frac{V_{OUT}}{R_{OUT}} (= M \cdot I_{OUT}) = (D + D_2) \frac{I_{Peak}}{2} \tag{10.58}$$

On solving (10.58), D_2 can be expressed in two forms,

$$D_2 = \frac{2MV_{OUT}}{R_{OUT} I_{Peak}} - D = \frac{2MI_{OUT}}{I_{Peak}} - D \tag{10.59}$$

The peak value of current can be written for ON and OFF cycle as

$$I_{Peak} = \frac{V_{IN}}{L_{CNV}} D \cdot T_{SW} = \frac{V_{OUT} - V_{IN}}{L_{CNV}} D_2 \cdot T_{SW} \tag{10.60}$$

Both the expressions of peak current in (10.60) will give a solution for M. However, for an easy solution, we can take the first expression of I_{Peak} and substitute this in (10.59) to get

$$D_2 = \frac{2M^2 L_{CNV}}{R_{OUT} D T_{SW}} - D = \frac{2MI_{OUT} L_{CNV}}{V_{IN} D T_{SW}} - D \tag{10.61}$$

The expressions of (10.61) can be substituted in (10.59) to give an expression for M in term of R_{OUT}

$$M^2 - M - \frac{R_{OUT} D^2 T_{SW}}{2L_{CNV}} = 0 \tag{10.62}$$

On solving quadratic expression (10.62) we will get

$$M = \frac{1}{2}\left(1 + \sqrt{1 + \frac{2R_{OUT} D^2 T_{SW}}{L_{CNV}}}\right) \tag{10.63}$$

Also, M can be expressed in terms of I_{OUT} as

$$M = 1 + \frac{V_{IN} D^2 T_{SW}}{2 I_{OUT} L_{CNV}} \tag{10.64}$$

The ripple voltage ΔV (pk – pk) is related to the load capacitor by expression [14]

$$\Delta V \geq \frac{I_{OUT} \cdot T_{SW}}{C_{OUT}} \left(1 - \sqrt{\frac{2 \cdot L_{CNV}}{R_{OUT} \cdot T_{SW}}}\right) \tag{10.65}$$

10.3.2 Maximum extraction of power from a low-voltage source V_{TEG}

Figure 10.13 shows that when using switches S_1 and S_2, the boost action will have the following effects:

- Input power = output power.
- Voltage increases, current decreases due to constant power from source.
- Equivalent to a transformer action.

Figure 10.13 also shows one method to boost the voltage V_{IN} to a high value V_{OUT} by a factor M. This requires an inductor that can store magnetic flux and discharge to the capacitor (C_{OUT}) charging it. The average value of current at C_{OUT} will determine the efficiency of the system and available power at the load R_{OUT}. In mesh I, considering a current I_{CNV} flows, when switch S_1 is closed, after considering initial boundary conditions can be written as

$$I_{CNV}(t) = \frac{V_{TEG}}{R_{SW} + R_{TEG} + R_L} \left(1 - \exp\left(-\frac{t\left(R_{SW} + R_{TEG} + R_L\right)}{L_{CNV}}\right)\right) \tag{10.66}$$

In (10.66), peak value of current will happen when $t = T_{ON}$, R_L is the internal resistance of the inductor. It is noteworthy that (10.52) is the approximate representation of (10.66).

In DCM the average value of current (I_{av}) is always less than half the peak value of current (I_P). Let the $I_{av} = I_{OUT}$ at the R_{OUT} of Figure 10.13. As the average value of current denotes the DC value of current to the load, from Figure 10.14 it can be given as

$$I_{av} = \frac{D_2 I_P}{2} = \frac{T_{OP} I_P}{2 T_{SW}} = \frac{D \cdot I_P}{2(M-1)} \tag{10.67}$$

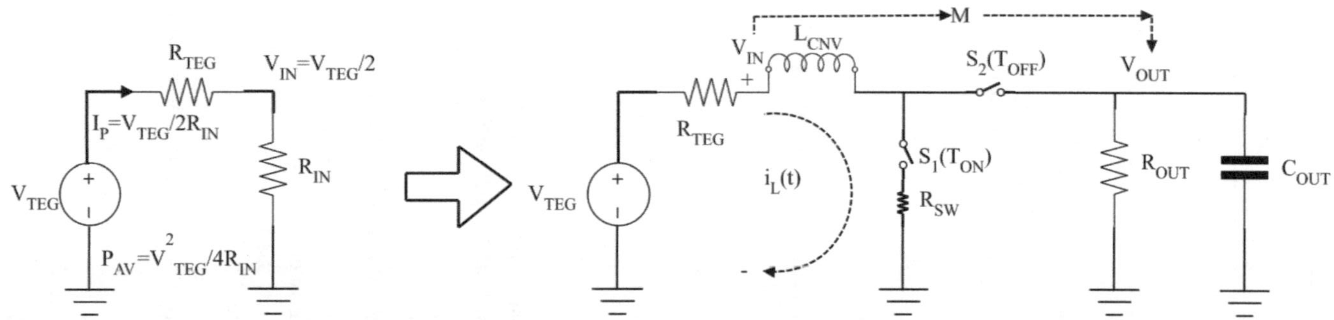

Figure 10.13 Depiction of maximum transfer theorem used in boosting voltage by factor M, R_{SW} is the switch S_1 resistance. T_{ON} and T_{OP} are the ON time of switches S_1 and S_2, and $R_{OUT} \cong M^2 \cdot R_{IN}$, where R_{IN} is the matched resistance [12].

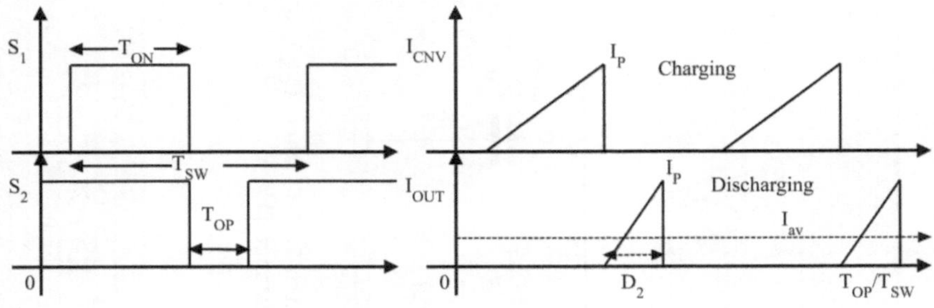

Figure 10.14 Waveform presents the availability of current from inductor L_{CNV}.

Substituting the value of D from (10.60) in (10.67) will result in the equation

$$I_{OUT} = \frac{I_P^2 \cdot L_{CNV}}{2(M-1)V_{IN}T_{SW}} \tag{10.68}$$

Equation (10.68) is the average value of current available obtained from one half of the cycle (T_{ON}). The average value of the current will be multiplied by 2, when both halves of the cycle are used. In (10.68), M >> 1 therefore output available power $P_{OUT} = I_{OUT}V_{OUT}$, can be expressed as in term of L_{CNV} as

$$P_{OUT} \approx \frac{I_P^2 \cdot L_{CNV}}{2 \cdot T_{SW}} \tag{10.69}$$

In (10.69), the I_P can be substitute by I_{CNV} of (10.66), to give an exact expression deciding the value of inductor that can deliver maximum value of the average load current.

10.3.3 Sizing auxiliary and main stage inductors

Table 10.2 shows the aspect ratio of switches used in our circuit design. The choice of inductor to power the V_{DDi} and V_{OUT} node is sized by using (10.69). The T_{SW} (CSRO-1) ≈ 10 µs, C_{DDi} = 1 nF; this ensures peak ripple is less than 20 mV (p). The *LPD* diode forward voltage drop is assumed as V_{DF} = 0.2 V. The values of parameters are: R_{TEG} = 5 Ω, R_L = 0.6 Ω, R_{SW} (NM0) at 0.5 V = 15 Ω, R_{SW} (NM1, 2) at 1 V = 0.45 Ω. This will give T_{SW} for CSRO-2 ≈ 42 µs.

Table 10.2 Switches aspect ratio

Low V_t devices	*NM0*	*NM1/NM2*	*PM1/PM2*	*PM3*
W/L (µm/µm)	(48×4)/0.48	(480×4)/0.12	(120×4)/0.12	(3×4)/0.12

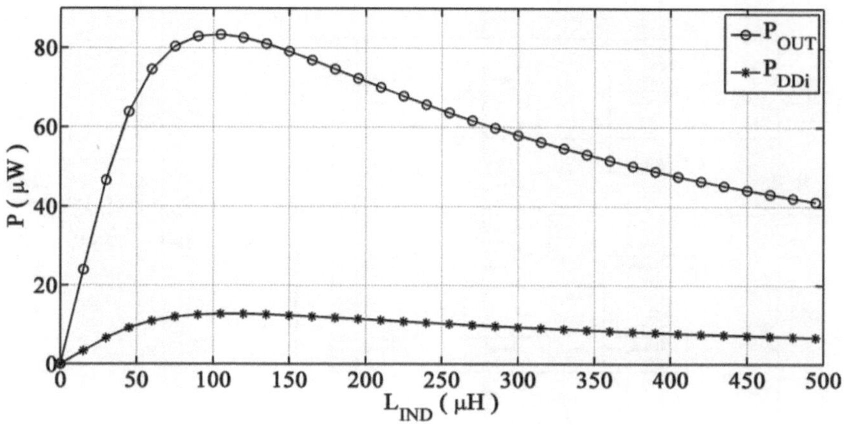

Figure 10.15 Plot of power using (3.2.4), i.e., average power available at the load (P_{OUT}) with two convertors [12].

The average power consumption by the peripherals is around 4 μW at 1V; choosing a L_{AUX} value lying between 100 μH and 500 μH will be sufficient to power the peripheral circuits. From Figure 10.15, the value of convertors: $L_{CNV1} = L_{CNV2} = 100$ μH was chosen to power the load. Maximum value of W/L of the NM1, 2 is estimated by considering the dynamic and static power consumption in (10.70) and (10.71).

$$P_{dyn} = C_{total} V_{DD}^2 f_{sw} \tag{10.70}$$

$$P_{Stat} = \frac{1}{T_{SW}} \int_0^{T_{SW}} R_{SW} I^2(t) \quad dt \cong \frac{R_{SW} I_P^2}{6} \tag{10.71}$$

For switch NM1-2: $V_{DD} = 1V$, $f_{SW} = 23.8$ kHz, $R_{SW} = 0.42\ \Omega$, $I_P = 6$ mA@ 50 mV. From this $P_{dyn} = 57.12$ nW, $P_{Stat} = 2.52$ μW.

10.4 PERIPHERAL CIRCUITS

10.4.1 Current starved ring oscillator

The schematic of current starved ring oscillator [21] is shown in Figure 10.16. To find the oscillation frequency of the circuit, we can follow some mathematical derivations as explained in this section. If I_D is the current flowing through R, then

$$I_D = \frac{V_{b1} - V_{b2}}{R} \tag{10.72}$$

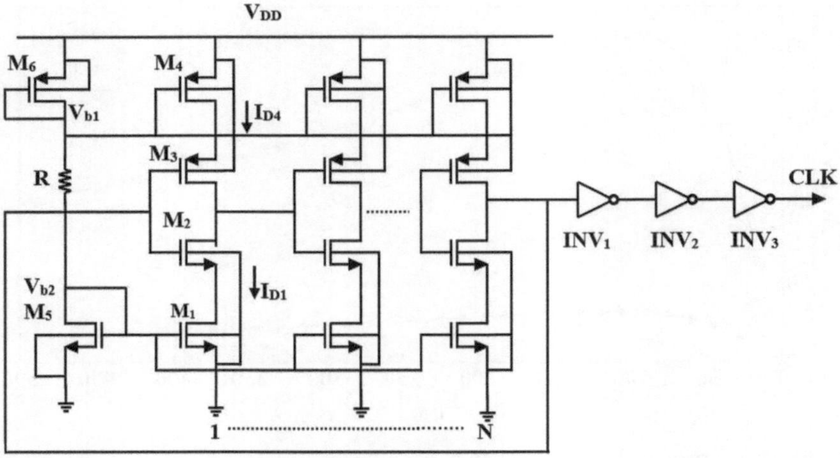

Figure 10.16 An arrangement of current starved ring oscillator used in our architecture. N is the number of stages. All the transistors are long channel devices; INV1, INV2 and INV3 are realized with a minimum width and length, and R = 8 MΩ is present on the chip.

In (10.72), V_{b1} and V_{b2} will be given by

$$V_{b1} = V_{DD} - U_T \ln\left(1 + \frac{I_D}{I_{S6}}\right) \tag{10.73}$$

$$V_{b2} = U_T \ln\left(1 + \frac{I_D}{I_{S5}}\right) \tag{10.74}$$

In (10.73) and (10.74), I_{S6} and I_{S5} are the reverse saturation current of diode-connected devices M_6 and M_5. On solving (10.72) to (10.74), we can get

$$R = \frac{V_{DD} - U_T\left[\ln\left(1 + \frac{I_D}{I_{S6}}\right) + \ln\left(1 + \frac{I_D}{I_{S5}}\right)\right]}{I_D} \tag{10.75}$$

Equation (10.75) is nonlinearly related between R and I_D. Hence the rule of thumb will be to choose one value of I_D, then calculate the value of R. Then frequency of oscillation of the ring oscillator can be determined. The current I_{D4} and I_{D1} will be given by

$$I_{D4} = I_{D04} \exp\left(\frac{V_{DD} - V_{b1}}{n_p U_T}\right) \tag{10.76}$$

$$I_{D1} = I_{D01} \exp\left(\frac{V_{b2}}{n_n U_T}\right) \tag{10.77}$$

In (10.76) and (10.77), n_p and n_n are the slope factors of PMOS and NMOS transistors determined from the weak inversion region of input characteristics; and I_{D04} and I_{D01} are the technological parameter of M_4 and M_1 transistors. The switching point V_{SP} for the inverting configuration (M_3 and M_2) is given by

$$V_{SP} = \frac{1}{2}\left[V_{DD} - U_T \ln\left(\frac{I_{D02}}{I_{D03}}\right)\right] \tag{10.78}$$

The time for charging and discharging of total capacitance C_{Total} can be given as

$$t_1 = C_{Total} \cdot \left(\frac{V_{SP}}{I_{D4}}\right) \tag{10.79}$$

$$t_2 = C_{Total} \cdot \left(\frac{V_{DD} - V_{SP}}{I_{D1}}\right) \tag{10.80}$$

From (10.79) and (10.80), the frequency of oscillation of the ring oscillator f_{osc} can be given as

$$f_{osc} = \frac{1}{N(t_1 + t_2)} \tag{10.81}$$

In (10.81), N is the number of oscillator stages. In (10.79) and (10.80), C_{Total} is the total gate to the bulk capacitance of the M_3 and M_2 transistors determined from the weak inversion region.

Table 10.3 The W/L of the transistors in the two ring oscillators

CSRO	*M_6 (μm/μm)*	*M_5 (μm/μm)*	*M_4 (μm/μm)*	*M_3 (μm/μm)*	*M_2 (μm/μm)*	*M_1 (μm/μm)*
1	4/4	4/4	4/4	4/2	2/2	4/4
2	4/4	4/4	4/4	8/4	4/4	4/4

Numerical calculation for CSRO-1.

1. $V_{O,DCP}$ = 0.5 V, $I_{S6} = I_{D04}$ = 0.175 nA, $I_{S5} = I_{D01}$ = 1.5 nA, I_D = 45.5 nA, V_{b2} = 89 mV, V_{b1} = 0.45 V, n_p = 1.436, n_n = 1.228, I_{D4} = 9.25 nA, I_{D1} = 24.8 nA, I_{D02} =1.5 nA, I_{D03} = 0.35 nA, V_{SP} = 0.28 V, $C_{oxn} = ((n_n - 1)/n_n)\times 6.3\times 10^{-14} = 1.165\times 10^{-14}$ F, $C_{oxp} = ((n_p - 1)/n_p)\times 12.55\times 10^{-14} = 3.81\times 10^{-14}$ F, $C_{Total} = C_{oxn} + C_{oxp}$ = 0.05 pF, $T_{AUX} \approx 10$ μs.
2. $V_{O,DCP}$ = 1 V, $I_{S6} = I_{D04}$ = 0.175 nA, $I_{S5} = I_{D01}$ = 1.5 nA, I_D = 91.5 nA, V_{b2} = 0.106 V, V_{b1} = 0.84 V, I_{D4} = 14 nA, I_{D1} = 42.6 nA, I_{D02} = 1.5 nA, I_{D03} = 0.35 nA, V_{SP} = 0.48 V, C_{Total} = 0.05 pF, $T_{AUX} \approx 11.62$ μs.

Selecting a high value resistance restricts the flow of current in the ring oscillator reducing, making the time period less sensitive to supply voltage variation. For the CSRO-2 the transistor parameters are same as CSRO-1, so the calculation method is the same; however the M_3 aspect ratio is taken as different, therefore C_{Total} = 0.2 pF, and $T_{main} \approx 42$ μs for CSRO-2 [22].

10.4.2 Reference generator and node sensing network

A diode-connected NMOS device NM-4 (Figure 10.18a) is used as a reference voltage generator. A reverse-biased NMOS device NM-3 limits the flow of current. The structure is simple, involving two transistors, and can give optimal performance with low current consumption.

If I_D is the drain leakage current through a gate to source-connected NM-3 device, then

$$I_D \cong I_{D03} = I_{spec3} \exp\left(-V_{T03}/nU_T\right) \tag{10.82}$$

For the forward-biased NM-4 device the reference voltage V_{RO} can be written as

$$I_D \cong I_{spec4} \exp\left(V_{RO}/U_T\right)\cdot\left(1 + \lambda_4 V_{RO}\right) \tag{10.83}$$

In (10.83), I_{spec} is the reverse saturation current and λ_4 is the channel length modulation of NM-4. On solving (10.82) and (10.83) we can derive an equation for the terminal voltage V_{RO} as

$$V_{RO} = \frac{\ln\left(I_{D03}/I_{spec4}\right)}{\lambda_4 + U_T^{-1}} \tag{10.84}$$

For the node sensing network shown in Figure 10.17b, c, the voltage at the output of V_{R1}, V_{R2} and V_{R3} can be given by

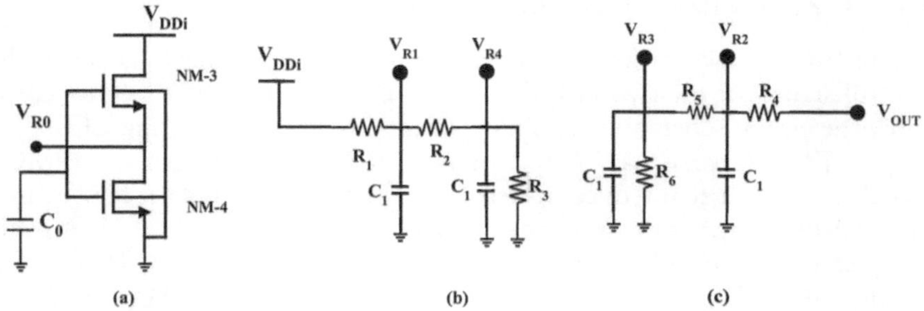

Figure 10.17 (a) A NMOS device voltage reference. NM-3 is long channel low Vt device and NM-4 is short channel standard Vt device. (b) Resistance ladder for sensing V_{DDi}. (c) Resistance ladder for sensing V_{OUT}.

$$V_{R1} = \frac{R_2 + R_3}{R_1 + R_2 + R_3} V_{DDi} \tag{10.85}$$

$$V_{R4} = \frac{R_3}{R_1 + R_2 + R_3} V_{DDi} \tag{10.86}$$

The scaling value of the resistance is in MΩ, to minimize voltage drop at the nodes. In Figure 10.17c, the value of $R_1 = R_4$, $R_2 = R_5$ and $R_3 = R_6$.

The plot in Figure 10.18 is with NM-3: W/L = 3 μm/1 μm, and NM-4: W/L = 0.16 μm/0.12 μm. The approximate value of V_{R0} is 0.262 V. The calculated values of resistances are: $R_1 = R_4 = 6.4$ MΩ; $R_2 = R_5 = 1.9$ MΩ; $R_3 = R_6 = 3$ MΩ.

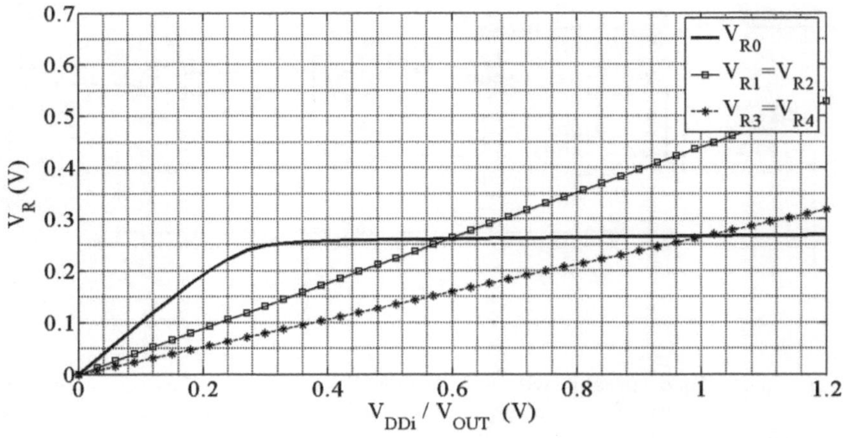

Figure 10.18 Plot of reference voltages that are used in the present work.

10.4.3 Zero current switching network

Figure 10.19 shows the zero current switching (ZCS) scheme to generate the control signal for the two PMOS switches PM1, 2. Figures 10.20 to 10.23 show the circuit schematics that have been used in implementing ZCS.

The (ZCS) scheme has been proposed in [10, 11] to open the PMOS switches when current reduces to zero. The concept presented in [11] inspired our ZCS design for both cycles of the clock as shown in Figure 10.19. A static complementary metal oxide semiconductor (CMOS)-based D flip-flop (Figure 10.20) is used to sense the change in the polarity of V_{SWD1}/V_{SWD2} at the rising edge of clock pulse $V_P/\overline{V_P}$. The D flip-flop runs a counter (Figure 10.21) to select one of the delays through the 16x1 multiplexer design shown in Figure 10.21. A logical NAND operation is performed with the main clocks to control the opening and closing of the PMOS switches. Due to the closed-loop condition, V_P and $\overline{V_P}$ is dynamically adjusted to a position when there is no more change in the polarity of $V_{SWD1,\,2}$. It has been found from simulation that a delay of 1.5 ps (between $V_P/\overline{V_P}$ and V_{SWD1}/V_{SWD2}) will be sufficient to sense the polarity at an appropriate instant of time. This is

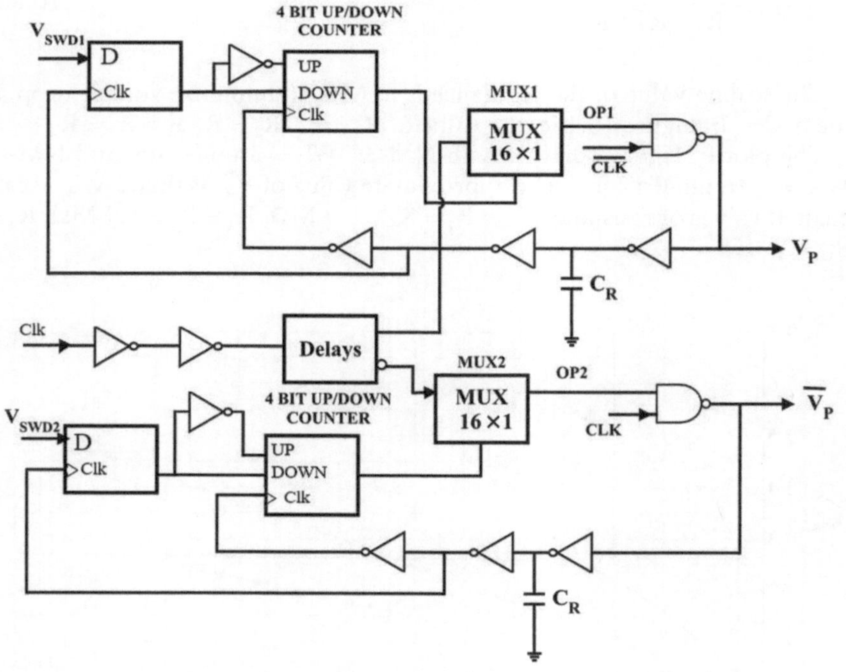

Figure 10.19 Zero current switching network to operate the switches for both cycles of a clock signal, C_R = 0.21 pF is used to introduce delay between $V_{SWD1,2}$ and $V_P/\overline{V_P}$ [12, 22].

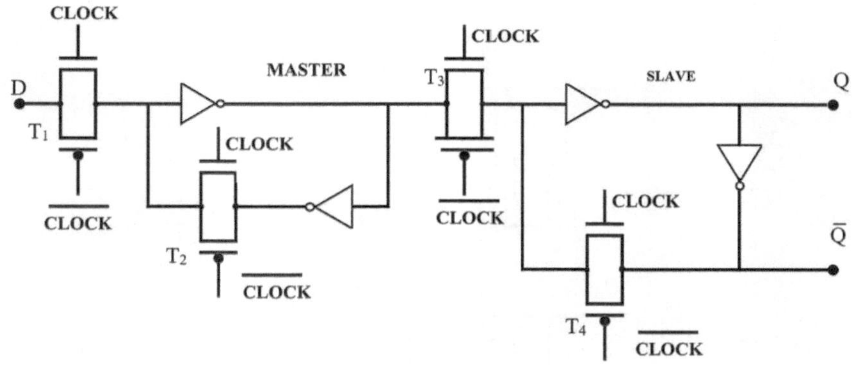

Figure 10.20 A static CMOS edge triggered D flip-flop [21].

also justified according to Figure 3 in [12]. Therefore, a C_R = 0.21 pF and inverter chain have been used in the delay block. To cover the 50–200 mV, we have selected delay range from 0.57 to 2.25 μs. In the delay block of Figure 10.20, a current starved arrangement shown in Figure 10.23 with low current consumption is used.

The delay circuit of Figure 10.23 works somewhere between moderate and strong inversion. For simplicity, we have considered strong inversion equations. Let I_b be the current flowing through R = 1 MΩ, and V_{DS4} is the voltage drop across the M_4, then

$$V_{DDi} - V_{DS4} = I_b \cdot R \tag{10.87}$$

On solving (10.87), the relation between R and I_b is as follows,

$$R = \frac{V_{DDi} - V_{TN4} - \sqrt{2 \cdot I_b/\beta_4}}{I_b} \tag{10.88}$$

The switching point V_{SP} for the inverting configuration (M_3 and M_2) is given by

$$V_{SP} = \frac{V_{TN2}\sqrt{\beta_{N2}/\beta_{P3}} + \left(V_{DDi} - V_{TP3}\right)}{1 + \sqrt{\beta_{N2}/\beta_{P3}}} \tag{10.89}$$

The delay time t_2 for discharging of total capacitance C_{Total} can be given as

$$t_2 = C_{Total} \cdot \left(\left(V_{DDi} - V_{SP}\right)/I_b\right) \tag{10.90}$$

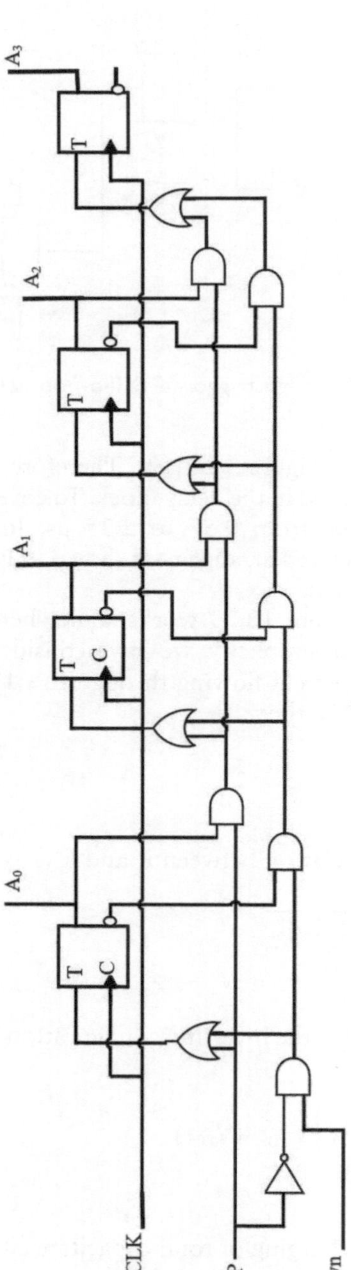

Figure 10.21 A 4-bit binary UP/DOWN synchronous counter [21].

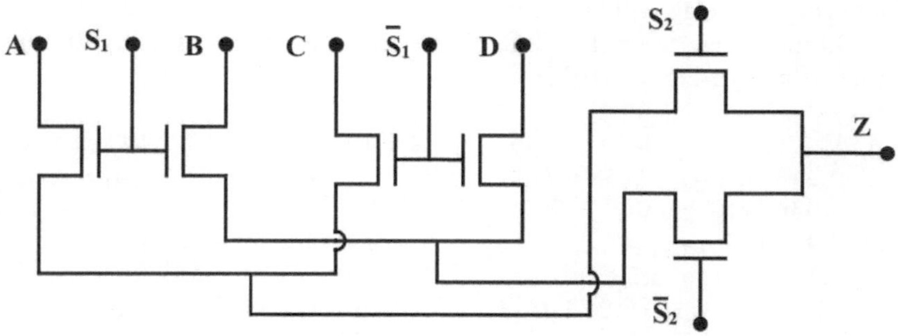

Figure 10.22 A 4x1 multiplexer; four such units are used to make a 16x1 multiplexer [21].

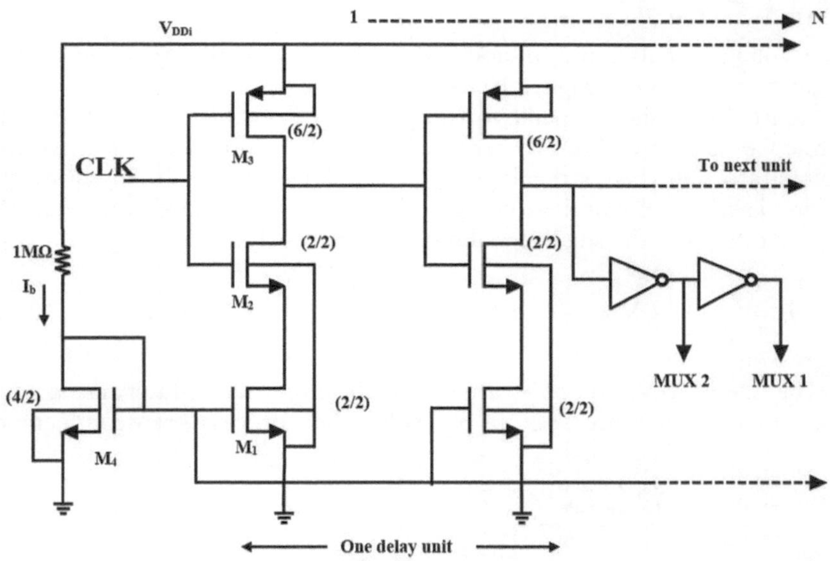

Figure 10.23 Schematic of the circuit arrangement used in delay block. The circuit is implemented with Tox = 52 Å transistors of the CMOS technology. N = 16 is the number of such delay circuits [12, 22].

The range of delay is calculated by considering: 50 mV $< V_{TEG} < 200$ mV, 25 mV $< V_{IN} <$ 100 mV, which implies $10 < M < 40$. From (10.54) the delay is expressed as

$$T_{Delay}(T_{OP}) = \frac{T_{SW}}{2(M-1)} \tag{10.91}$$

Therefore, the delay will be between: 0.57 μs < T_{Delay} < 2.25 μs, covering the 50–200 mV range from the TEG. Numerical calculation of delay obtained from one unit of Figure 10.23:

1. $V_{DDi} = 1$ V, $n_n = 1.4$, $n_p = 1.51$, $\beta_4 = 2 \times 0.143 \times 10^{-3}$ A/V^2, $V_{TN4} = V_{TN2} =$ 0.348 V, $V_{TP3} = 0.389$ V, $\beta_{N2} = (2/2) \times 0.143\times10^{-3}$ A/V^2, $\beta_{P3} = (6/2) \times$ 0.0368×10^{-3} A/V^2, $I_b = 587$ nA, $I_{b1} = I_d/2 = 293.5$ nA, $V_{b2} = 0.413$ V, $V_{SP} =$ 0.4365 V, $C_{Total} = C_{oxn} + C_{oxp} = (6\times2+2\times2) \times 6.6405\times10^{-3} \times 10^{-12} =$ 0.106 pF, $t_2 \approx 0.203$ μs.
2. $V_{DDi} = 0.6$ V, $I_b = 0.213$ μA, $I_{b1} = 0.106$ μA, $V_{b2} = 0.387$ V, $\beta_{N2}/\beta_{P3} = 1.3$, $V_{SP} = 0.284$ V, $C_{Total} = 0.106$ pF, $t_2 \approx 0.316$ μs.

The above calculation agrees with the observed simulation result.

10.4.4 Control logic

The voltage regulation and clocks generation are realized by the control logic circuit shown in Figure 10.24.

Figure 10.24 shows that two comparators, Comp 1, 3 have hysteresis, while Comp 2 is without hysteresis. Hysteresis is introduced in these two comparators for better regulation of the voltage. Figure 10.25 shows the internal architecture of these comparators. In Figure 10.25 we can relate the input voltages to the small signal output currents (I_3 and I_4) by

$$I_3 = \frac{g_m}{2}(V_+ - V_-) + \frac{I_b}{2} = I_b - I_4 \tag{10.92}$$

The size of MN_1 and MN_2 are set by diff-amp transconductance g_m and the input capacitance. Let us consider that process transconductance

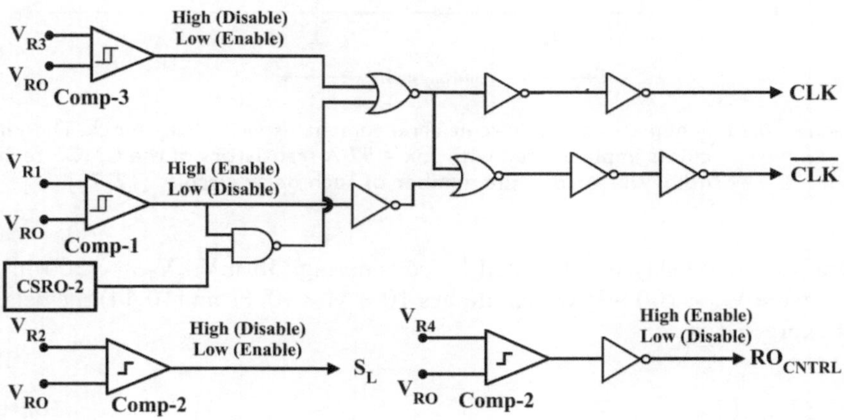

Figure 10.24 Circuit to implement the control logic for regulating voltage [12].

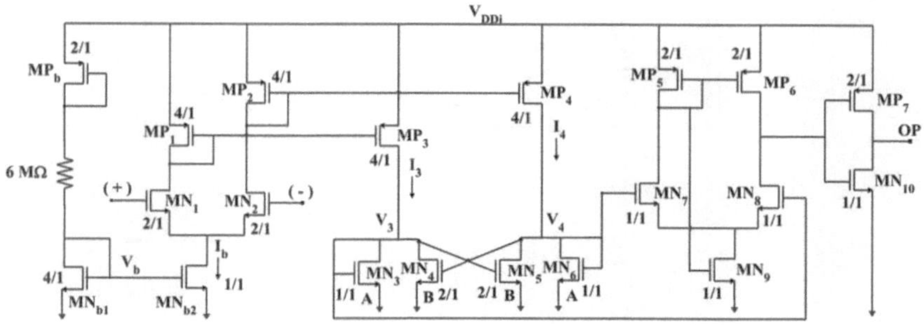

Figure 10.25 The circuit diagram of Comp 3 used in the control logic [21].

parameter of MN_4–MN_5 is β_B and MN_3–MN_6 is β_A. The switching point high (SPH) and switching point low (SPL) are related as

$$V_{SPH} = -V_{SPL} = \frac{I_b}{g_m} \frac{(\beta_B/\beta_A) - 1}{(\beta_B/\beta_A) + 1} \tag{10.93}$$

for $\beta_B > \beta_A$. Eq. (10.93), will set the hysteresis in this comparator.

Total power consumption by the peripherals: the auxiliary stage is driven by the low-voltage starter coupled with RO-1, therefore for the peripheral circuit to start working, the V_{DDi} should be at a potential greater than 0.6 V. At present the peripherals consume around 1.5 μA at 0.6 V, and 4 μA at 1.0 V.

10.5 TEST AND MEASUREMENT

Figure 10.26 shows the circuit diagram along with pinout, and Figure 10.27 shows the photo of the die. The circuit occupies an area of 0.8 × 0.82 mm^2. A TEG element from Tellurex reported with a sensitivity of 25 mV/K is used. The measurement set-up to measure the extracted power by the chip is shown in Figure 10.28. The transient response is recorded using a digital oscilloscope.

The start-up circuit is crucial in circuit design, therefore the inductor response in the ESRO configuration is presented.

10.5.1 Inductors for ESRO coupled with DCP

In this section we have to choose pairs of inductors, which can start oscillation from minimum voltage V_{IN}. For the value of W/L in $MN_{1,2}$, the extracted value of g_0 at input voltage 10–50 mV is close to the 4.08 mA/V. Figure 10.29 shows the plot of the g_{ms}/g_0 ratio versus input voltage for the ZVT transistor in ESRO.

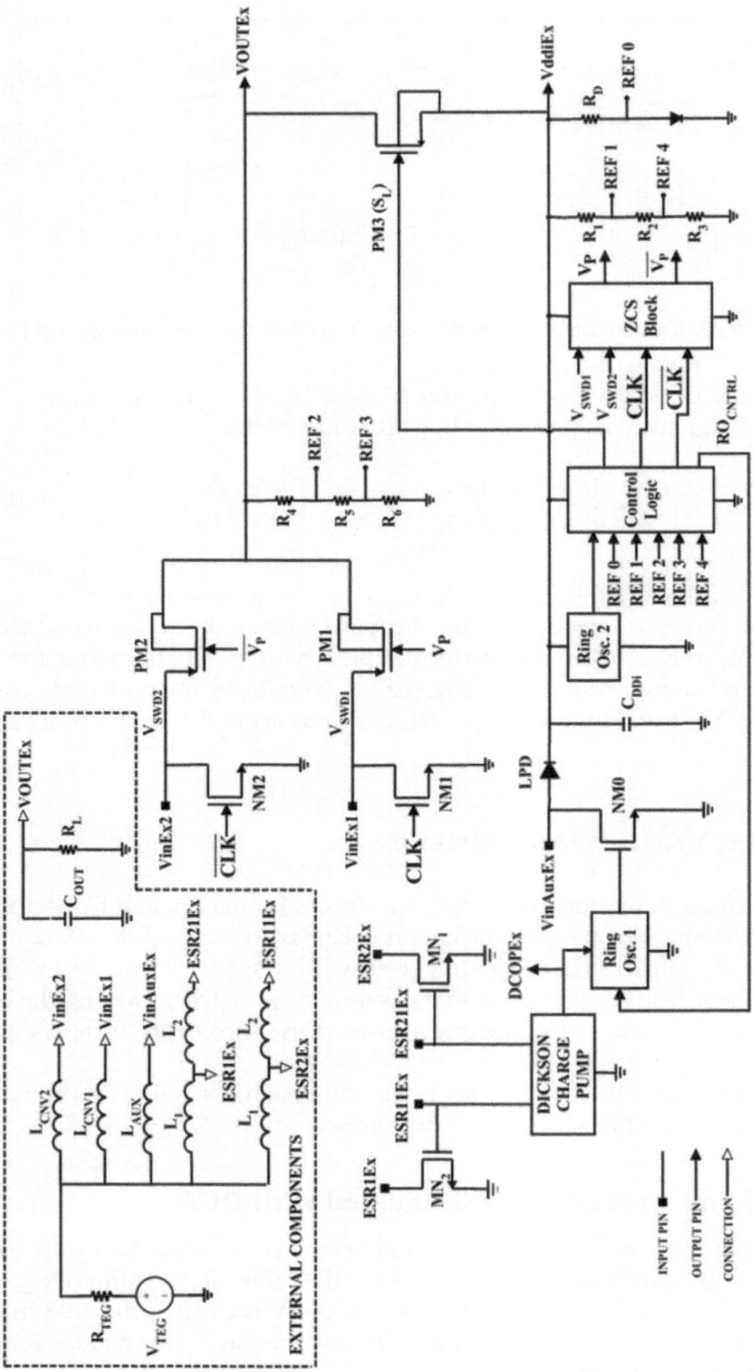

Figure 10.26 Circuit diagram of the energy harvesting circuit with pinout.

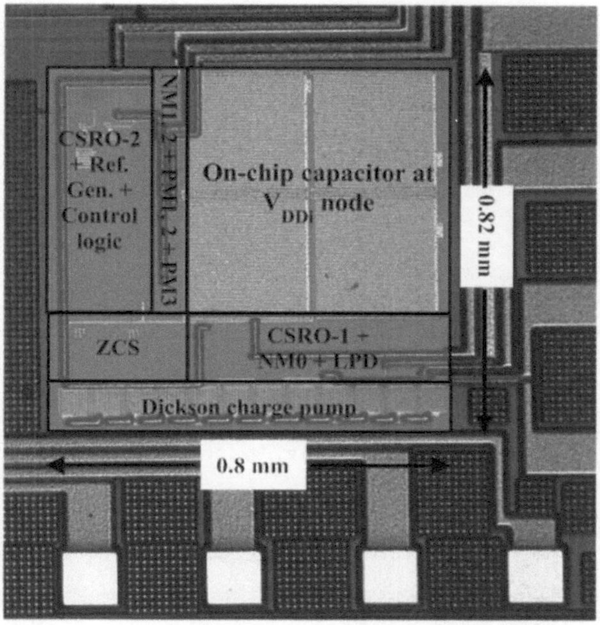

Figure 10.27 Photo of the die implementing the circuit shown in Figure 10.26 [12].

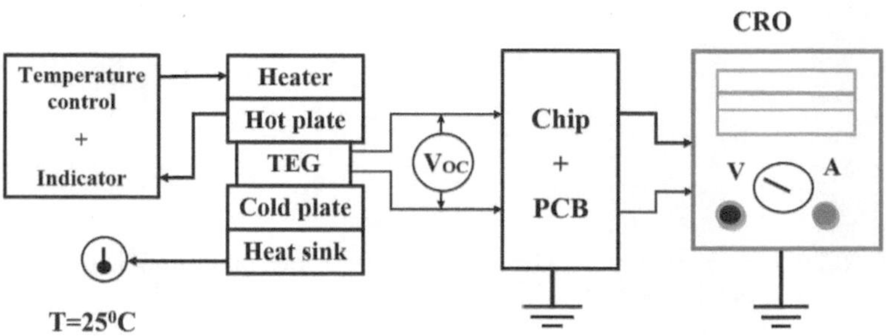

Figure 10.28 Measurement set-up to measure performance of the prototype.

The condition given in (10.29) was plotted for three different values of inductors shown in Figure 10.30. Using an LCR meter, the parallel resistance R_p have been measured at 1 MHz and 500 kHz for the inductors. In Figure 10.29, the value of g_{ms}/g_0 is taken as 1.57, because if ESRO oscillates at 20 mV it will also oscillate at 50 mV. In Figure 10.30, it can be observed that the larger value inductors are more likely to oscillate, because they strongly satisfy the condition given by (10.48). However, large value inductors will also decrease the oscillation frequency (f_0) according to (10.23); this is shown by the plot in Figure 10.31.

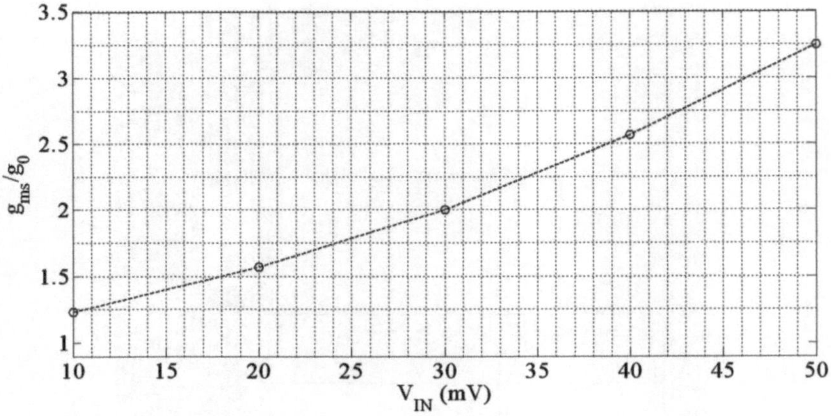

Figure 10.29 Plot of V_{IN} versus extracted g_{ms}/g_0 for ZVT transistor 4.2 μm/0.42 μm.

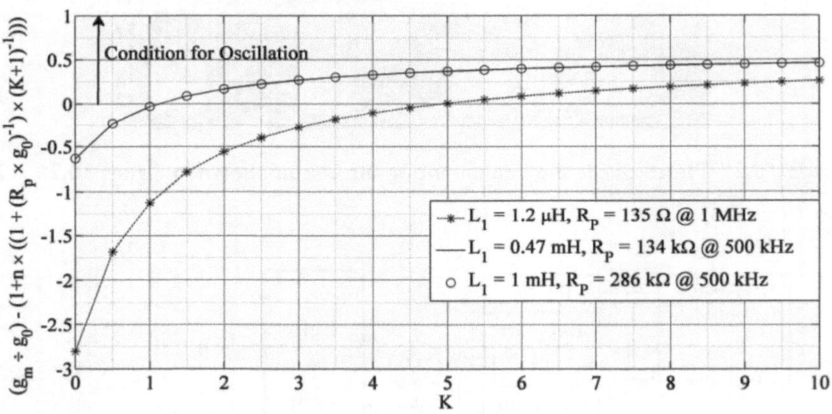

Figure 10.30 Plot shows that with inductor 0.47 mH and 1 mH, oscillation is more likely than 1.2 μH, for the greater value of K.

As given by (10.49), decreasing frequency will decrease the charge pump output shown by Figure 10.32, for the three values of inductor. In Figure 10.31, it is clear that the Dickson output falls with the decreasing frequency, which in turn depends on the inductor values (after C_{net} is fixed). Minimum start-up can be achieved only by increasing the inductor values: it is a trade-off in a low-voltage oscillator coupled with charge pump. For our circuit a DCP output ≈ 0.45 V is enough to oscillate CSRO-1, and this in turn can drive the auxiliary inductor. The measurement reported in Figures 10.33 to 10.50, shows the result from ESRO and the charge pump.

1. L_1 = 1.2 μH, L_2 = 2.2 μH.
2. L_1 = 1.2 μH, L_2 = 4.7 μH.

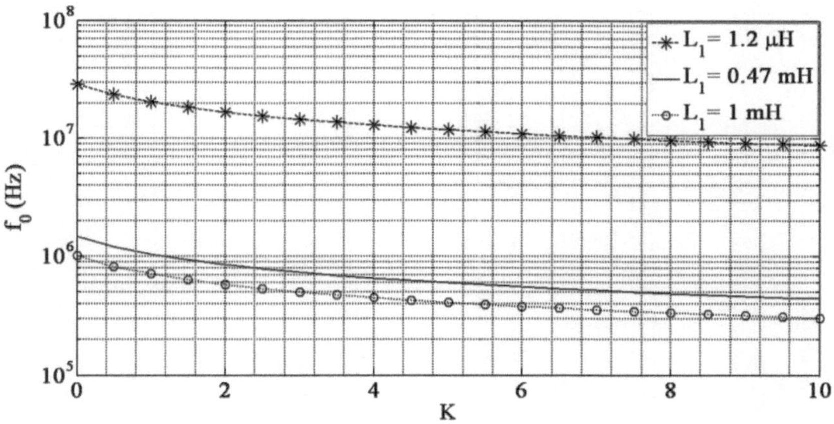

Figure 10.31 Plot shows that f_0 decreases with increasing K, and also with increasing value of L_1. The C_{net} = 25 pF and $L_P \approx L_1$.

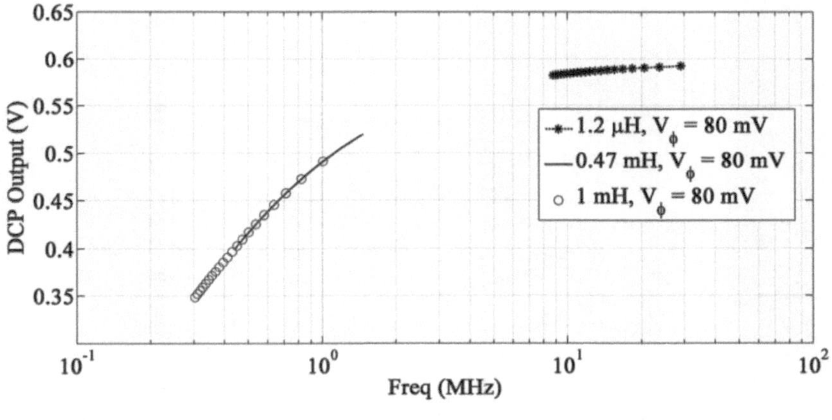

Figure 10.32 Plot shows that the charge pump output drops with an increase in value of inductors and decrease in frequency. The charge pump output depends to a large extent upon swing peak voltage V_ϕ.

3. L_1 = 0.1 mH, L_2 = 0.2 mH.
4. L_1 = 0.1 mH, L_2 = 0.47 mH.
5. L_1 = 0.2 mH, L_2 = 2.2 mH.
6. L_1 = 0.47 mH, L_2 = 2.2 mH
7. L_1 = 1 mH, L_2 = 2.2 mH.

It can be concluded from Figures 10.33 to 10.50 that an ESRO can start oscillating from 25 mV; however to get the output voltage = 1 V at load, the input should be near to 50 mV. Table 10.4 compiles the measurements of voltage from oscillator and charge pump.

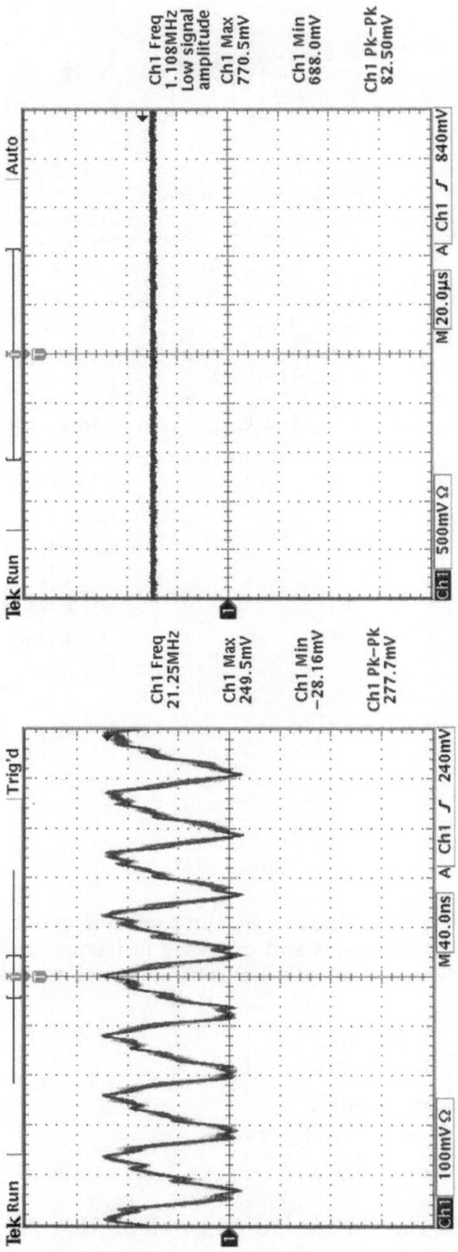

Figure 10.33 For V_{TEG} = 80 mV, (Left) ESRO oscillation. (Right) Charge pump output.

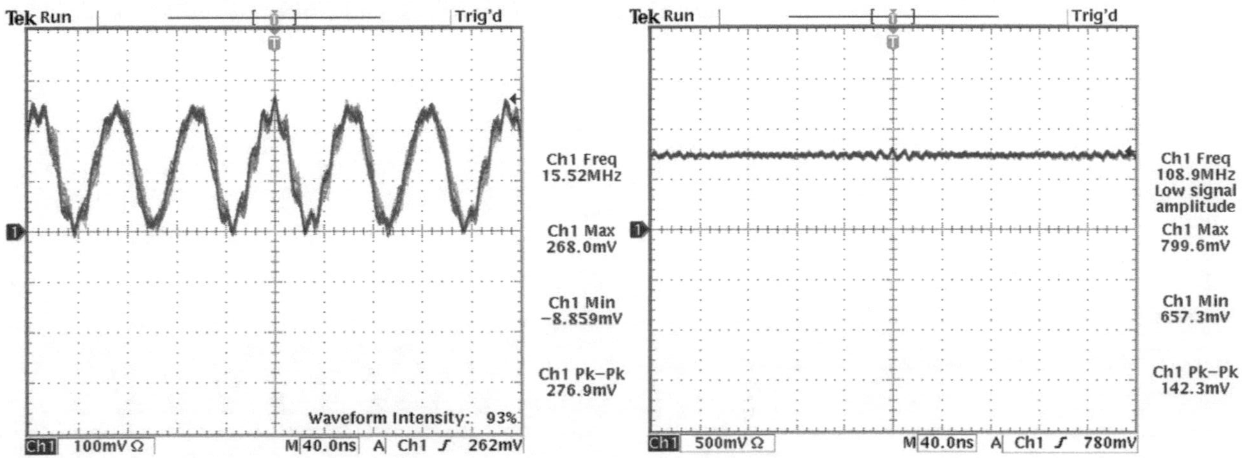

Figure 10.34 For V_{TEG} = 90 mV, (Left) ESRO oscillation. (Right) Charge pump output.

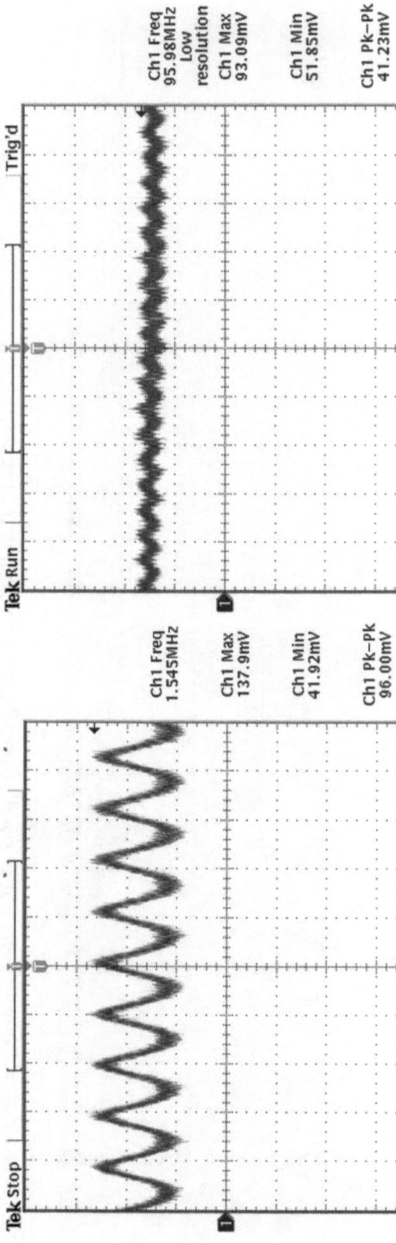

Figure 10.35 For V_{TEG} = 40 mV, (Left) ESRO oscillation. (Right) Charge pump output.

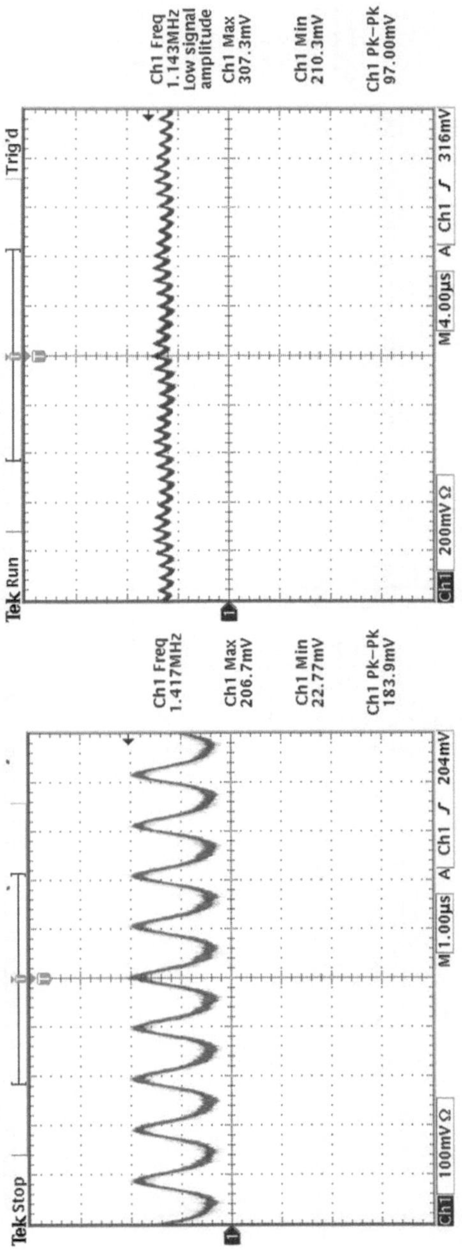

Figure 10.36 For V_{TEG} = 50 mV, (Left) ESRO oscillation. (Right) Charge pump output.

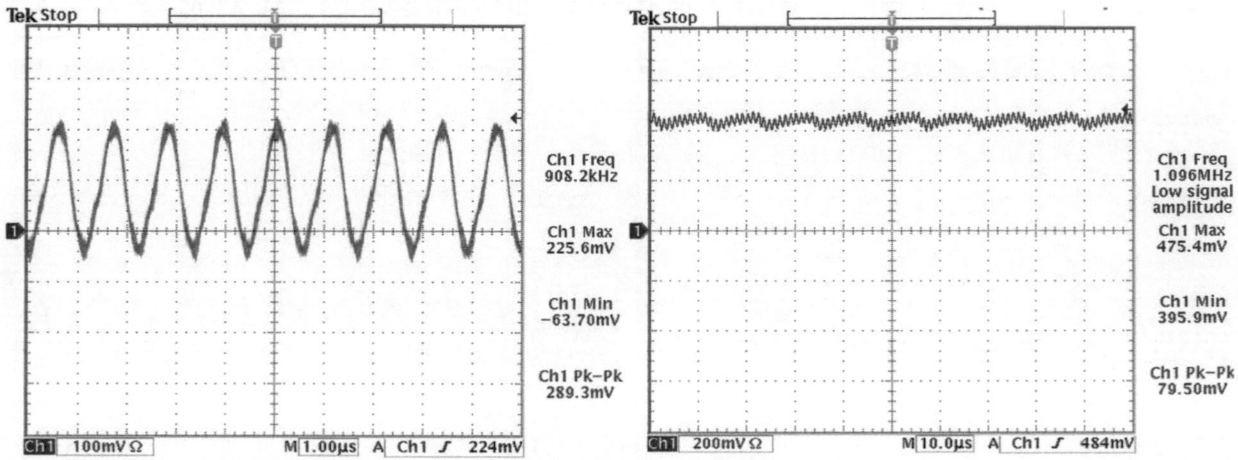

Figure 10.37 For V_{TEG} = 40 mV, (Left) ESRO oscillation. (Right) Charge pump output.

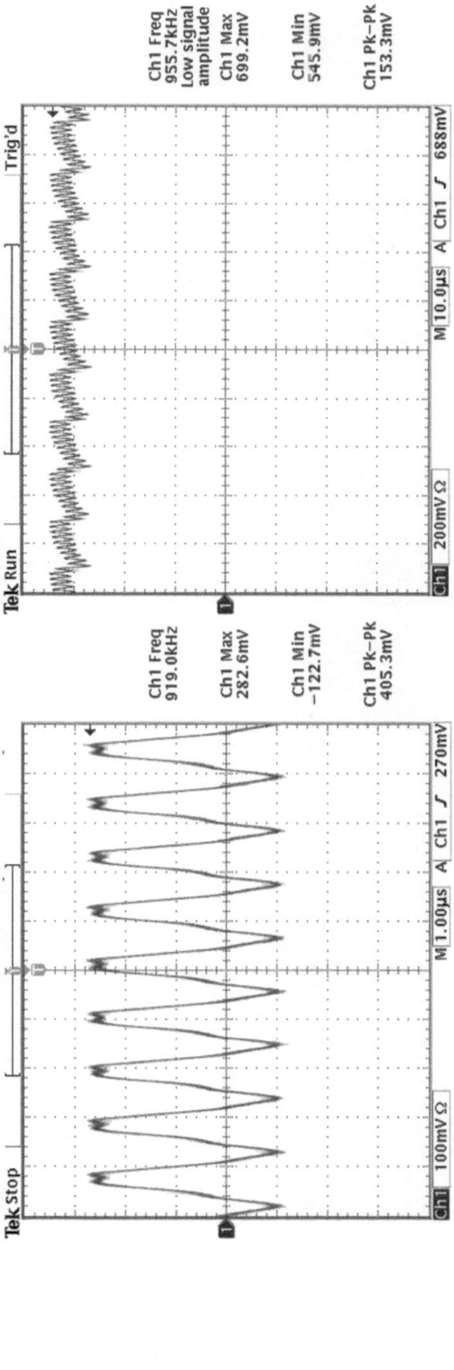

Figure 10.38 For V_{TEG} = 50 mV, (Left) ESRO oscillation. (Right) Charge pump output.

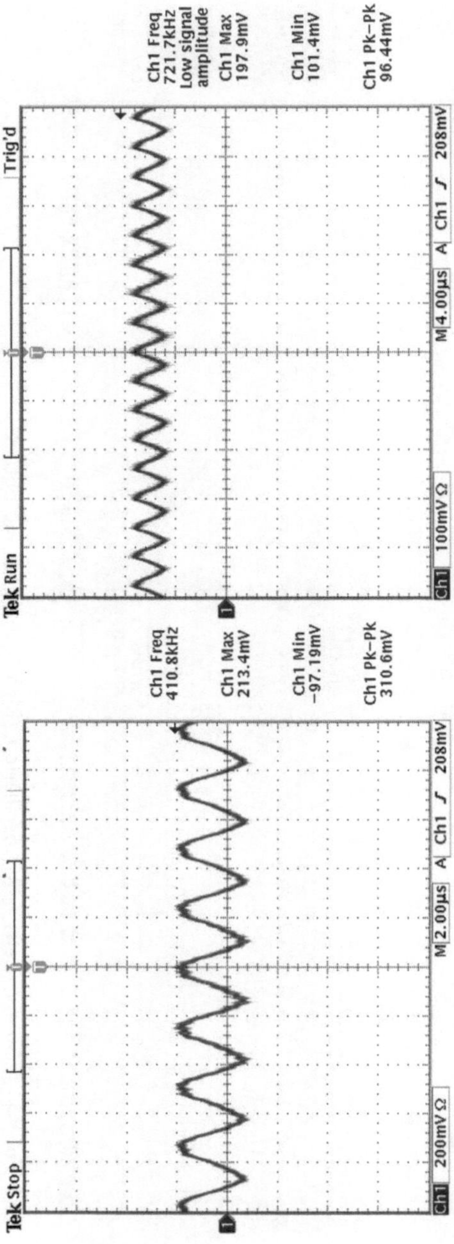

Figure 10.39 For V_{TEG} = 25 mV, (Left) ESRO oscillation. (Right) Charge pump output.

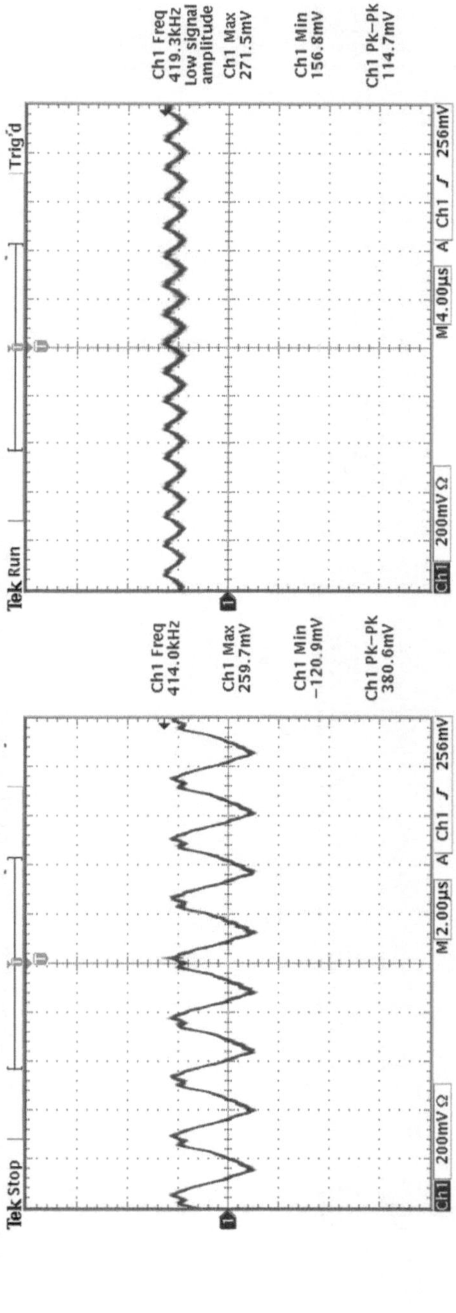

Figure 10.40 For V_{TEG} = 30 mV, (Left) ESRO oscillation. (Right) Charge pump output.

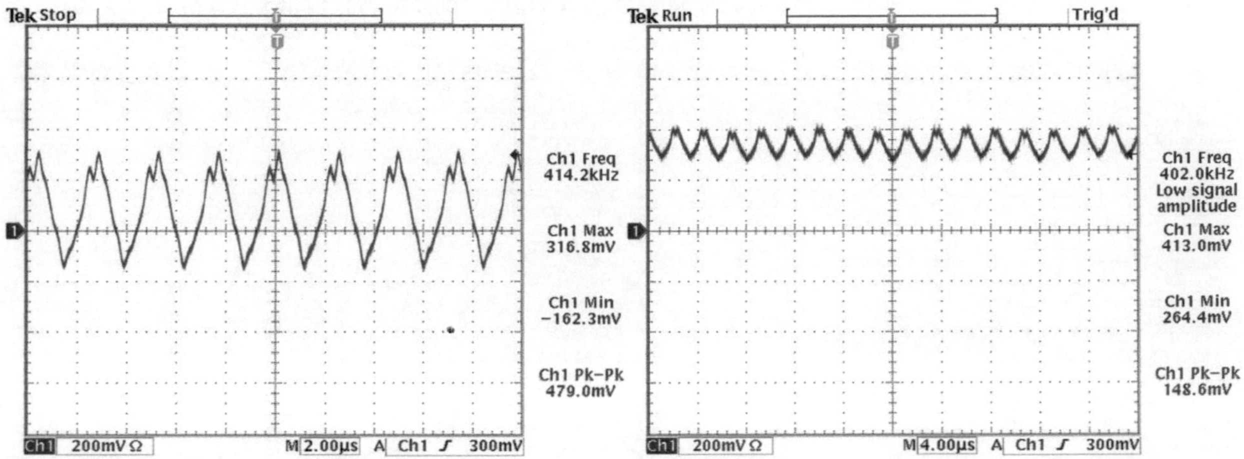

Figure 10.41 For V_{TEG} = 40 mV, (Left) ESRO oscillation. (Right) Charge pump output.

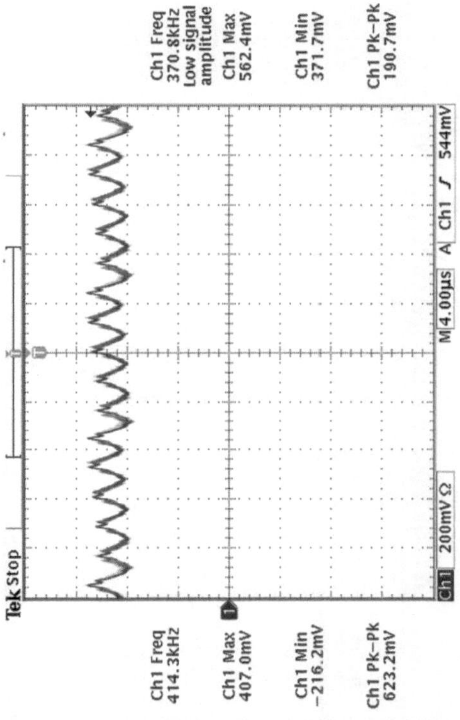

Figure 10.42 For V_{TEG} = 50 mV, (Left) ESRO oscillation. (Right) Charge pump output.

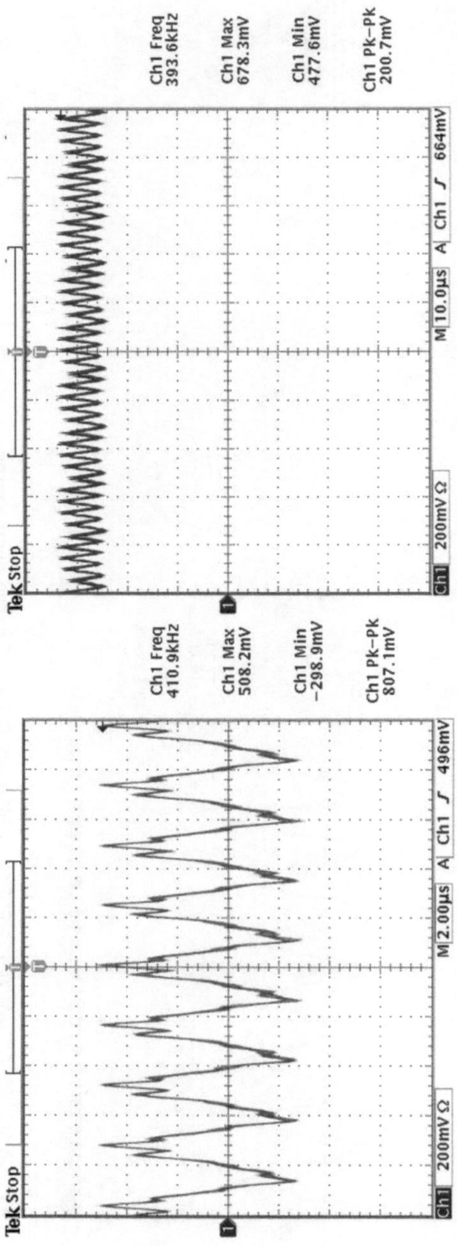

Figure 10.43 For V_{TEG} = 60 mV, (Left) ESRO oscillation. (Right) Charge pump output.

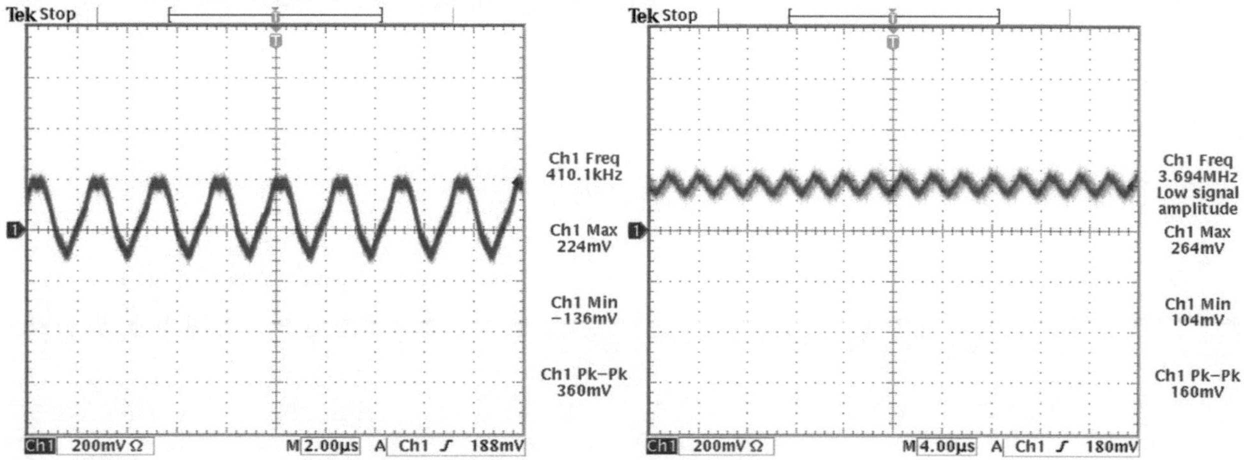

Figure 10.44 For V_{TEG} = 25 mV, (Left) ESRO oscillation. (Right) Charge pump output.

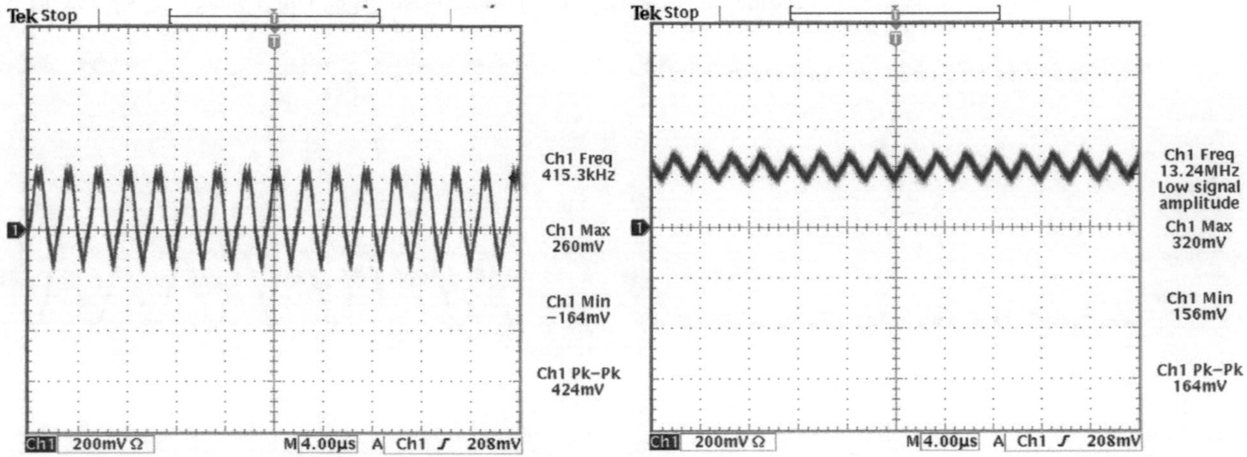

Figure 10.45 For V_{TEG} = 30 mV, (Left) ESRO oscillation. (Right) Charge pump output.

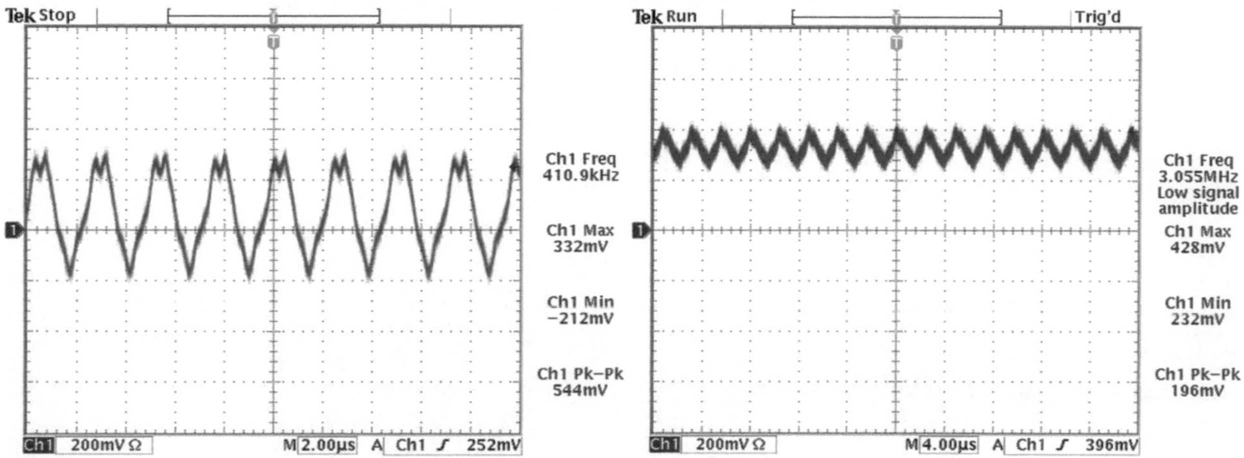

Figure 10.46 For V_{TEG} = 40 mV, (Left) ESRO oscillation. (Right) Charge pump output.

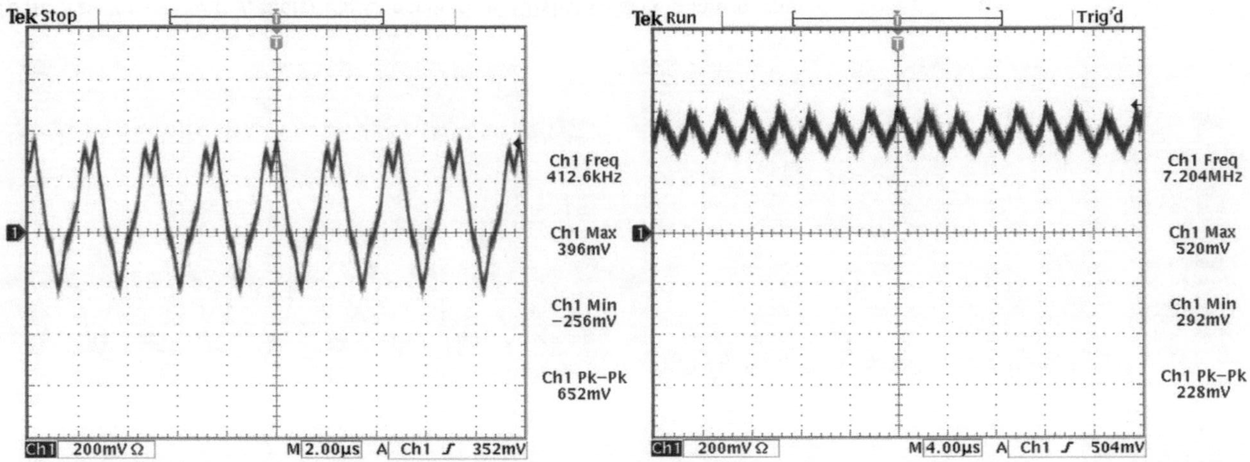

Figure 10.47 For V_{TEG} = 50 mV, (Left) ESRO oscillation. (Right) Charge pump output.

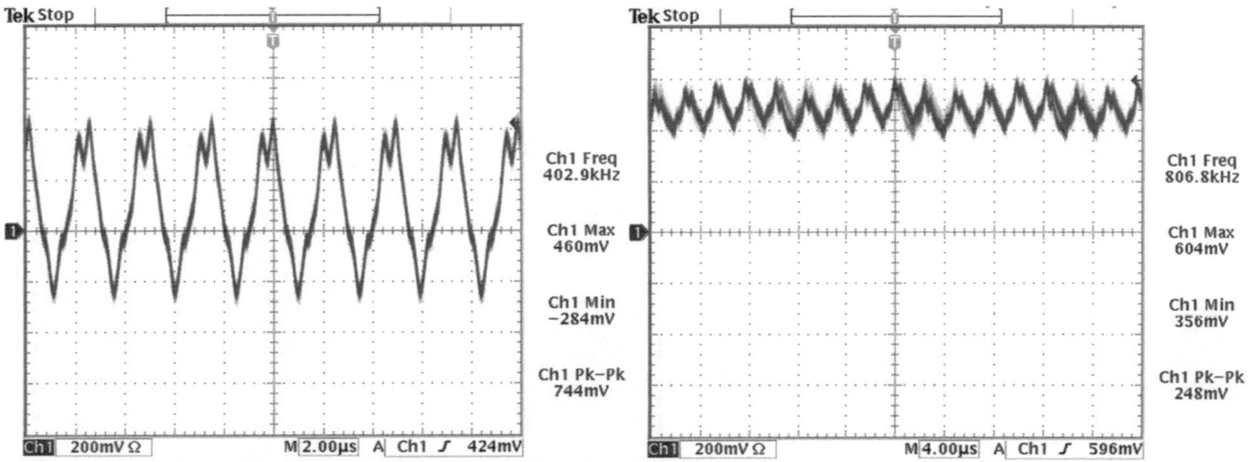

Figure 10.48 For V_{TEG} = 60 mV, (Left) ESRO oscillation. (Right) Charge pump output.

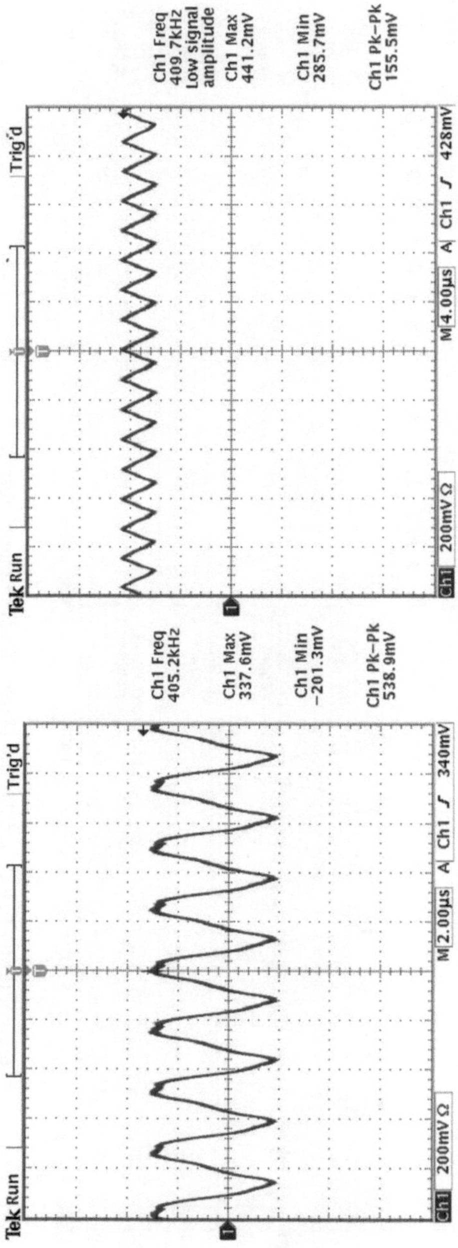

Figure 10.49 For V_{TEG} = 40 mV, (Left) ESRO oscillation. (Right) Charge pump output.

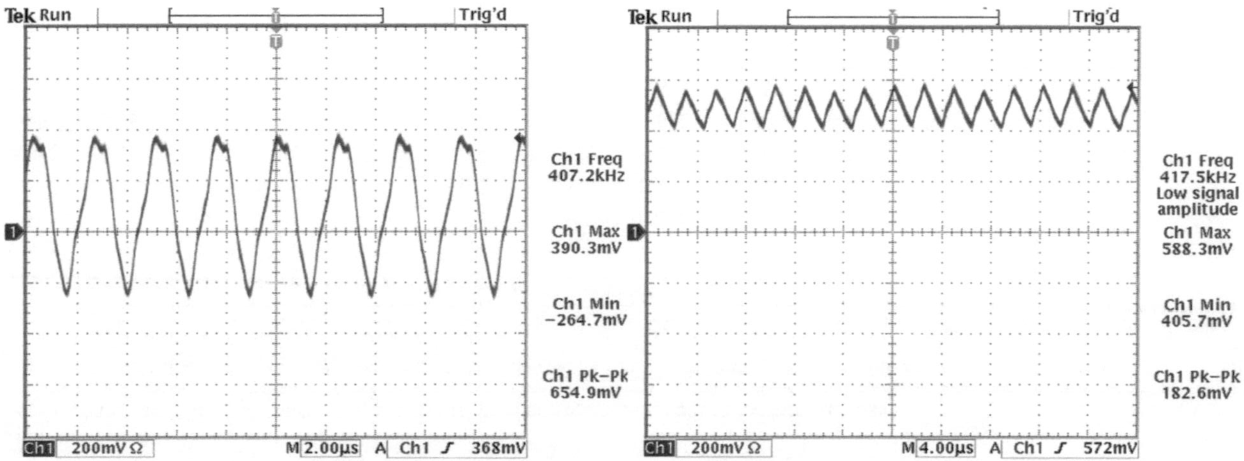

Figure 10.50 For V_{TEG} = 50 mV, (Left) ESRO oscillation. (Right) Charge pump output.

Table 10.4 Oscillator and Dickson charge pump outputs at different start-up voltage for inductor pairs

	25 mV		*30 mV*		*40 mV*		*50 mV*		*60 mV*		*80 mV*		*90 mV*	
L_1, L_2	*Osc.*	*CP*	*Osc.*	*CP*	*Osc.*	*CP*	*Osc.*	*CP*	*Osc.*	*CP*	*Osc.*	*CP*	*Osc.*	*CP*
1.2, 2.2 μH	x	x	x	x	x	x	x	x	x	x	0.277	0.77	–	–
1.2, 4.7 μH	x	x	x	x	x	x	x	x	x	x	x	x	0.277	0.8
0.1, 0.2 mH	x	x	x	x	x	x	0.096	0.0725	0.184	0.26	–	–	–	–
0.1, 0.47 mH	x	x	x	x	0.29	0.434	0.405	0.623	–	–	–	–	–	–
0.2, 2.2 mH	0.31	0.15	0.38	0.214	0.48	0.338	0.623	0.467	0.807	0.578	–	–	–	–
0.47, 2.2 mH	0.36	0.184	0.424	0.238	0.544	0.33	0.652	0.406	0.744	0.48	–	–	–	–
1, 2.2 mH	x	x	x	x	0.539	0.363	0.655	0.497	–	–	–	–	–	–

Table 10.5 External components and their values

S. No	*Components*	*Values (Units)*
1.	C_{OUT}	100 nF
2.	R_{OUT} (a potentiometer)	0–50 KΩ
3.	$L_{CNV1,2}$	100 μH
4.	L_{AUX}	330 μH
5.	ESRO: L_1 and L_2	0.47 mH, 2.2 mH

Table 10.4 reports the measurement values from ESRO and DCP. In Table 10.4, "Osc." refers to pk-pk oscillator voltage swing, and "CP" refers to average voltage of the DCP output. Double dash means the measurement was not recorded for $V_{TEG} > 60$ mV.

A load potentiometer was connected to the output of the circuit, and the measurements were recorded. The definition of efficiency (i.e., η) is given by

$$\eta = \frac{P_{OUT}|_{V_{OUT}=1V}}{P_{AV}} \times 100\% \tag{10.94}$$

In (10.94), V_{OUT} = 1V is the voltage across the load; however, this voltage will depend on the susceptibility of the reference generator to the process variation. The Table 10.5 shows the external components value used to test the chip performance. Figures 10.51 and 10.52 present some transient responses at the output of the chip with varying input voltage.

Figure 10.51a, shows the response of the chip for when V_{TEG} = 50 mV, showing that the V_{DDi} and V_{OUT} closes at 600 mV and the output is boosted to 1 V. In our circuit the auxiliary stage starts as soon the ESRO + DCP powers the CSRO-1; this considerably reduces the circuit transient settling time compared with [10, 11]. This is possible because in ESRO the peak-peak swing is greater than V_{IN} [18]. This demonstrates that the prototype chip is working according to our design specifications. In Figure 10.52 we have shown another transient response at 100 mV; in this measurement the load is decreased. Figure 10.51b shows the regulated output in the form of ripples and the average value of voltage stays around 1 V. Figure 10.52a shows the voltage drop across LPD, Figure 10.52b shows the $V_{SWD1,\ 2}$, and Figure 10.53a shows the voltage across PM1 switch.

Figure 10.53b shows the plot of available power and extracted power at different values of the V_{TEG}. The V_{TEG} value represents a temperature difference of 2°K–8°K around the TEG surface. The plot shows efficiency varying between 60% and 65% at 50 mV and 200 mV respectively. These are close enough to our derivation on the extraction of maximum power from available power. The power estimation using circuit simulation to match the measurement results at 50 mV are as follows: total available power = 125 μW (100%); extracted power (measured) = 74 μW (≈ 60%); power lost in the inductor and switching resistance = 9 μW; circuit power consumption = 4 μW; parasitic and switching loss = 7 μW; and total power lost 16 μW (≈ 13%).

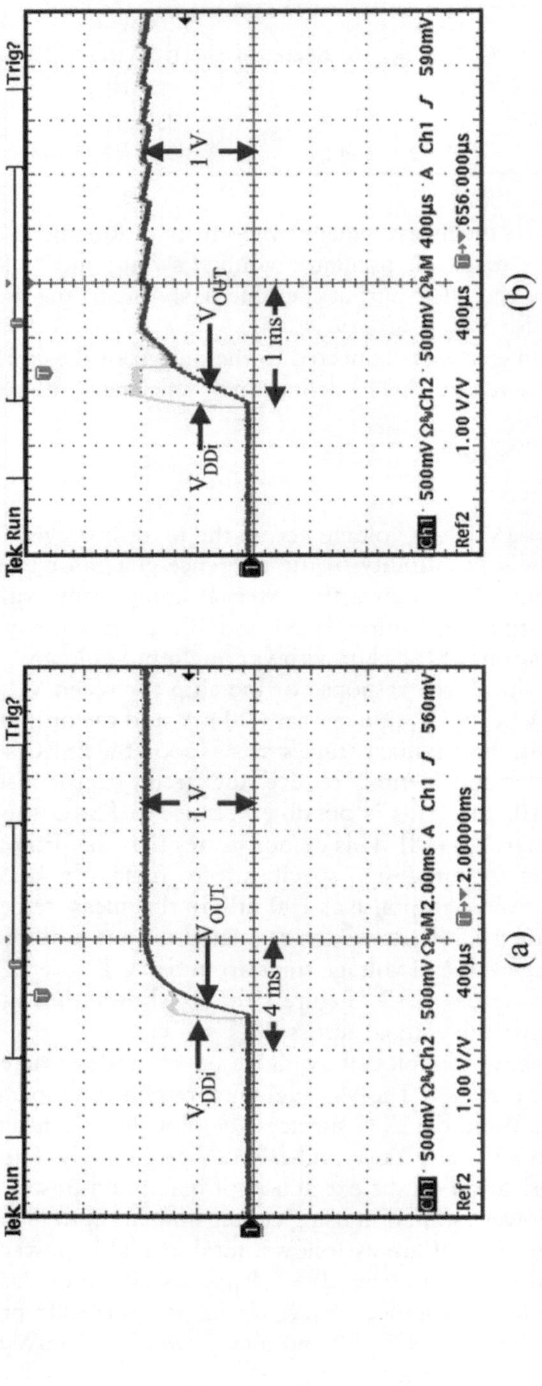

Figure 10.51 (a): Transient response when V_{TEG} = 50 mV, (b) Transient response when V_{TEG} = 100 mV, shows the regulated output.

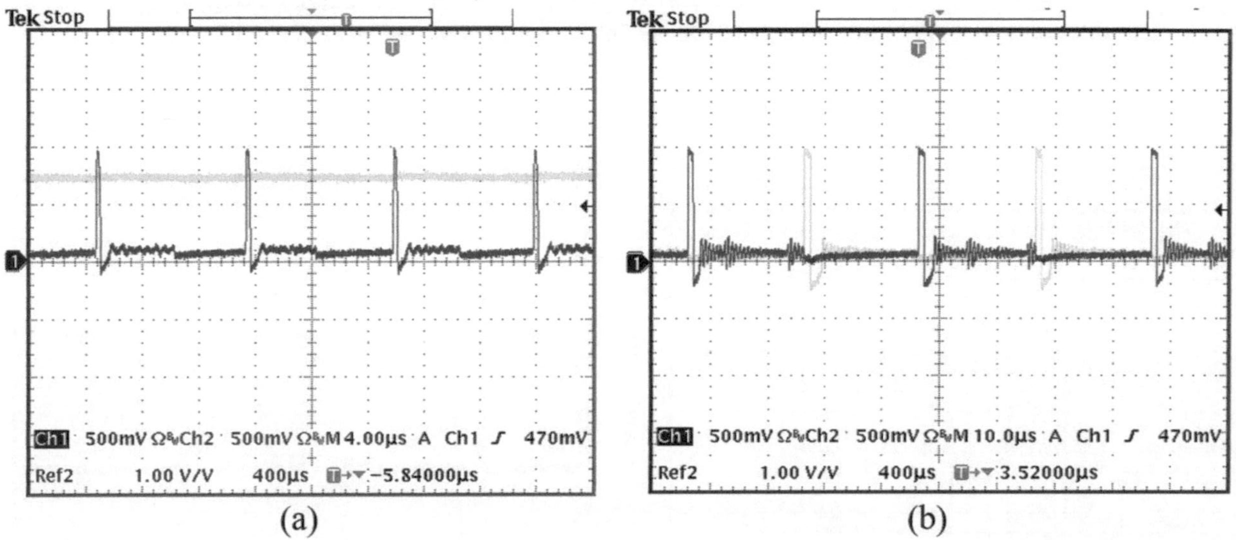

Figure 10.52 (a) Voltage drop across LPD to show transient at auxiliary stage, V_{TEG} = 100 mV, (b) Transient measurement shows voltage response at $V_{SWD1,2}$, the V_{TEG} = 100 mV.

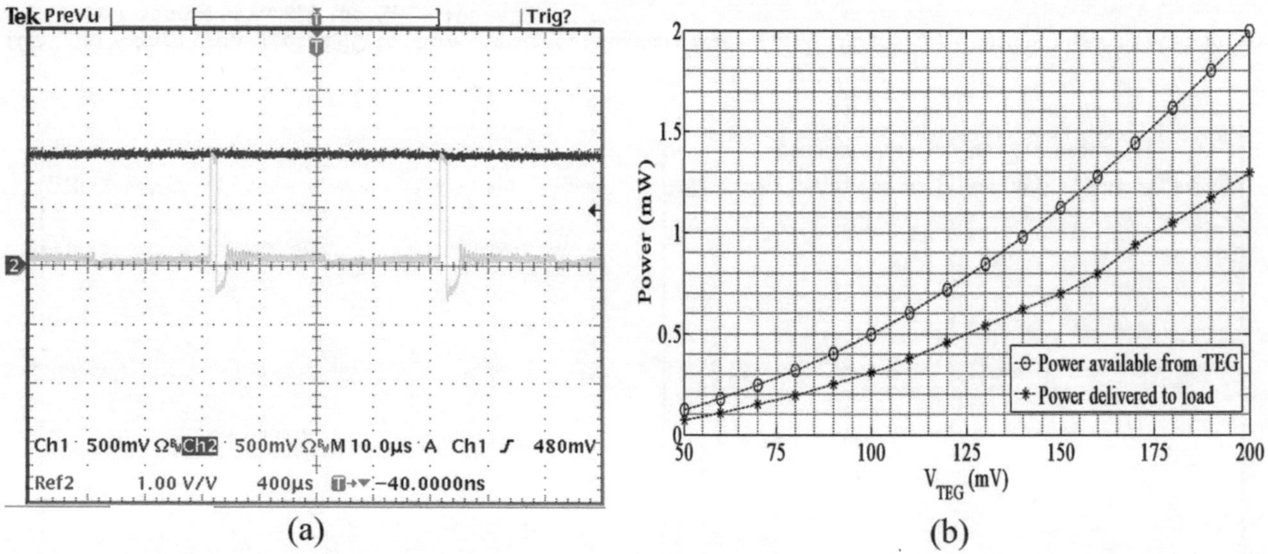

Figure 10.53 (a) Transient response across PM1. No drop is visible compared to Figure 10.52a, V_{TEG} = 100 mV, (b) Diagram showing the available power and the extracted power from TEG [12].

10.6 CONCLUSION

In this chapter we presented a complete design methodology of a thermal energy harvester, including mathematical derivation and calculation. The oscillator and charge pump design were presented in detail along with measurement results. The auxiliary and main converter values and the DCM mode of operation was explained and estimated. The overall system works with peak 65% efficiency.

ACKNOWLEDGMENTS

The authors would like to thank MOSIS for providing fabrication services for the MEP program. The author at Federal University of Santa Catarina, Brazil, carried out this work while receiving a CNPq fellowship grant no.-164911/2013-8. The author gratefully acknowledges the assistance of Ms Sangam Sri Lakshmi (VIT-AP University) with the writing of the chapter.

REFERENCES

[1] Y. Yao, Y. Chen, H. Yao, Z. Ni and M. Motani, "Multiple task resource allocation considering QoS in energy harvesting systems," *IEEE Internet of Things Journal*, vol. 10, no. 9, pp. 7893–7908, 2023.

[2] B. Vikrant, and O. Philip, "Energy harvesting for assistive and mobile applications," *Energy Science & Engineering*, vol. 3, no. 3, pp. 153–173, 2015.

[3] Y. Jung et al., "Soft multi-modal thermoelectric skin for dual functionality of underwater energy harvesting and thermoregulation," *Nano Energy*, vol. 95, pp. 1–12, 2022.

[4] N. Yujin et al., "Energy harvesting from human body heat using highly flexible thermoelectric generator based on Bi2Te3 particles and polymer composite," *Journal of Alloys and Compounds*, vol. 924, pp. 1–8, 2022.

[5] T. T. Kim, N. V. Toan, and T. Ono, "Self-powered wireless sensing system driven by daily ambient temperature energy harvesting," *Applied Energy*, vol. 311, pp. 1–10, 2022.

[6] T. Amid, G. Mohammadreza, and D. Samer, "A novel thermoelectric approach to energy harvesting from road pavement," in Proceedings of the International Conference on Transportation and Development, American Society of Civil Engineers Reston, VA, pp. 174–181, 2020.

[7] A. S. Soliman, R. Ali, Li Xu, J. Dong, and P. Cheng, "Energy harvesting in diesel engines to avoid cold start-up using phase change materials," *Case Studies in Thermal Engineering*, vol. 31, pp. 1–20, Jan. 2022.

[8] N. Jaziri et al., "A comprehensive review of thermoelectric generators: technologies and common applications," *Energy Reports*, vol. 6, no. 7, pp. 264–287, 2020.

[9] E. J. Carlson, K. Strunz, and B. P. Otis, "A 20 mV input boost convertor with efficient digital control for thermoelectric energy harvesting," *IEEE Journal of Solid-State Circuits*, vol. 45, no. 4, pp. 741–749, 2010.

[10] Y. K. Ramadass, and A. P. Chandrakasan, "A batteryless thermoelectric energy harvesting interface circuit with 35 mV startup voltage," *IEEE Journal of Solid-State Circuits*, vol. 46, no. 1, pp. 333–341, 2011.

[11] P. S. Weng, H. Y. Tang, P. C. Ku, and L. H. Lu, "50 mV-input Batteryless boost convertor for thermal energy harvesting," *IEEE Journal of Solid-State Circuits*, vol. 48, no. 4, pp. 1031–1041, 2013.

[12] A. K. Sinha, and M. C. Schneider, "Short startup, batteryless, self-starting thermal energy harvesting chip working in full clock cycle," *IET Circuit Devices and Systems*, vol. 11, no. 6, pp. 521–528, 2017.

[13] A. K. Sinha, and M. C. Schneider, "Efficient, 50mV startup, with transient settling time < 5 ms, energy harvesting system for thermoelectric generator," *IET Electronics Letters*, vol. 52, no. 8, pp. 646–648, 2016.

[14] C. P. Basso, *Switched-Mode Power Supplies SPICE Simulations and Practical Designs*. Mc-Graw Hill Publication, New York, NY, 2008.

[15] Understanding boost power stages in switch mode power supplies, Application report, March 1999.

[16] M. B. Machado, M. C. Schneider, and C. G. Montoro, "On the minimum supply voltage for MOSFET oscillators," *IEEE Transactions on Circuits and Systems I: Regular Papers*, vol. 61, no. 2, pp. 347–357, 2014.

[17] J. Dickson, "On-chip High-voltage Generation in NMOS integrated circuits using an improved voltage multiplier technique," *IEEE Journal of Solid-State Circuits*, vol. 11, no. 6, pp. 374–378, 1976.

[18] C. G. Montoro, M. C. Schneider, and M. B. Machado, "Ultra-low-voltage operation of CMOS analog circuits: Amplifiers, oscillators, and rectifiers," *IEEE Transactions on Circuits and Systems—II*, vol. 59, no. 12, 2012.

[19] K. K. Clarke, and D. T. Hess, *Communication Circuits: Analysis and Design*. Krieger Publishing Company, Malabar, Florida, USA, 1992.

[20] F. Pan, and T. Samaddar, *Charge Pump Circuit Design*. McGraw-Hill, New York, NY, 2006, ch 9.

[21] R. J. Baker, *CMOS Circuit Design, Layout, and Simulation*. John Wiley & Sons, New Jersey, NJ, 2005.

[22] A. K. Sinha, "A self-starting 70 mV-1 V, 65% peak efficient, TEG energy harvesting chip with 5 ms startup time," *Journal of Circuit System and Computers*, vol. 26, no. 3, pp. 1750040-1–1750040-22, 2017.

Chapter 11

Managing concept drift in IoT health data streams

A dynamic adaptive weighted ensemble approach

M. Geetha Pratyusha and Arun Kumar Sinha
VIT-AP University, Near Vijayawada, India

11.1 INTRODUCTION

The widespread integration of internet of things (IoT) technology across various sectors has transformed the face of healthcare [1]. It has introduced innovative possibilities for continuous, real-time patient monitoring beyond traditional clinical settings. The hallmark of these IoT-based health monitoring systems is the generation of voluminous data streams that, when leveraged correctly, significantly enhance the accuracy and timeliness of medical interventions. According to a recent report by Mordor Intelligence, the IoT healthcare market is anticipated to reach over 11.6% CAGR by 2028, demonstrating the scale and speed at which this data is generated [2]. Data streams, as conceptualized in our research, are essential as an ordered series of instances that occur in real time. These are supported by various digital sources, each with distinctive features. These sources include IoT devices such as wearables, sensors, actuators, and other emerging technologies. Today's digital ecosystem is defined by the ceaseless evolution of this high-dimensionality data generation, which has the potential for infinite expansion. The advent of IoT technology and the rapid digitization of our world have magnified the complexity of data streams, challenging traditional data handling and storage methods [3]. Innovative approaches, such as online learning/adaptive learning, have evolved in response to these constraints [4]. These approaches allow data to be processed as it arrives, effectively addressing memory constraints and the immediacy demanded by data streams. Data streams are characterized by their immediate processing requirements, their inability to be stored in their entirety due to their volume, and their frequent temporal pattern shifts. As such, they demand a specific set of management techniques that can accommodate these features. The dynamic adaptive weighted ensemble (DAWE) model, introduced in our research, addresses these demands and combines the models with ensemble methods with drift detection mechanisms [5], to manage concept drift observed in streaming health data. This approach enhances the overall predictive accuracy of modern health monitoring systems that can ensure

DOI: 10.1201/9781032628059-11

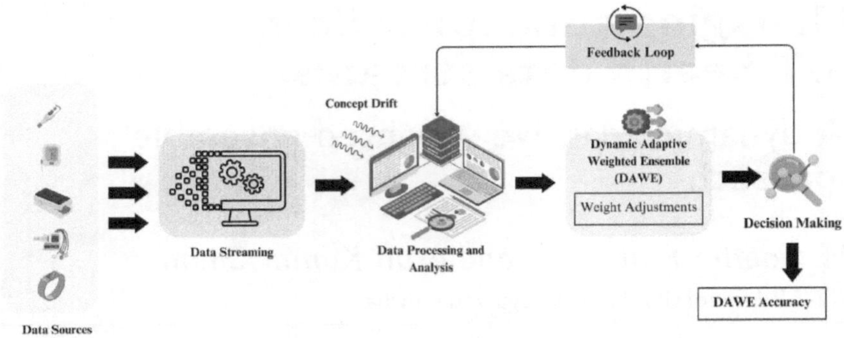

Figure 11.1 Illustration of concept drift in real-time IoT health monitoring systems.

timely medical interventions, in contrast to the inherent inadequacies of traditional data mining methods in the face of continuous streams of data [6]. However, these vast data streams bring with them the complex issue of concept drift (see Figure 11.1).

Concept drift refers to the changes in the statistical properties of the target variable that can occur over time in data streams [5]. In the healthcare context, these drifts might result from several factors, including changes in a patient's health condition, variations in diagnostic techniques, new disease trends, and shifts in population health patterns. For instance, a machine learning model trained to predict patient health status based on certain features might become less accurate over time if patients' health conditions evolve or if medical practices and technologies change. This concept drift can significantly affect the model's prediction accuracy and, thus, the quality of healthcare interventions. Therefore, identifying and managing these concept drifts is crucial. Adaptive learning model, such as DAWE offer a promising solution. Our proposed DAWE model represents the practical application of these paradigms. By integrating a multitude of models, including the Hoeffding tree (HT), stochastic gradient trees (SGT), and streaming random patches (SRP), with adaptive windowing (AWSRP) and the drift detection method (DDMSRP), the DAWE model can dynamically adjust to evolving data patterns [7]. By dynamically adjusting model weights based on prediction errors using the exponential weighted moving average (EWMA) technique, these models can continually refine their predictive performance in the face of concept drift. This approach enables the model to swiftly respond to changes, ensuring reliable, high-accuracy predictions in real time.

This research provides detailed data streaming and adaptive learning concepts, focusing on real-world implementation in IoT-based health monitoring systems. The DAWE model was evaluated using the MIMIC-III dataset [8], a popular choice in healthcare informatics, with results illustrating a substantial improvement in prediction accuracy compared to traditional models. By demonstrating how adaptive machine learning techniques can effectively manage high-velocity, high-volume data, this work offers a promising path forward for the future of healthcare service delivery and patient outcomes.

11.1.1 Concept of drift

The concept of drift, a significant challenge in IoT data streams, refers to changes in the underlying statistical properties of the target variable, which are functions of input data features over time [5]. Mathematically, if we denote the joint probability distribution at time t as $P_t(X, Y)$ (where X represents the feature vector and Y represents the target variable), concept drift occurs when $P_t(X, Y)$ differs from $P_s(X, Y)$ for $t \neq s$. Given the temporal nature of IoT data, these changes may occur suddenly or gradually and could be recurrent or incremental [9]. IoT data streams exhibit non-stationary environments due to changes in devices, users, context, and other factors affecting the sensors. For instance, if we consider a binary classification problem in IoT data stream mining, where $Y \in \{0, 1\}$, the posterior probability $P_t(X, Y)$ would essentially describe the concept at time t. Concept drift would thus occur when the conditional probability $P_t(Y|X) \neq P_s(Y|X)$ for $t \neq s$. Adapting to these changes is crucial to allow predictive models applied to IoT data streams to maintain their performance over time. This is often addressed using adaptive learning techniques, which update the model based on the latest data, accommodating the new "concept" or the changed data distribution.

There are four fundamental types of concept drift: Sudden drift is an abrupt change in the concept, such as when a sensor fails and begins sending erroneous readings, significantly altering the data stream. Incremental drift represents a gradual transition from an old concept to a new one, much like the slow change in sensor readings as a machine component wears out over time. Gradual drift is characterized by oscillation between the old and new concepts until the new one becomes established. This pattern can be observed when a new heating system is installed and adjusted, causing room temperature readings to fluctuate before settling into a new pattern. Recurring drift happens when previous concepts re-emerge, similar to how an outdoor weather sensor might record very similar readings in the same season across different years, reflecting recurring environmental conditions. Mathematically, we can consider a data stream as a sequence of instances $(x_1,y_1),(x_2,y_2),\ldots,$ where each x is an input vector and y is the corresponding output, denoted as the target concept at time t as $y(t) = f(x(t))$, where f is an unknown target function. Concept drift can then be formally defined as a situation where f changes over time. In IoT data streams, a time-dependent probabilistic model would describe the data generation process better, given by Equation (11.1).

$$(X_t, Y_T) \sim P(X, Y \mid \Theta_t) \tag{11.1}$$

where Θ_t is the set of parameters of the joint distribution at time t. When Θ_t changes over time, it induces the concept drift.

11.1.2 Significance of adaptive learning in IoT with machine learning for health data

Adaptive learning and batch learning are two pivotal methodologies used in machine learning (ML), and their respective applications have significant implications, particularly in the healthcare sector when applied to internet of things (IoT) systems. Batch learning, also known as offline learning, is a traditional ML approach where the model is trained using a complete dataset [10]. In batch learning, once the model is trained and deployed, it no longer learns from or adapts to new incoming data. Therefore, the model's performance is determined by its ability to generalize from the initial training set. While batch learning can produce effective models, it may be less suitable for dealing with IoT healthcare data streams due to its lack of adaptability to new data. Adaptive learning, also known as online learning, on the other hand, involves training models incrementally on each new instance of data or small batches of new data. This allows the model to continually adapt and update its understanding of the problem over time. Adaptive learning models are especially well suited to handling high-speed, high-volume data streams like those encountered in IoT health monitoring systems. This approach makes it possible to handle concept drift, where the statistical properties of the target variable, which the model is trying to predict, change over time. Figure 11.2 vividly illustrates the contrast between batch and adaptive learning approaches in the context of IoT data streams in healthcare. By showcasing how the two methods respond to continuous

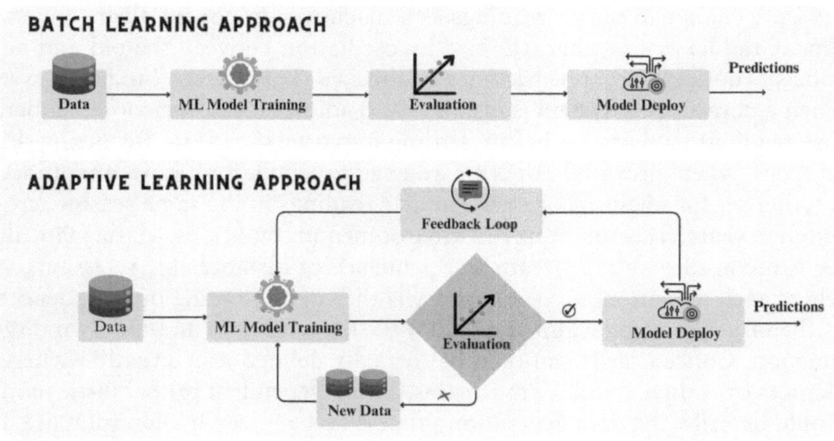

Figure 11.2 Batch and adaptive learning approaches in IoT-based healthcare data stream.

data flow and evolving trends, it highlights the superior flexibility and performance of adaptive learning models in the face of concept drift.

Suppose θ as the parameters of our model and $L(\theta; x, y)$ as the loss function for a data instance (x, y), an online learning algorithm may update the parameters based on the new instance using the rule: $\theta_{new} = \theta_{old} - \eta \nabla L(\theta_{old}; x, y)$, where η is the learning rate and $\nabla L(\theta_{old}; x, y)$ is the gradient of the loss function with respect to the parameters at the old values. The significance of this capability in health applications is multifold. First, it enables real-time patient monitoring and disease prediction, allowing for more proactive and personalized patient care. For instance, an adaptive model could learn to detect anomalous patterns in an individual patient's vital signs data, potentially catching early signs of disease progression or treatment side effects. Second, it accommodates the non-stationary nature of health data, as a patient's health status and thus data patterns can change over time due to factors like aging, lifestyle changes, disease progression, or treatment effects. The model can adapt to these changes and maintain high performance. Third, by continually learning from new data, adaptive learning can accommodate the evolving nature of healthcare, where new treatments, procedures, or health guidelines can change health data patterns.

Adaptive learning techniques, such as ensemble methods, can also effectively handle such data characteristics. For example, our proposed DAWE model leverages ensemble learning [11] and drift detection mechanisms to handle high-dimensional, imbalanced, and drifting data effectively. The ability of adaptive learning to accommodate concept drift-enhancing data streams is crucial in health data analysis. By constantly updating the models with new data, adaptive learning can effectively track and adapt to these changes.

11.2 LITERATURE SURVEY

The literature survey discusses two interconnected domains connected with IoT data streams: the concept of drift and the dynamic role of adaptive learning in magnifying model performance.

11.2.1 Drift phenomenon

The drift phenomenon in IoT data streams represents the significant, often challenging, alterations in data over time that are intrinsic to the dynamic environments these systems monitor. As IoT systems continuously collect and analyze data, understanding these shifts is crucial for effective decision-making, predictive maintenance, and more. The literature reveals various strategies to manage these drifts, each addressing sudden, incremental, gradual, or recurring drifts and offering solutions designed for the real-time nature of IoT applications. Gama *et al.* [5] presented a broad survey of methodologies used for dealing with the concept drift problem in data streams, where the foundations of taxonomy were structured around several factors such as learning paradigms, nature of the change, and method operational mode. Gama *et al.* suggested that one effective way of dealing with concept drift is to combine active learning and ensemble models. Active learning is a learning paradigm where the model actively chooses the data from which it learns, while ensemble models involve using multiple learning models and aggregating their predictions. This combination allows the model to continuously update its understanding of the concept and remain robust to concept drift. Zliobaite [12] proposed the development of adaptive IoT systems to handle concept drift. These systems are designed to recognize when concept drift has occurred and adjust their behavior in response. By making systems adaptive in this way, the author ensures that the IoT devices continue to function effectively even in the face of frequent and unpredictable changes in the data they are handling.

Agrahari and Singh conducted an extensive literature review to understand the categorization of various concept drift detectors, highlighting their key points, limitations, and advantages [13]. Their research also proposes adaptive mechanisms for detecting and adapting to concept drift, which is crucial in real-time applications that deal with semi-supervised and unsupervised data streams. Yang and Shami, in their two different works, have made significant strides towards addressing concept drift in IoT data streams [14]. In their first work, they conducted a review of AutoML approaches for IoT data analytics, recognizing the dynamic nature of IoT data, which often introduces concept drift issues. The research proposes ways to automatically select, construct, tune, and update ML models for optimal performance. In their second work, they proposed a novel, lightweight drift detection and adaptation framework designed specifically for IoT data streams [15]. This work offers an adaptive LightGBM model for anomaly detection, enhancing the efficiency and accuracy of handling IoT data streams. Liu *et al.* presented a comprehensive active learning method specifically designed for multiclass imbalanced streaming data with concept drift [16]. Their proposed method, CALMID, includes an ensemble classifier, a drift detector, a label sliding window, sample sliding windows, and an initialization training sample sequence. The authors demonstrated that CALMID is more effective and efficient than several state-of-the-art learning algorithms. Bayram, Ahmed, and Kassler [17] further deepened the discourse by examining the impact of concept drift on the performance degradation of predictive ML models. The authors proposed a consolidated taxonomy for the field by grouping concept drift types by their mathematical definitions. They also reviewed and classified performance-based concept drift detection methods, providing valuable insights into managing the performance of ML models in non-stationary environments. The authors in [14–17] have provided substantial evidence to explore adaptive mechanisms, providing efficacy in managing concept drift. The next subsection will discuss the value of adaptive learning methods, which is further emphasized by their inherent ability to continuously update the models' understanding of data concepts, making them robust to changes and enabling effective real-time decision-making.

11.2.2 Adaptive learning models towards IoT

Adaptive learning models have emerged as a focal point of exploration and development owing to their inherent ability to dynamically update and refine their predictive algorithms based on incoming data streams. Adaptive learning models have recently been used in various applications such as agriculture [18], transportation [19], health [20], etc. These models improved the capabilities of these sectors by enhancing real-time data analysis, ensuring optimal decision-making processes and fostering predictive capabilities. Codispoti *et al.* [21] discuss K-Active Neighbors (KAN), a new interactive and adaptable learning model for identifying what machines are in smart houses. This method uses the principles of stream-based active learning (SAL) and takes into account the user's access and actions during the labeling process. Notably, the suggested KAN algorithm changes the searching approach based on how available the user is and how good the device signatures are. By using real device data collected through a low-cost Arduino-based smart outlet and the ECO smart home dataset, KAN has been able to obtain good accuracy with little data. Jeong *et al.* [22] demonstrated the importance of traffic control systems in real time. They proposed a new way to predict short-term traffic flow called online learning weighted support vector regression (OLWSVR). The OLWSVR made more accurate predictions than other models, such as artificial neural networks and locally weighted regression. A pivotal study proposed by Gomes *et al.* [23] demonstrated the successful implementation of adaptive random forests (ARF) for IoT data streams, thereby significantly improving data processing efficiency. Their model capitalized on the random subspace method and drift detection to dynamically adjust to changes in data streams, yielding impressive results.

With STREAMDM-C++, a new system that implements decision trees for data streams in C++, Bifet *et al.* [24] showed how important it is to have fast and efficient ways to process big, changing data streams. The STREAMDM-C++ system was very flexible, easy to understand, and efficient. It was also much faster and used fewer resources than the current state-of-the-art C version, VFML. In their thorough study, Gaber *et al.* [25] also looked at the difficulties of adapting traditional data mining techniques to data streams, with a focus on classification techniques. They stress the need to change the way traditional methods are made to account for limited resources and changing ideas, which are common in the data stream area. The writers give an overview of the most up-to-date techniques for classifying data streams, pointing out their pros and cons.

From this literature survey, it is apparent that adaptive learning models are essential to address the unique challenges posed by real-time data processing and streaming in various domains, from smart homes to traffic management and beyond. The methods are uniquely positioned to adapt to changing environments and requirements, thereby enabling efficient and accurate decision-making and prediction processes in dynamic contexts.

11.3 METHODOLOGY

This research was designed to evaluate the efficacy of different adaptive learning models in handling concept drifts in health data streams, as shown in Figure 11.3. The process follows standard practice in ML research, providing a robust mechanism for unbiased model evaluation. The dataset comprises an array of features such as gender, age, admission type, admission location, admission diagnosis, insurance, and admission procedure, among others, that provide a comprehensive view of a patient's health profile. The initial step in the methodology involves data preprocessing. In this phase, the data is ingested, and label encoding is applied to categorical columns to convert them into a machine-understandable format. The dataset used is further transformed to ensure all remaining object columns are converted to numeric types. Once the preprocessing step is completed, the training and testing proceed. The base learning models include the streaming random patches classifier (SRP) with adaptive windowing (AWSRP) and drift detector method (DDMSRP) as drift methods, along with the tree-based models *stochastic gradient trees* (SGT) and the Hoeffding tree classifier (HT). Each model is initially trained on the training data, following which its performance is assessed on the testing data. Performance accuracy is quantitatively measured, providing a concrete metric for model evaluation. The proposed DAWE approach integrates an ensemble of adaptive models, and uses the exponential weighted moving average (EWMA) technique to dynamically adjust the weights of individual models based on their prediction errors. This innovative approach allows for a flexible model that is highly adaptive to concept drift, effectively enhancing the overall predictive performance over time.

11.3.1 Dataset description

The MIMIC-III (Medical Information Mart for Intensive Care III) Clinical Database is a significant, widely accessible dataset commonly used to train predictive medical data models [26]. It incorporates comprehensive, anonymized, health-related data extracted from a diverse population of patients admitted to intensive care units, covering a wide variety of demographics, admission details, and clinical data, and making it an irreplaceable resource for creating models aimed at enhancing clinical decision support systems and patient care outcomes. The dataset holds a total of 46,520 observations. For this work, our model tailors its methodology to focus exclusively on

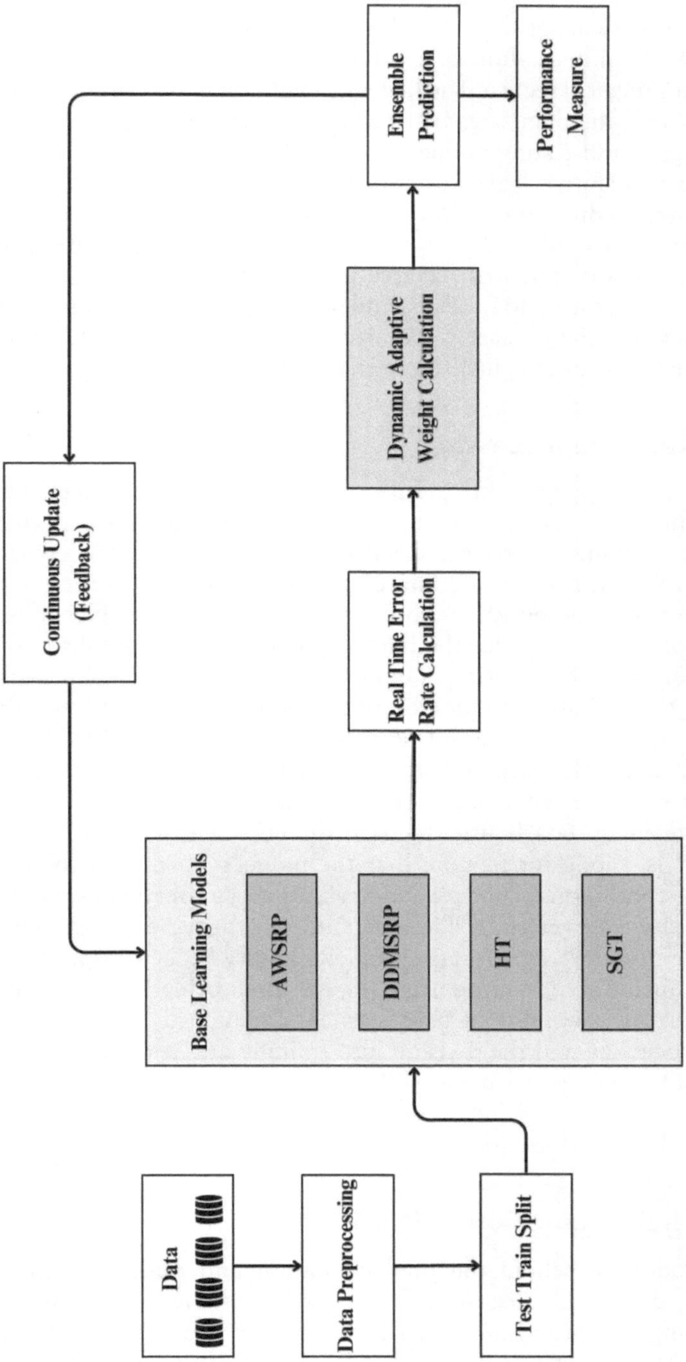

Figure 11.3 Methodology of proposed model dynamic adaptive weight calculation.

patients within the MIMIC-III dataset suffering from cardiovascular disease [27]. This subset encapsulates roughly 9,990 individual instances, each represented by 24 unique columns, signifying distinct features or characteristics of a patient's medical record. The high granularity and real-time nature of the data, mirroring the data streams from IoT devices, offer excellent grounds for developing and testing online learning algorithms. With the continuous flow of data, adaptive learning models can incrementally learn from new data instances, adjust their parameters, and improve their accuracy over time. In the context of cardiovascular disease, this could potentially enable early detection and personalized treatments, demonstrating the significant potential of integrating IoT, EHRs, and adaptive learning to improve health outcomes. While the dataset is expansive and multifaceted, it is relatively space efficient, occupying little more than 1.8MB of storage.

11.3.2 Data preprocessing

To ensure the integrity and relevance of our findings, data processing plays a significant role in our research methodology. The data procured from IoT health data streams is first subjected to a preprocessing stage where any noise or irrelevant features are filtered out. This stage involves cleaning the data, addressing missing values, and standardizing the variables to create a more uniform dataset. Once the data is preprocessed, it is split into training and test sets. The splitting is done in a time-ordered manner since the sequence of data matters in the case of data streams [28]. Typically, the first 80% of the data is allocated for training, and the remaining 20% is reserved for testing. This split ensures that the models are trained on a substantial portion of the data while still leaving a sizable test set for model validation. The training data is used to train the DAWE ensemble and the base learners. It is important to note that the model's learning is incremental, meaning it continuously updates and refines its parameters as it processes new data chunks over time. The test data, on the other hand, is used to evaluate the model's performance and its ability to generalize to unseen data. This includes measuring accuracy for monitoring the real-time error rate to adjust the weights of base learners dynamically. This set-up, with data preprocessing and the division into training and test sets, allows us to ensure that the models are being evaluated in a manner consistent with their intended use: processing and predicting outcomes in streaming data, where the availability of information changes over time.

11.3.3 Implementation of DAWE

The methodology behind the implementation and evaluation of DAWE is predicated on the integration of two foundational concepts: building base learning and proposing an ensemble DAWE strategy. Essential to this strategy is the use of robust base models and the amalgamation of these models using an innovative, dynamic ensemble method. Initially, four base models, streaming random patches classifier (SRP) [29] with adaptive windowing [30];, drift detection method [31]; and the tree-based models stochastic gradient trees (SGT) [32] and Hoeffding tree classifier (HT) [33], are trained. Each of these provides diverse, independent predictions based on their unique learning algorithms. Subsequently, these base learners are combined in a novel way using the DAWE with ensemble strategy. Unlike conventional ensemble techniques, which use static weights, DAWE employs adaptive weights that adjust according to the real-time performance of each model. This ensures that the ensemble can rapidly adapt to concept drifts in the data stream, a critical aspect of IoT data stream analytics. This dynamic approach offers enhanced accuracy, adaptability, and robustness in handling various concept drift scenarios, providing an advanced methodology for IoT data stream analytics.

11.3.3.1 Base learners

The four base models of the DAWE approach each have unique characteristics and mathematical intuitions. Importantly, all the models are designed to support data streaming, which is a crucial requirement for real-time IoT data handling, making them ideal for use in the DAWE approach, aimed at efficiently managing high-speed, high-volume data streams.

a) ***Streaming random patches with adaptive windowing (AWSRP)***: Streaming random patches with adaptive windowing (AWSRP) is an advanced adaptive ensemble learning technique that has been designed specifically to deal with the challenges presented by evolving data streams. The streaming random patches (SRP) model [34], is a type of ensemble learning method based on random patches, which are subsets of the original data features and instances. The adaptive windowing component [35], was derived to handle concept drift or shifts in the data distribution over time. It involves dynamic adjustment of the data window size w for training, based on drift detection. Mathematically, if a drift is detected at time t, the window size w decreases, focusing the model on more recent data to facilitate quick adaptation to the new concept. Conversely, in the absence of detected drift, w can increase to gather more data, promoting model stability and performance. For every data stream, each learner in the ensemble is trained on a patch $P_i = \{x'_i\}$, where x'_i is a random subset of X. The window size w_i is adaptively adjusted using the formula with Equation (11.2).

$$w_i = w_i + \left(1 - \delta_i\right) - \theta\delta_i \tag{11.2}$$

where δ_i is the drift measure and θ is a control parameter.

b) ***Streaming random patches with drift detection method (DDMSRP)***: Streaming random patches with drift detection method (DDMSRP) is an adaptive ensemble learning strategy designed specifically for evolving data streams with possible concept drift. The strength of SRP lies in this diversification, which makes the overall ensemble model more robust and generalizable. DDMSRP builds upon SRP by integrating a drift detection method (DDM) [36]. DDM works by tracking the error rate of each individual model within the ensemble. If there is a significant increase in the error rate, this may suggest a concept drift, that is, a change in the underlying distribution of the data stream.

c) ***Hoeffding tree (HT)***: The HT model is a type of decision tree that uses the Hoeffding bound principle [37]. For a given node and two features A and B, the Hoeffding bound H is computed to decide whether A is significantly better than B or more information is needed. The Hoeffding bound is computed with Equation (11.3).

$$H = sqrt\left(R^2 * \ln\left(1/\delta\right)\right)/2n, \tag{11.3}$$

where R is the range of the output variable, 0.05 is the confidence level. When H is smaller than the difference in the errors of A and B, A is chosen as the splitting feature.

d) ***Stochastic gradient trees (SGT)***: SGTs are online decision tree models that use gradient-based splitting criteria. For a node t and a feature, A, the gradient $G_{t,a}$of the loss function with respect to the model parameters is computed [37]. The feature that provides the maximum absolute gradient is chosen as the splitting feature. This can be mathematically expressed as Equation (11.4):

$$A = argmax\left|G_{t,a}\right|. \tag{11.4}$$

These mathematical intuitions underline the operations of each base model used in DAWE. The combination of these base models and their dynamic weighting strategy using the exponential weighted moving average (EWMA) technique enable DAWE to effectively manage concept drift in IoT health data streams.

11.3.3.2 Proposed ensemble strategy for DAWE

The DAWE ensemble strategy introduces a dynamic method of estimating the target class, employing a weighted combination of predictions from base learners. For a given data instance x, the estimated class, $\bar{y}$, is calculated as Equation (11.5).

$$\bar{y} = argmax\left(y\right)S_m\left[W_m\left(t\right) * P_m(y \mid x)\right], \tag{11.5}$$

where $P_m(y|x)$represents the prediction probability of class y for instance x by model m. The sum operates over all base models, m. The weights, $W_m(t)$ are determined based on each learner's real-time performance, updated dynamically to swiftly respond to concept drift. To calculate the weights, the model considers each base learner's real-time error rate, $E_m(t)$, determined as the proportion of incorrect predictions in recent data instances. This is calculated using the exponential weighted moving average (EWMA) [38], a prominent feature of the DAWE strategy. The EWMA method applies a smoothing factor, α, yielding with the error rate defined as Equation (11.6).

$$E_m(t) = \alpha * E_m(t-1) + (1-\alpha) * l\left(y_t, P_m(y_t \mid x_t)\right), \tag{11.6}$$

where l represents the loss function, y_t is the true class of instance t, and $P_m(y_t|x_t)$is the predicted class by model m. This formula ensures that more recent data instances have a larger impact on the error rate calculation, hence enabling swift adaptation to concept drift. The weights, $W_m(t)$ are then determined as $W_m(t) = 1 - E_m(t)$, resulting in models with lower error rates having higher weights, and therefore a more significant influence on the ensemble's decision. In weight calculation, DAWE also includes a mechanism to avoid computational issues arising from a zero denominator. To this end, a small constant ε is added to the denominator, ensuring the stability of the calculation. The updated weight equation will then be by Equation (11.7).

$$W_m(t) = \left(1 - E_m(t)\right) / \left(\Sigma_m\left(1 - E_m(t)\right) + \varepsilon\right), \tag{11.7}$$

where ε is the small constant to avoid computational loss and the denominator normalization ensures that the sum of all weights is 1, which is essential for the weighted average in the ensemble prediction. In all these equations. (11.5) to (11.7), m is the index over the base models in the ensemble. The DAWE technique provides several significant advantages over existing ensemble strategies, and the incorporation of the EWMA forms a key part of these benefits, enabling more responsive weighting of individual models based on their real-time performance. Instead of relying on static or fixed weights, EWMA places more emphasis on recent data points, allowing for quicker detection and adaptation to concept drift. This dynamism is vital in the rapidly changing context of IoT health data streams, which often feature evolving trends and patterns. Moreover, by using real-time error rates for weight calculations [39], EWMA ensures that the ensemble strategy remains flexible and adaptive. Models that perform well under the current concept have their weight increased, thus contributing more significantly to the ensemble's final prediction. This strategy results in an ensemble system that is constantly learning, evolving, and adapting to deliver optimal performance. The DAWE method shows much promise for the future. Because its design is flexible, DAWE could easily add newer and better drift adaptation methods as they become available. The adaptability and flexibility of the DAWE system are ideally suited to the dynamic, data-driven landscape of IoT health data streams. As technological progress brings new models and methods, DAWE's dynamic weight adjustment and real-time error monitoring provide a solid foundation for continuous improvement and enhancement, making it an effective solution for managing concept drift in IoT health data streams analysis.

11.4 EXPERIMENTAL RESULTS

The experiments were performed in a Python environment on a Jupyter Notebook on a machine with an Intel Core i5 processor, 16.0GB of RAM, and a 4GB NVIDIA GeForce RTX 3850. To simulate the concept of drift in the data stream, the River library was used, a Python library dedicated to online machine learning and stream data mining. The data stream was analyzed and processed by DAWE, with the ensemble learners' real-time performance compared against the base learners' and recorded throughout the experiment.

The proposed DAWE model demonstrated a highly competitive performance, as shown in Figure 11.4, outperforming all the individual base learners with a performance metric accuracy [40] of 96.49%. DAWE's enhanced performance can be attributed to the synergistic integration of the strengths of different base learners and the effective mechanism for handling concept drift provided by the EWMA technique. The base learner adaptive windowing

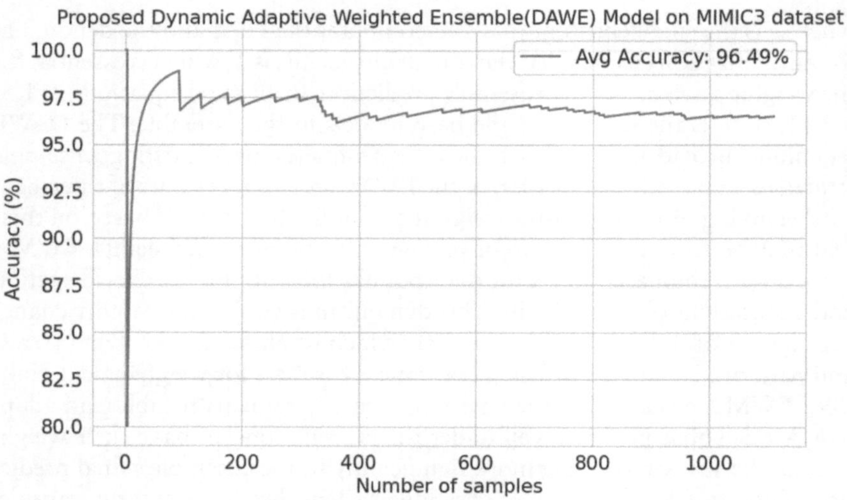

Figure 11.4 Average accuracy observed for the data streams from proposed DAWE model.

streaming random patches AWSRP achieved an average accuracy of 95.50%. It performed commendably, indicating a strong ability to learn from streaming data and adapt to concept drifts. DDMSRP was assessed and reported an average accuracy of 95.23%. Although slightly lower than AWSRP, this score still signals that DDMSRP performed robustly in detecting and adapting to concept drifts in the data stream. The SGT scored an average accuracy of 91.53%. While this was the lowest of the four base learners, it still represented a substantial degree of accuracy, especially considering that this model was applied to a challenging IoT data stream with inherent concept drift. The Hoeffding tree (HT) learner on evaluation yielded the highest average accuracy of all the base learners at 95.68%.

The accuracy achieved underscores DAWE's ability to handle IoT health data streams proficiently. It shows that the ensemble approach in DAWE significantly improves on the accuracies achieved by its individual base learners. The average accuracies are compared in Figure 11.5. DAWE's improved performance validates its design goal of creating a system capable of accurately classifying health data in dynamic and evolving environments, a characteristic inherent in IoT health data streams. The exceptional performance of DAWE also highlights its robust adaptability to the concept drift common in IoT data streams. Its dynamic weighting strategy, powered by the EWMA technique, allows it to respond rapidly and accurately to changes in the underlying data distribution, contributing to its impressive accuracy. DAWE's drift adaptation mechanisms demonstrate superior efficiency and responsiveness in real-world applications, crucial for time-sensitive IoT health data streams.

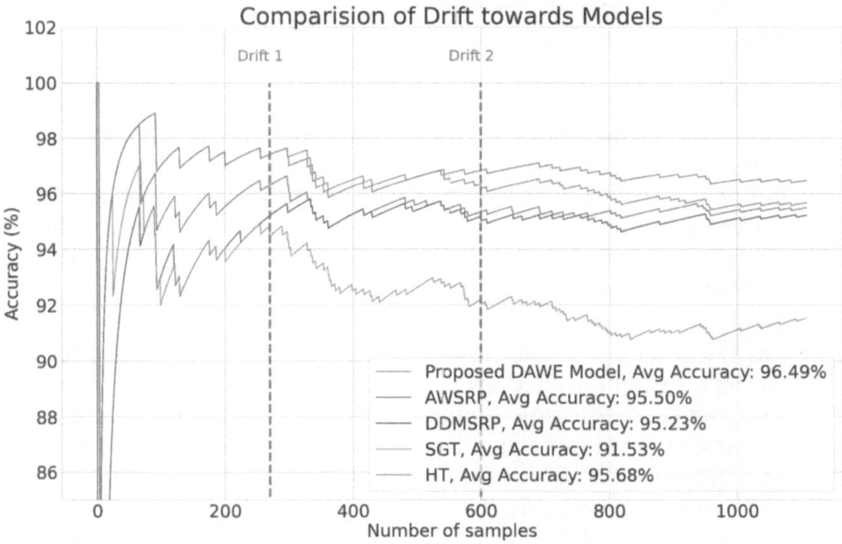

Figure 11.5 Comparison of drift across models.

11.5 DISCUSSION AND FUTURE SCOPE

DAWE's exceptional performance validates its design goal of creating a system capable of accurately classifying health data in dynamic and evolving environments, a characteristic inherent in IoT health data streams. The following aspects of its performance were highlighted, along with the future scope of this research:

- *Superior integration of multiple models.* Health data from IoT devices are often heterogeneous, complex and can involve multiple data streams. The ability of DAWE to integrate different base models such as SRP, AWSRP, DDMSRP, SGT, and HT provides it with the capability to handle a wide variety of data patterns. The DAWE model leverages the unique strengths of these varied ML models to deliver robust and versatile predictive capabilities. This diversity in model integration allows DAWE to effectively handle the high-dimensional, multifaceted health data, enhancing the accuracy of predictions and making the system more resilient to outliers and anomalies.
- *Dynamic adaptability to concept drift.* Healthcare data streams can vary over time due to a range of factors such as changes in patient health conditions, advances in medical technology, and alterations in healthcare policies. Traditional ML models might struggle with such concept drift, resulting in reduced performance over time. However, DAWE prioritizes adaptability, dynamically adjusting model weights in response to data changes. This adaptability makes DAWE more resilient to changes in the data, ensuring the model's predictive accuracy remains high even as underlying data patterns change.
- *Effective real-time weight adjustment.* The ability to make effective real-time adjustments is crucial in healthcare data management. DAWE uses the EWMA technique to dynamically adjust the weights of individual models based on their recent prediction errors. This means that as soon as a model's performance begins to degrade, its influence on the overall prediction is reduced. Conversely, models that perform better on the new data have their weights increased, maintaining the overall accuracy of the ensemble.
- *Augmented predictive accuracy.* Healthcare predictions need to be highly accurate due to the potential implications for patient treatment and care. By using multiple base learners and adjusting their weights dynamically, DAWE improves its predictive performance. This dynamic ensemble approach provides a level of accuracy that surpasses that of individual base learners, improving the reliability of healthcare decisions based on these predictions.
- *Future-proof design.* As the field of healthcare evolves, so too does the need for models that can adapt to new advances. The architecture of the DAWE model allows for the integration of new and more

effective drift adaptation methods as they become available. This forward-compatible design makes the DAWE model adaptable to future advances, ensuring its continued effectiveness in managing healthcare data streams.

These factors establish the DAWE model as a robust and adaptable system for handling data streams, particularly in healthcare, where data is continuously evolving. By maintaining high prediction accuracy, quickly adapting to changes, and offering a flexible design for future advances, the DAWE system presents a promising solution for managing concept drift in data stream analysis. The future scope of real-time data processing and streaming in the DAWE model is extensive and rich with potential. Enhanced adaptive learning capabilities can be further explored, refining the model's adaptability and efficiency in handling concept drift and other non-stationary environments, thereby increasing its responsiveness and predictive accuracy. The DAWE model's integration abilities can also be expanded to incorporate newly developed advanced algorithms, ensuring that it remains at the forefront of predictive modeling techniques.

As the complexity and volume of data streams increase, an important focus area is enhancing the model's real-time anomaly detection capabilities. This will allow the model to promptly identify unusual data patterns and outliers as they occur, crucial in domains like healthcare or finance. With the advent of the IoT and the rise of edge computing, optimizing the DAWE model for these applications could significantly improve the real-time processing and analysis of IoT data streams. Data privacy and security issues are also of paramount importance, inviting exploration into improving the DAWE model's features concerning data anonymization, encryption, and secure transmission. The potential for industry-specific customization of the DAWE model is another promising area, providing room for refinement according to the unique needs of various sectors. Finally, as data volumes escalate, attention must be paid to enhancing the scalability and performance of the DAWE model, optimizing computational efficiency and resource usage to handle larger data streams more effectively. By addressing these future scopes, the DAWE model can continue to evolve, becoming an increasingly powerful tool in managing real-time data streams.

11.6 CONCLUSIONS

In this research, a dynamic adaptive weighted ensemble (DAWE) model was developed as a means of addressing the inherent difficulties associated with IoT health data streams. Our proposed DAWE model is a significant improvement in this IoT domain. By incorporating multiple base learners, the exponential weighted moving average (EWMA) technique enables the DAWE model to achieve superior predictive performance by dynamically adjusting weights based on prediction errors. The experimental results demonstrated that DAWE obtained an accuracy of 96.5%, strongly outperforming the average accuracy score of the base learners. This remarkable performance exhibits DAWE's robust adaptability to concept drift and demonstrates the method's efficacy and responsiveness in real-world applications, particularly for time-sensitive IoT health data streams. As IoT networks continue to expand and evolve, the significance of dynamic adaptability to concept distribution will grow. With its superior performance and drift management capabilities, the DAWE model functions as a benchmark for future research and development in this area. As a result, it is anticipated that our findings will considerably contribute to the ongoing efforts to optimize the management of IoT health data streams and, consequently, improve the efficacy of health interventions.

ACKNOWLEDGMENTS

The authors would like to acknowledge the support and encouragement received from VIT-AP University.

REFERENCES

[1] J. T. Kelly, K. L. Campbell, E. Gong, and P. Scuffham, "The Internet of Things: Impact and implications for health care delivery," *Journal of Medical Internet Research*, vol. 22, no. 11, p. e20135, 2020.

[2] Mordor Intelligence, "Internet of Things in Healthcare Market – Growth, Trends, COVID-19 Impact, and Forecasts (2021–2026)," [Online]. Available: https://www.mordorintelligence.com/industry-reports/internet-of-things-in-healthcare-market/market-size. [Accessed: 5 June 2023].

[3] S. Kumar, P. Tiwari, and M. Zymbler, "Internet of Things is a revolutionary approach for future technology enhancement: a review," *Journal of Big Data*, vol. 6, no. 1, pp. 1–21, 2019.

[4] D. J. Foster, A. Rakhlin, and K. Sridharan, "Adaptive online learning," *Advances in Neural Information Processing Systems*, vol. 28, 2015.

[5] J. Gama, I. Žliobaitė, A. Bifet, M. Pechenizkiy, and A. Bouchachia, "A survey on concept drift adaptation," *ACM Computing Surveys (CSUR)*, vol. 46, no. 4, pp. 1–37, 2014.

[6] A. Bifet, J. Read, I. Žliobaitė, B. Pfahringer, and G. Holmes, "Pitfalls in benchmarking data stream classification and how to avoid them," in *Machine Learning and Knowledge Discovery in Databases: European Conference, ECML PKDD 2013*, Prague, Czech Republic, September 23–27, 2013, Proceedings, Part I 13, pp. 465–479, Berlin, Heidelberg: Springer, 2013.

[7] S. Wares, J. Isaacs, and E. Elyan, "Data stream mining: methods and challenges for handling concept drift," *SN Applied Sciences*, vol. 1, pp. 1–19, 2019.

[8] A. E. Johnson, T. J. Pollard, L. Shen, L. W. H. Lehman, M. Feng, M. Ghassemi, B. Moody, P. Szolovits, L. A. Celi, and R. G. Mark, "MIMIC-III, a freely accessible critical care database," *Scientific Data*, vol. 3, no. 1, pp. 1–9, 2016.

[9] R. Elwell and R. Polikar, "Incremental learning of concepts drifted in nonstationary environments," *IEEE Transactions on Neural Networks*, vol. 22, no. 10, pp. 1517–1531, 2011.

[10] M. Sayed-Mouchaweh, (Ed.), "*Learning from Data Streams in Evolving Environments: Methods and Applications*," 1st ed., vol. 41 of Studies in Big Data. Cham: Springer, 2018. [Online]. Available: https://doi.org/10.1007/978-3-319-89803-2

[11] B. Krawczyk, L. L. Minku, J. Gama, J. Stefanowski, and M. Woźniak, "Ensemble learning for data stream analysis: A survey," *Information Fusion*, vol. 37, pp. 132–156, 2017.

[12] I. Žliobaitė, "Learning under concept drift: An overview," arXiv preprint arXiv:1010.4784, 2010.

[13] S. Agrahari and A. K. Singh, "Concept drift detection in data stream mining: A literature review," *Journal of King Saud University-Computer and Information Sciences*, vol. 34, no. 10, pp. 9523–9540, 2022.

[14] L. Yang and A. Shami, "IoT data analytics in dynamic environments: From an automated machine learning perspective," *Engineering Applications of Artificial Intelligence*, vol. 116, p. 105366, 2022.

[15] L. Yang and A. Shami, "A lightweight concept drift detection and adaptation framework for IoT data streams," *IEEE Internet of Things Magazine*, vol. 4, no. 2, pp. 96–101, 2021.

[16] W. Liu, H. Zhang, Z. Ding, Q. Liu, and C. Zhu, "A comprehensive active learning method for multiclass imbalanced data streams with concept drift," *Knowledge-Based Systems*, vol. 215, p. 106778, 2021.

[17] F. Bayram, B. S. Ahmed, and A. Kassler, "From concept drift to model degradation: An overview on performance-aware drift detectors," *Knowledge-Based Systems*, p. 108632, 2022.

[18] F. Kyriazi, D. D. Thomakos, and J. B. Guerard, "Adaptive learning forecasting, with applications in forecasting agricultural prices," *International Journal of Forecasting*, vol. 35, no. 4, pp. 1356–1369, 2019.

[19] L. Pozueco et al., "Adaptive learning for efficient driving in urban public transport," in *2015 International Conference on Computer, Information and Telecommunication Systems (CITS)*, pp. 1–5, July 2015.

[20] C. A. Figueroa et al., "Adaptive learning algorithms to optimize mobile applications for behavioral health: guidelines for design decisions," *Journal of the American Medical Informatics Association*, vol. 28, no. 6, pp. 1225–1234, 2021.

[21] J. Codispoti et al., "Learning from non-experts: an interactive and adaptive learning approach for appliance recognition in smart homes," *ACM Transactions on Cyber-Physical Systems (TCPS)*, vol. 6, no. 2, pp. 1–22, 2022.

[22] Y. S. Jeong, Y. J. Byon, M. M. Castro-Neto, and S. M. Easa, "Supervised weighting-online learning algorithm for short-term traffic flow prediction," *IEEE Transactions on Intelligent Transportation Systems*, vol. 14, no. 4, pp. 1700–1707, 2013.

[23] H. M. Gomes et al., "Adaptive random forests for evolving data stream classification," *Machine Learning*, vol. 106, pp. 1469–1495, 2017.

[24] A. Bifet, J. Zhang, W. Fan, C. He, J. Zhang, J. Qian, G. Holmes, and B. Pfahringer, "Extremely fast decision tree mining for evolving data streams," in *Proceedings of the 23rd ACM SIGKDD International Conference on Knowledge Discovery and Data Mining*, pp. 1733–1742, August 2017.

[25] M. M. Gaber, A. Zaslavsky, and S. Krishnaswamy, "A survey of classification methods in data streams," *Data Streams: Models and Algorithms*, pp. 39–59, 2007.

[26] A. Goldberger et al., "PhysioBank, PhysioToolkit, and PhysioNet: Components of a new research resource for complex physiologic signals," *Circulation* [Online], vol. 101, no. 23, pp. e215–e220, 2000.

[27] A. Johnson, T. Pollard, and R. Mark, "MIMIC-III Clinical Database Demo (version 1.4)," *PhysioNet*, 2019.

[28] A. Shaker and E. Hüllermeier, "Recovery analysis for adaptive learning from non-stationary data streams: Experimental design and case study," *Neurocomputing*, vol. 150, pp. 250–264, 2015.

[29] H. M. Gomes, J. Read, and A. Bifet, "Streaming random patches for evolving data stream classification," in *2019 IEEE International Conference on Data Mining (ICDM)*, pp. 240–249, November 2019.

[30] I. Khamassi, M. Sayed-Mouchaweh, M. Hammami, and K. Ghédira, "Self-adaptive windowing approach for handling complex concept drift," *Cognitive Computation*, vol. 7, pp. 772–790, 2015.

[31] F. Pinagé, E. M. dos Santos, and J. Gama, "A drift detection method based on dynamic classifier selection," *Data Mining and Knowledge Discovery*, vol. 34, pp. 50–74, 2020.

[32] H. Gouk, B. Pfahringer, and E. Frank, "Stochastic gradient trees," in *Asian Conference on Machine Learning*, pp. 1094–1109, October 2019.

[33] X. C. Pham, M. T. Dang, S. V. Dinh, S. Hoang, T. T. Nguyen, and A. W. C. Liew, "Learning from data stream based on random projection and Hoeffding tree classifier," in *2017 International Conference on Digital Image Computing: Techniques and Applications (DICTA)*, pp. 1–8, November 2017.

[34] H. M. Gomes, J. Read, A. Bifet, and R. J. Durrant, "Learning from evolving data streams through ensembles of random patches," *Knowledge and Information Systems*, vol. 63, no. 7, pp. 1597–1625, 2021.

[35] Y. Sun, Z. Wang, H. Liu, C. Du, and J. Yuan, "Online ensemble using adaptive windowing for data streams with concept drift," *International Journal of Distributed Sensor Networks*, vol. 12, no. 5, p. 4218973, 2016

[36] A. Kumar, P. Kaur, and P. Sharma, "A survey on Hoeffding tree stream data classification algorithms," *CPUH-Research Journal*, vol. 1, no. 2, pp. 28–32, 2015.

[37] G. G. Moisen, E. A. Freeman, J. A. Blackard, T. S. Frescino, N. E. Zimmermann, and T. C. Edwards Jr., "Predicting tree species presence and basal area in Utah: a comparison of stochastic gradient boosting, generalized additive models, and tree-based methods," *Ecological Modelling*, vol. 199, no. 2, pp. 176–187, 2006.

[38] J. S. Hunter, "The exponentially weighted moving average," *Journal of Quality Technology*, vol. 18, no. 4, pp. 203–210, 1986.

[39] Q. Fan, C. Liu, Y. Zhao, and Y. Li, "Unsupervised Online Concept Drift Detection Based on Divergence and EWMA," in *Web and Big Data: 6th International Joint Conference, APWeb-WAIM 2022*, Nanjing, China, November 25–27, 2022, Proceedings, Part I, pp. 121–134. Cham: Springer Nature Switzerland, February 2023.

[40] G. P. Miriyala and A. K. Sinha, "Voting Ensemble Learning Technique with Improved Accuracy for the CAD Diagnosis," in *ICDSMLA 2021: Proceedings of the 3rd International Conference on Data Science, Machine Learning and Applications*, pp. 477–Es487. Singapore: Springer Nature Singapore, February 2023.

Chapter 12

GraLSTM

A distributed learning model for efficient IoT resource allocation in healthcare

M. Geetha Pratyusha and Arun Kumar Sinha
VIT-AP University, Near Vijayawada, India

12.1 INTRODUCTION

The healthcare sector has experienced significant transformation in recent years due to the rapid acceleration of digital technologies. Two emerging technologies, the internet of things (IoT) and artificial intelligence (AI) [1], are leading the digital revolution and creating a paradigm shift known as "digital healthcare." IoT enables interconnectivity between various healthcare devices, increasing the wealth of data that can be used for patient monitoring, preventive care, and diagnosis. At the same time, AI offers robust data processing capabilities, learning from the patterns in the data to deliver superior decision-making support. In healthcare, many IoT devices contain embedded systems that provide the intelligence behind their operation. These embedded systems collect data from the healthcare device, process it, and either make a decision or send that data to another machine for analysis [2]. The synergy with embedded systems is paving the way for intelligent solutions that are significantly enhancing operational efficiency, productivity, and the overall quality of life. The healthcare sector's global IoT market is expected to witness considerable growth, and to rise from USD26.5 in 2021 to reach USD94.2 billion by 2026 [3]. This expansion represents a compound annual growth rate (CAGR) of 28.9%. Concurrently, the AI segment within healthcare is also set to experience robust growth, projected to rise from USD14.6 billion in 2023 to an impressive USD102.7 billion by 2028, demonstrating an exceptional CAGR of 47.6% during this forecast period [4]. The combination of IoT devices, data transmission, and AI is transforming patient outcomes, offering a promising future where healthcare is proactive rather than reactive, making data driven decisions, and supporting patients and physicians as partners in managing health. In this work, our aim is to explore how the combined technologies of IoT and AI are reshaping the clinical decision-making process. Challenges and potential solutions for the digital transformation in healthcare are also discussed, highlighting the path towards a more efficient, effective, and equitable healthcare system.

DOI: 10.1201/9781032628059-12

12.1.1 How is IoT useful for the healthcare sector?

Patients use various IoT devices as nodes, such as wearables, smart inhalers, and implanted devices, to monitor their health parameters, and the device sensors collect continuous real-time data. This data is then transmitted over the electronic health record (EHR) under the complete healthcare administration node and sent to servers for storage and analysis by AI algorithms. These algorithms can identify patterns, predict health trends, and provide decision-making support. The insights generated from the analysis are shared with doctors, acting as end users, who can make informed decisions about a patient. As doctors apply these insights and monitor patient outcomes, feedback is provided to the AI system, allowing it to learn more and improve its predictions and recommendations. This feedback loop ensures continuous refinement of the AI algorithms and enhances the quality of care provided. The overall structure of IoT and AI in healthcare is shown in Figure 12.1.

12.1.2 Open discussion and challenges in IoT- AI with EHR

An electronic health record (EHR) is a digital version of a patient's medical history, maintained over time by health care providers. These records can include a range of data, including demographics, medical history, medication and allergies, immunization status, laboratory test results, radiology images, and personal stats like age and weight. This invaluable information can significantly aid in predicting and preventing cardiac events, adjusting treatment strategies, and ultimately improving patient outcomes. Moreover, when EHRs are combined with real-time information from IoT devices [5], it can assist healthcare experts in identifying small yet crucial alterations in a patient's state. IoT devices can play a key role in this scenario by continuously monitoring a patient's vital signs, such as heart rate or blood pressure, and updating the EHR in real time. This real-time data can be a lifesaver, alerting medical professionals to any significant changes that may indicate a need for immediate intervention. This not only helps patients have more control and involvement in their healthcare but also allows healthcare professionals to make well-informed and timely decisions. Further, AI can augment the use of EHRs [6] and IoT by providing advanced data analytics, predictive modeling, and decision-making support. AI can process vast amounts of data from IoT devices and EHRs to identify patterns, predict health outcomes, and suggest preventive measures, leading to improved patient care. Despite the vast potential, the integration of IoT, AI, and EHRs comes with a set of challenges. The main challenges relating to the combination of IoT and AI with EHR are (Figure 12.2) the following:

- *Unpredictable patient behavior*. AI algorithms struggle with the unpredictability and diversity of patient behavior due to varying genetic, lifestyle, and environmental factors influencing how health conditions present themselves.

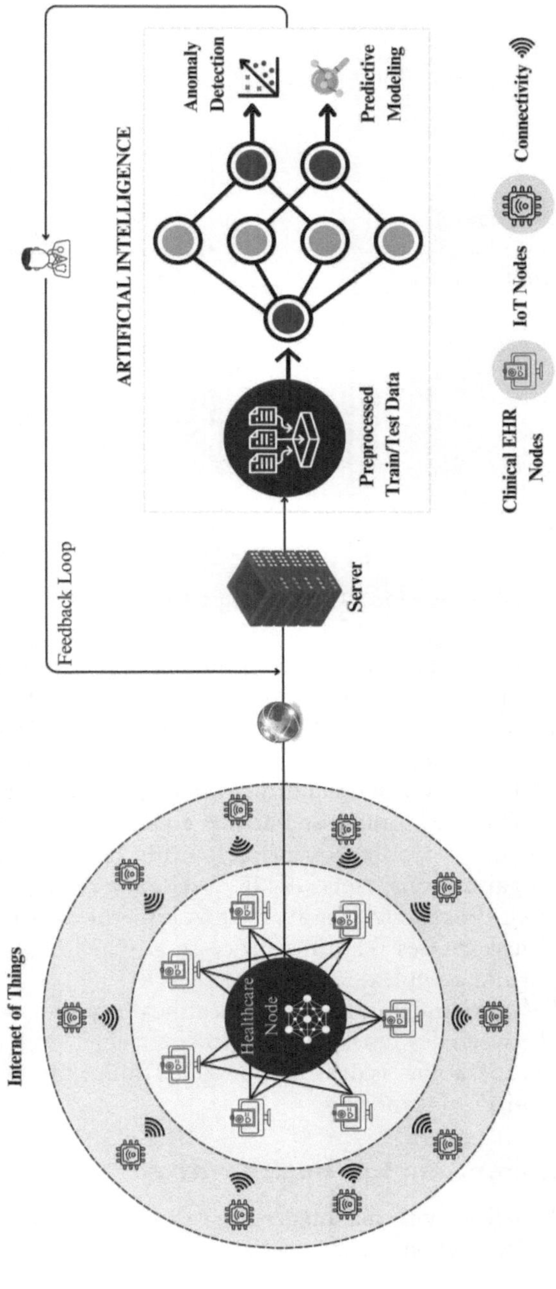

Figure 12.1 Overall use of IoT and AI in Health Care Sector.

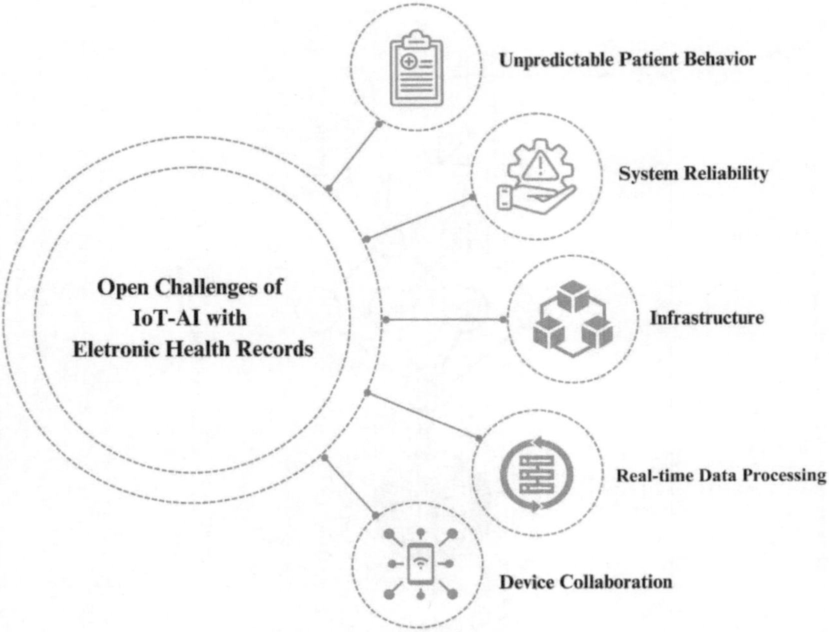

Figure 12.2 Open challenges of IoT-AI observed towards EHR.

- *System reliability*. The need for continuous, accurate functioning of IoT and AI systems in healthcare raises concerns about system reliability, as any errors or downtime can lead to serious health consequences.
- *Infrastructure*. Building a reliable, secure IT infrastructure to integrate IoT, AI, and EHRs is a major hurdle due to the need for high-speed connectivity, massive storage, and advanced data processing capabilities, which could be too costly or technically challenging for many healthcare organizations, especially in developing regions.
- *Real-time data processing*. The massive volume of data generated by IoT devices necessitates real-time processing, which is computationally intensive and complex.
- *Device collaboration*. In an IoT-based healthcare system, seamless collaboration between devices is critical. However, achieving standardization and interoperability is difficult due to the multitude of devices and the lack of universal standards.

12.1.3 Recent trends in IoT health with AI

Observing the challenges with the integration of IoT, AI, and EHRs, emerging trends in AI like anomaly detection and distributed learning present innovative solutions. Anomaly detection, a trend gaining momentum within AI, can offer significant value in handling the unpredictable and diverse nature of patient behavior. It can identify outliers in data obtained from EHRs and IoT devices, potentially revealing crucial health insights that might have otherwise been overlooked. These unexpected patterns could signal the early onset of disease or alterations in a patient's health condition, further enhancing proactive patient care. The author, N. Han, informed us about the importance of anomaly detection as a central concern in modern healthcare [7], with applications spanning from early disease detection to monitoring treatment effectiveness. To address the anomalies, our model has integrated graph neural networks (GNNs) [8] and long short-term memory (LSTM) networks to bring a new level of sophisticated model, GraLSTM, to data management and interpretation. The unique learning processes of GNN offer exceptional capability in data representation, enabling the capture and analysis of complex patterns and anomalies in healthcare data [8]. Meanwhile, LSTM networks address the heterogeneity of IoT devices by processing temporal and spatial data, providing valuable insights into real-time health status and trends [9]. However, given the variety and volume of data processed in the healthcare sector, the effectiveness of advanced technologies also hinges on the efficiency of resource allocation [10]. On the other hand, resource allocation and distributed learning address system reliability and infrastructural issues faced in the implementation of AI and IoT in healthcare. Distributed learning decentralizes the learning process across numerous IoT devices, permitting individual devices to learn from local data and contribute to the overall model, addressing real-time data processing challenges. This also ensures privacy, as raw data can remain on local devices. Additionally, efficient resource allocation optimizes usage, balancing computational requirements and power consumption—a critical aspect in IoT health systems that often work on limited resources.

In our research with IoT-AI, anomaly detection was prioritized, and our model combines model parallelism [11], data parallelism [12], and dynamic resource allocation [13] to optimally manage computational resources. The model parallelism divides an AI model across several processors, reducing the computation time, while the data parallelism splits the data across multiple processors, allowing simultaneous model training. And the dynamic resource allocation further enhances the system's efficiency by adapting to workload changes and ensuring that resources are used judiciously without compromising the system's performance. The integration of AI, IoT, and sophisticated resource management techniques was used to optimize healthcare delivery, particularly in the realm of cardiac care. Our research question looks at these emerging intersections, examining the potential of AI and IoT in restructuring healthcare systems for enhanced clinical decision-making and better patient outcomes.

12.2 LITERATURE SURVEY

This literature survey explores key areas related to the effective management of EHR data. Section 12.2.1 focuses on the role of IoT devices in decision-making processes, specifically examining their contribution to data analysis and decision support in healthcare settings. Section 12.2.2 considers the paradigm shift in anomaly detection in EHR data, exploring the evolving approaches and techniques for identifying anomalies or irregular patterns within EHR datasets. This shift reflects the need for more advanced and sophisticated methods that can effectively detect anomalies in complex and vast healthcare data, enabling timely interventions and improved patient care. Section 12.2.3 focuses on resource allocation with data and model parallelization and distributed learning techniques. It investigates how these techniques can optimize the utilization of resources, such as computational power and data processing capabilities, in handling EHR data. This subsection examines the potential benefits of parallelization and distributed learning for improving computational efficiency and scalability in EHR data management.

12.2.1 IoT towards decision-making

Cardiovascular diseases, particularly ischemic heart disease or coronary artery disease (CAD), are a leading cause of death and illness worldwide [14]. Patients with heart conditions receive treatment in various healthcare settings, generating a large volume of medical data. In current clinical practice, it is important to document every medical encounter. Healthcare professionals gather patient information, resulting in text-based documents such as admission notes, medical histories, physical examinations, progress notes, test results, discharge summaries, and recommendations. These documents are stored in an EHR system, which categorizes diseases and procedures using International Classification of Diseases (ICD) codes [15]. However, there are limits to how accurately the costs and outcomes of chronic diseases can be modeled using traditional methods such as cost-effect models, regression models, and simulation models [16]. To address these limitations, IoT was linked with the EHR in order to integrate vital monitoring of the patient with the help of sensors [18]. To date, advanced EHR functionality, such as clinical decision-making with the data including medical history, complications, medicine usage, physical reports, and sensor measurements, is lacking. Market awareness and use of smart sensors [19], wearable devices [20, 21], and medical devices have risen. The global market size of the IoT in healthcare is expected to grow from USD3.8 billion in 2019 to USD8.9 billion by 2024, at a CAGR of 18.7% during the forecast period [17]. Market potential data are shown in Figure 12.3.

According to [18], eliminating the traditional system of preserving reports, scans, prescriptions, and medical history with the help of an EHR will make

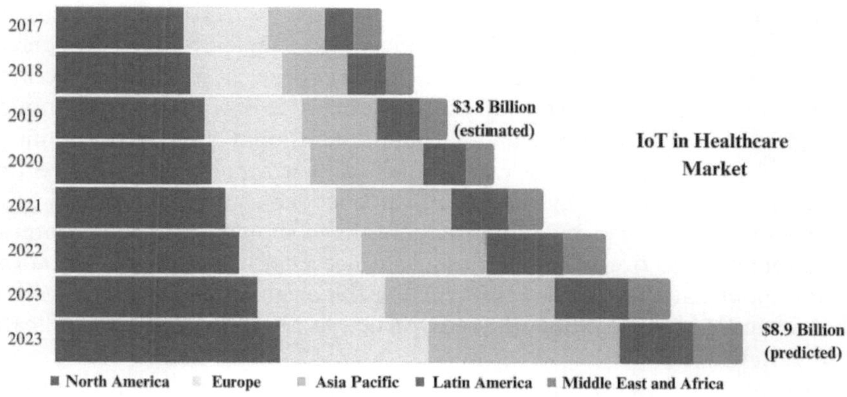

Figure 12.3 IoT in the healthcare market, varied by region [17].

the system paperless, with access to the system controlled completely by IoT technology. The authorization used by EHR device records from an individual device to be transferred to personal health records (PHR) is observed by Nai-Wei Lo [19]. EHR is mostly perceived as a portable tool, as proposed by G. Ma [20] and M. Lubell [21]. Within the IoT landscape, Hamidi [22] discussed the inherent challenges associated with hardware algorithms, software, networks, and data processing in a smart decision-making health system, some of which could be overcome by integrating biometric technology, thus ensuring secure access to smart health device data. The connection of heterogeneous IoT devices is delicate and complex, as highlighted by Mandel [23], who proposed an Android application for EHR devices to manage this heterogeneity effectively. However, this heterogeneity also raises hardware-related problems, necessitating efficient strategies for handling diverse IoT devices. To reduce hardware-related challenges and manage the heterogeneity of IoT devices, our research model seeks to provide new insights into leveraging EHR and IoT for improving decision-making and healthcare outcomes.

12.2.2 Paradigm shift in anomaly detection

The concept of anomaly detection has sparked a series of research efforts. Traditional anomaly detection techniques primarily employed statistical methods and machine learning-based techniques. Statistical approaches, such as regression models, clustering methods, outlier detection algorithms, and multivariate Gaussian models, are well suited for simple data distributions and for cases where a clear pattern of normal behavior can be defined by the author [24]. Reif [25] used machine learning techniques that have also been applied to anomaly detection, with support vector machines (SVMs) and decision trees being among the most popular. These methods require feature engineering and are typically sensitive to the quality and relevance of the features used [25]. The advent of deep learning has triggered a shift towards the application of neural networks in anomaly detection. Chalapathy and Chawla showed that the ability of deep learning to model complex data distributions, handle high-dimensional data, and eliminate the need for manual feature extraction makes it a compelling choice for anomaly detection [26]. By extension, autoencoders, a type of artificial neural network developed by Sakurada [27], have shown significant promise for anomaly detection. They work by learning a compressed representation of the input data and then reconstructing the original data. Anomalies are then identified by examining the reconstruction error; larger errors indicate anomalous data.

Convolutional neural networks (CNNs) were also applied to anomaly detection, specifically for image data, by Xu et al. [28]. CNNs have demonstrated strong performance in detecting anomalies due to their ability to automatically learn and extract features from images. According to Malhotra et al., recurrent neural networks (RNNs), along with their variants like LSTM networks, have been effective in detecting anomalies in time-series data as they can model temporal dependencies [29]. And generative models, such as variational autoencoders (VAEs) and generative adversarial networks (GANs), have also been used for anomaly detection by Schlegl et al. These models learn the underlying data distribution and can generate new data similar to the input data, making them effective at detecting anomalous instances [30]. Despite the progress made in anomaly detection using neural networks, several challenges persist. These include interpretability issues, the need for large amounts of data, and difficulties in determining an appropriate threshold for anomaly scores.

As graph mining advanced, attention gradually shifted towards anomaly detection within graphs. However, early work in this field relied heavily on domain knowledge and statistical methods, requiring handcrafted features for detecting anomalies. This approach was both time-consuming and labor-intensive. Furthermore, real-world graphs usually comprise a substantial number of nodes and edges, often labeled with numerous attributes. Consequently, these graphs are large scale and high dimensional. In an attempt to bypass these limitations, recent studies have shifted focus onto deep learning approaches for detecting anomalies in graphs. This integration operates by allowing GNNs to encapsulate the distinctive characteristics of a graph, while deep learning approaches manage other types of data, particularly temporal information. One of the first attempts to use GNNs for anomaly detection was proposed by Li et al., who used a graph convolutional autoencoder to learn representations and detect anomalies [31]. This demonstrated the ability of GNNs to use relational structures for identifying anomalous instances, leading to improved detection performance. Ding et al. [32] proposed a method based on GNNs for unsupervised anomaly detection in attributed graphs. The proposed model, named Graph CNN

(GCN), successfully uses node attributes and graph topology to detect anomalies. Further, Duta et al. introduced a method that uses GNNs for anomaly detection in time-evolving graphs. This model, known as DyReG, employs a recurrent graph autoencoder to capture dynamic patterns in graph structure and node attributes [33].

For managing large volumes of spatial-temporal data, deep learning architectures such as GNNs have made significant strides. By combining GNNs with other deep learning models, it is possible to efficiently handle spatial and temporal dimensions simultaneously. GNNs are excellent at modeling spatial dependencies. They leverage the graph structure of data, allowing them to propagate and aggregate information through the graph's nodes and edges, capturing spatial relationships [34]. To handle the temporal dimension, a variety of deep learning methods can be integrated with GNNs. RNNs, including LSTM units and gated recurrent units (GRUs), have been extensively used to handle sequential data. First, a GNN captures the spatial dependencies between nodes at each time step. The features extracted by the GNN are then processed by the deep learning model to account for temporal dependencies. Scaling these models to handle large data volumes requires efficient learning algorithms and training techniques. For GNNs, methods like GraphSAGE, proposed by Hamilton [35], generate node embeddings efficiently for large graphs. Despite the progress, challenges such as high computational costs, difficulties in model interpretation, and potential overfitting remain. Many IoT devices for EHR have followed the parallelization route, and distributed learning techniques have been implemented with data and a model for resource allocation to speed up training.

12.2.3 Resource allocation with parallelization and distributed learning

Maintaining speed and efficiency while balancing the workload to handle conventional machine learning algorithms is often challenging in view of the vast volume and complex nature of EHR data. This complexity is largely attributed to the need for high scalability and computational efficiency, which are demanding requirements for traditional algorithms to meet. The limitations of these algorithms have led to a need for innovative solutions that can more effectively manage the enormous volumes of EHR data while maintaining computational speed and time. One of the authors, Victor Toporkov [36], works at the problem of resource co-allocation for the safe execution of parallel jobs on heterogeneous hosts in high-performance computing systems. Confusion can be caused by a number of things, such as the way jobs are carried out, how they are run locally, and other steady and dynamic utilization events. Even with all of these unknowns, there is a big need for solid computing services that guarantee a good level of quality. The authors show that there needs to be a balance between available scheduling services, like securing resources, and the general efficiency of resource use in order to improve computing efficiency.

Recently, researchers have proposed parallelization—performing multiple tasks concurrently—and distributed learning techniques to overcome the challenges of scalability and computational efficiency [37]. With the IoT, each device can process a distinct subset of data independently, and the overall speed of data analysis is significantly increased. Distributed learning involves dividing not only the data but also the learning model across several machines or devices, an approach that alleviates the pressure on a central server and speeds up the learning process. This is of particular significance for large-scale GNNs [38]. Despite these promising advances, challenges remain with the integration of parallelization and distributed learning techniques in IoT devices specifically for EHR data management. A major concern is ensuring data security while maintaining the locality of EHR data. The next section focuses on the methodological aspects of the GraLSTM model, showcasing its ability to deliver high-performance capabilities while maintaining the lowest training loss and batch loading time.

12.3 METHODOLOGY

The methodology for this research revolves primarily around the use of the MIMIC-III dataset, a critical database of real-world patient data, and the choice of instances related to cardiovascular disease. A total of 9,990 patient records were identified as relevant, each record comprising a broad spectrum of data encompassing demographics, admission data, and clinical specifics over a span of 24 distinct columns. The first step in our unique approach was to represent this dataset in the form of a graph for a more intuitive understanding of patient interactions and to facilitate anomaly detection on a larger scale. Each patient data record is depicted as a node, with the edges outlining relationships between these nodes, capturing the intricate network of patient interactions. The graph thus created consists of 9,990 nodes and a staggering 1,552,447 edges determining the relationships and complexities of cardiovascular disease instances. This intricate graphical representation aids identification of abnormal data points, which helps early-stage anomaly detection. The graph, paired with the 23 feature columns and length of stay (LOS) days as a target, is then transformed through label encoding, a crucial step for ensuring data compatibility. The transformed data served as the input to our specially engineered GNN model, integrated with an LSTM layer in a configuration referred to as GraLSTM. The GraLSTM model is a complex structure composed of two graph convolutional network (GCN) layers, an LSTM layer that allows for temporal analysis, an attention mechanism for differential weighting of LSTM outputs, and a final fully connected layer responsible for generating the output. The attention mechanism employed here highlights unusual patterns in the data. The model was trained and optimized using the L1-Loss function and the stochastic gradient descent (SGD) optimizer, respectively. The training process consisted of 100 epochs, in which at each epoch, the forward pass was executed, the loss was computed, gradients were backpropagated, and the model's parameters were updated accordingly This process ensured optimal usage of computational resources while maintaining model performance. In a performance comparison with the GCN and GNN models, the model had the lowest loss values and significantly lower batch loading times, showing GraLSTM's high computational efficiency in terms of speed. This result confirms the efficiency of our resource allocation strategy, facilitating faster processing while reducing the computational burden.

To further enhance our aim of IoT scalability and adaptability for providing clinical decision-making systems by aligning the model with real-world healthcare systems, our model performs data and model parallelism with distributed learning. This implies an efficient allocation of resources among multiple IoT devices. The data were broken down into dynamic chunks that were processed by ten distinct worker models, or IoT devices, in parallel. Each of these worker models was equipped with a GraLSTM model, allowing simultaneous independent data processing. The results showed a further reduction in loss values, validating our hypothesis about the efficacy of a distributed learning approach for handling large-scale healthcare data in an IoT ecosystem. This approach not only optimizes resource allocation but also allows for more efficient and comprehensive anomaly detection across a larger dataset. The methodology developed is shown in Figure 12.4.

12.3.1 Dataset description

The MIMIC-III (Medical Information Mart for Intensive Care III) Clinical Database is a comprehensive and publicly accessible resource frequently used for training models that handle EHR [39–41]. It includes detailed, deidentified health-related data derived from a diverse number of patients admitted to intensive care units. This rich collection of data covers a broad range of demographics, admission information, and clinical data, making it an invaluable resource for developing models aimed at improving clinical decision-making support and patient care outcomes. The total number of observations in this dataset is 46,520. The structured format enhances both the accessibility and usability of the data, facilitating seamless integration into various data analysis and machine learning workflows. For the specific purpose of this work, our model develops methodology focused only on those patients suffering from cardiovascular diseases within the MIMIC-III dataset. This specific subset consists of approximately 9,990 individual instances. These instances have 24 distinct columns, each representing a unique feature or characteristic of the patient's medical record. Despite the size and complexity of the dataset, it is relatively compact in terms of storage requirements, taking up slightly more than 1.8MB of space. A detailed

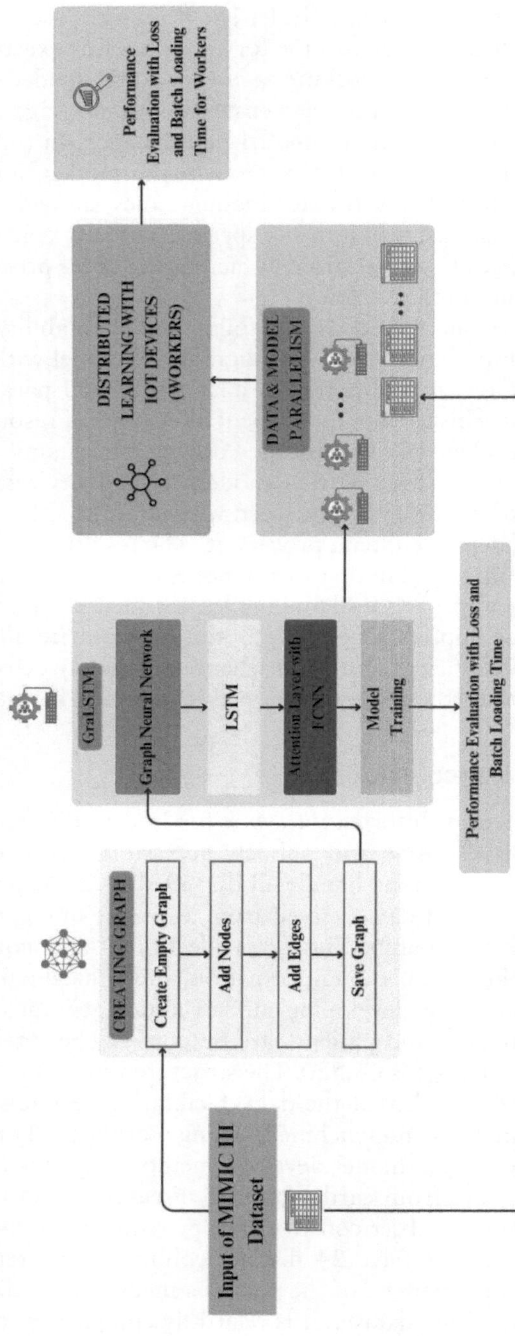

Figure 12.4 Methodology of the developed work.

Table 12.1 Dataset description

Column name	*Description*
gender	Patient's gender
age	Age of the patient
LOSdays	Length of stay in the hospital in days
admit_type	Type of admission (e.g., emergency, elective, urgent)
admit_location	Location from which the patient was admitted
AdmitDiagnosis	The diagnosis that was made when the patient was admitted
insurance	Type of insurance coverage of the patient
NumCallouts	Number of callouts during the patient's stay
NumDiagnosis	Number of diagnoses during the patient's stay
NumProcs	Number of procedures the patient underwent
AdmitProcedure	Procedure that was done when the patient was admitted
NumCPTevents	Number of Current Procedural Terminology (CPT) events
NumInput	Number of input events during the patient's stay
NumLabs	Number of laboratory tests conducted
NumMicroLabs	Number of microbiology lab tests conducted
NumNotes	Number of notes made during the patient's stay
NumOutput	Number of output events during the patient's stay
NumRx	Number of prescription medications given
NumProcEvents	Number of procedure events during the patient's stay
NumTransfers	Number of times the patient was transferred (either within the hospital or to another facility)
NumChartEvents	Number of charted events during the patient's stay
ExpiredHospital	Indicates whether the patient died in the hospital (Yes/No)
TotalNumInteract	Total number of interactions during the patient's stay
LOSgroupNum	Group number based on the length of stay, typically used for stratification or classification of patients based on length of stay

description of the dataset, including the list and description of the 24 columns (features), is provided in Table 12.1.

12.3.2 Structured graph for anomaly detection

In this work, an undirected graph G is expressed as $G = (v, \varepsilon)$ where v signifies the set of nodes and ε the set of edges. Each node, represented as $v_i = V, i = 1, 2, \ldots ..., N$, corresponds to a single patient record from our chosen subset of cardiovascular disease cases within the MIMIC-III dataset. Meanwhile, the edges forming ε shows the relation between these nodes, forming interactions, similarities, or dependencies that could exist between various patient records. To form these node connections, an adjacency matrix A was formed, with dimensions $N \times N$, where N represents the total number of nodes in graph G [40]. Each element within this matrix, denoted as A_{ij}, signifies the relationship between nodes v_i and v_j. These elements, interpreted as binary values in particular situations, define the presence or absence of an edge between v_i and v_j. However, in this work, these elements assume real values, encoding the similarity between v_i and v_j, based on a predetermined similarity measure. This adjacency matrix A, then, aids us in defining the degree matrix D, which is a diagonal matrix where each element corresponds to the sum of the rows in the adjacency matrix, showing the total degree or connections of each node within the graph. Moreover, each node v_i in the graph is linked with a feature vector matrix $x_i \in R^{N \times D}$, symbolizing the patient data record of the corresponding cardiovascular disease case. This data record comprises multiple features or parameters such as demographics, admission information, and clinical data, etc., establishing a feature space of D dimensions. These feature vectors for each node are then consolidated to generate the feature vector matrix X. In instances where a feature vector for each node is not readily accessible, node embedding techniques can be employed to learn vector-based node representations. These methods transform the nodes of the graph into a low-dimensional space, retaining the structural properties of the graph, thus simplifying subsequent machine learning tasks like anomaly detection [42].

12.3.3 GraLSTM model

Our research utilizes GraLSTM, a model combining graph neural networks (GNNs) [43] and long short-term memory (LSTM) [44] networks, to analyze structured graph data (see Figure 12.5). GraLSTM captures complex dependencies in graph data with its GCN layers, effectively addressing spatial dependencies. To cover temporal-spatial dependencies, the GCN [45] outputs were fed into an LSTM layer. An attention mechanism then assigns weights to the LSTM outputs, focusing on nodes most likely linked to anomalies. The final step uses a fully connected layer to evaluate performance based on attention-weighted LSTM outputs. These evaluations provide insights into the data structure and behavior, assisting in anomaly detection and optimization of high-dimensional data processing, crucial in fields like real-time analysis of electronic health records for clinical decision support. By monitoring training loss and batch loading times, GraLSTM ensures efficient resource management and accurate representation of data. Comparisons with traditional GCN and GNN models underscore GraLSTM's superior performance in handling high-dimensional data and detecting anomalies.

The GraLSTM model initially used a GNN, an innovative approach that extends the application of deep learning techniques to structured data, specifically, graph data. Unlike conventional neural networks that assume that instances are independently and identically distributed, GNNs focus on graph-structured data where the entities (nodes) and their connections (edges) exhibit complex dependencies. In GNN, a specific type of two-GNN

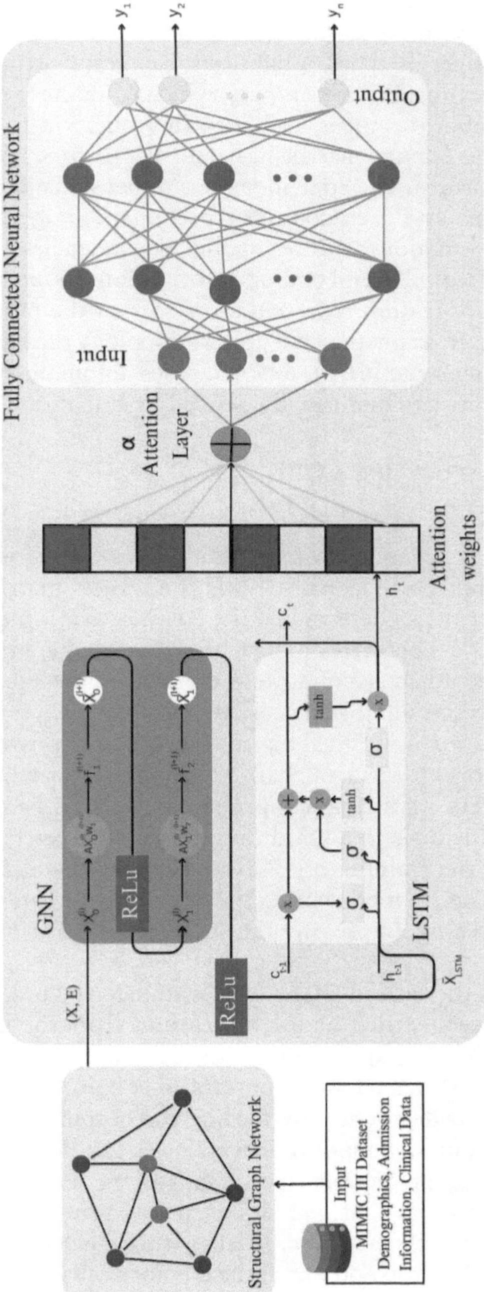

Figure 12.5 GraLSTM model architecture.

layer called graph convolutional network (GCN) was implemented using the PyTorch geometric function [46]. The GCN layer takes a feature matrix of nodes and an edge index as inputs. The GCN is particularly effective at performing semi-supervised learning tasks on graph-structured data. It works by learning a function of its inputs, which could be the features of the node and its neighboring nodes and uses this function to transform the node's features. The feature matrix has dimensions $[\nu \times \varepsilon]$, with F as the number of node features. The edge index has dimensions $[2 \times \varepsilon]$, where ε is the number of edges and each column represents an edge connecting two nodes. This transformation captures spatial dependencies [47] within the graph structure, essentially embedding information about a node's neighborhood into its new feature representation, so that the model can discern relationships and patterns within the graph's topology that might be indicative of an underlying structure or, in some cases, anomalies. The operation of a single GCN layer can be formally defined by Equation (12.1).

$$\bar{X} = \sigma\left(\left(D^{(-1/2)}AD^{(-1/2)}\right)(X \times W)\right) \tag{12.1}$$

where A is the adjacency matrix of the graph, D is the degree matrix of the graph, X is the input node features, W is the weight matrix of the GCN layer, σ in Equation (12.1) determined the ReLu activation function. In the equation, $D^{(-1/2)}AD^{(-1/2)}$ performs a normalization on the adjacency matrix. Each element of the adjacency matrix is divided by the square root of the degrees of the corresponding nodes. This is known as symmetric normalization. The reason for using the square root (or –1/2 power) comes from spectral graph theory. It ensures that the resulting normalized adjacency matrix maintains nice mathematical properties, such as being symmetrical and having its eigenvalues in a good range. This symmetric normalization is known to make the training of GCNs more stable and successful. This normalization step helps in preventing the exploding or vanishing gradients problem during training, improving the stability and efficiency of learning in GCNs.

The output from the second GCN layer is transferred to the LSTM layer. Generally, GCN layers extract high-level features from the nodes and their neighboring nodes in the graph but may not capture the temporal dependencies among nodes [48], which are important in certain applications such as health records and medical data. By feeding the output of the GCN layers into an LSTM layer, these temporal dependencies can be captured and considered when making predictions. Long short-term memory (LSTM) is a type of RNN architecture that was designed to address the issue of long-term dependencies in sequence data. It has gating mechanisms that control the flow of information into and out of the memory cells of the network. The LSTM layer operates on sequences. Each node's output from the GCN layers forms part of the sequence that the LSTM layer processes. The LSTM layer updates its hidden state H_t and cell state c_t at each time step t according to the following input gate (i_t), forget gate (f_t), cell candidate gate (g_t), and output gate (o_t). The input gate in LSTM determines how much of the incoming information should be stored in the memory cell and is a function of the current input and the previous hidden state. The forget gate decides what portion of the previous memory cell should be forgotten or kept. Meanwhile, the cell candidate generates new candidate values that could be added to the memory cell. The weights and biases involved in this process are parameters that are learned by the input gate, forget gate, and cell candidate gate during training. The output gate then determines how much of the information in the memory cell should be output at this time. Concurrently, the memory cell, representing the internal memory of the LSTM cell, gets updated with each time step. And the hidden state shows the output of the LSTM cell for this time step. This process in time t is defined by Equations (12.2) to (12.7).

$$i_t = \sigma\left(W_{ii}X_t + b_{ii} + W_{hi}H_{(t-1)} + b_{hi}\right) \tag{12.2}$$

$$f_t = \sigma\left(W_{if}X_t + b_{if} + W_{hf}H_{(t-1)} + b_{hf}\right) \tag{12.3}$$

$$g_t = \tanh\left(W_{ig}X_t + b_{ig} + W_{hg}H_{(t-1)} + b_{hg}\right) \tag{12.4}$$

$$o_t = \sigma\left(W_{io}X_t + b_{io} + W_{ho}H_{(t-1)} + b_{ho}\right) \tag{12.5}$$

$$c_t = f_t C_{t-1} + i_t g_t \tag{12.6}$$

$$H_t = o_t \tanh(C_t) \tag{12.7}$$

where σ in Equation (12.3) denotes the sigmoid activation function, tanh denotes the hyperbolic tangent activation function. After the LSTM layer, an attention mechanism is applied over the LSTM outputs. The purpose of the attention mechanism is to weigh the importance of each node's features based on the output from the LSTM layer. The attention mechanism can be formally defined by Equations (12.8) and (12.9), where W and b denote weight matrices and bias vectors, respectively.

$$e_i = w_v{}^T \tanh(Wh_i + b) \tag{12.8}$$

$$\alpha_i = \frac{\exp(e_i)}{\Sigma \exp(e_j)} \tag{12.9}$$

where from Equation (12.8) and Equation (12.9), the 'exp' is the exponential function and the denominator $\sum \exp(e_j)$ is the sum of the exponential of all alignment scores for all nodes. Essentially, α_i is the SoftMax of e_i, which normalizes the alignment scores into a probability distribution. This means that the sum of all α_i for all nodes equals 1.

In the final step of GraLSTM, a fully connected (linear) layer was used, which takes the attention-weighted LSTM outputs as input and produces the final performance evaluations. This final layer can capture the non-linear relationships between the learned graph features and the target variable. In our model, the main goal is not to predict a specific outcome but to understand the underlying structure and behavior of the data; therefore, testing loss was not the primary focus. In this case, understanding anomalies in the data and optimizing the processing speed of the high-dimensional data represented as a graph were vital. This model, GraLSTM, constructed with GNN and LSTM, helps in this by modeling the complex, non-Euclidean structure of the graph data and the potential temporal relationships between the nodes, which may reveal the anomalies. The training loss is useful as it provides a measure of how well the model is capturing the structure of the data. A decreasing training loss indicates that the model is learning to represent the data effectively. As the dimensionality and size of the data grows, the time required to load the data and the computational resources required to process it can increase significantly. Thus, keeping track of batch loading times helps optimize the process and manage resources effectively, which is crucial for this proposed model with high-dimensional data.

12.3.4 GraLSTM-IoT resource allocation

The developed GraLSTM model is now applied to the primary application of the IoT for resource allocation by using distributed learning principles to enhance the efficiency and scalability of resource usage [49]. The design of the model shown in Figure 12.6 allows for a multitude of workers, each acting as a virtual IoT device, thereby ensuring a decentralized and parallel processing approach. In an IoT environment, a large number of devices continuously generate data, therefore resource allocation becomes a challenging task. IoT devices are usually resource constrained and work in highly dynamic environments. The ability to optimize resource allocation comes with implications, such as computational capacity, network bandwidth, power, and memory, which are crucial for maintaining the system's responsiveness and efficiency. The GraLSTM model was developed to handle these challenges and enable optimal resource allocation.

In our work, the GraLSTM model training process is divided among different workers, each responsible for processing a distinct portion of data. This strategy uses the concept of data parallelism: each worker operates independently and simultaneously rather than sequentially, mitigating the risk of system-wide failures and privacy concerns and enhancing the system's

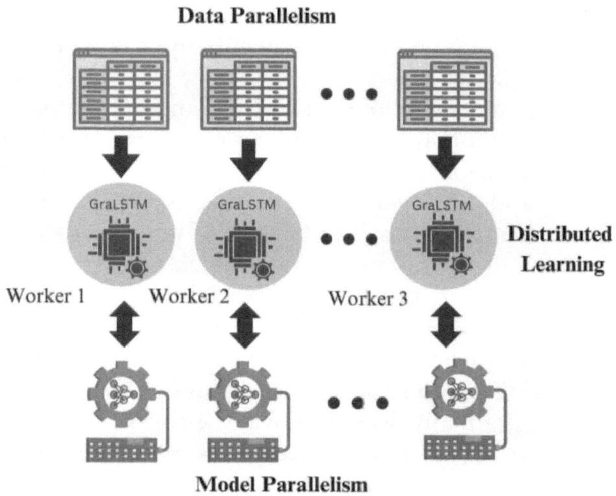

Figure 12.6 GraLSTM-IoT distributed model.

reliability and resilience. Another vital aspect of this approach is its adaptability to dynamic IoT environments. Given that IoT devices often work under changing network conditions and data patterns, the distributed learning mechanism using the GraLSTM model can also allocate more resources to workers dealing with larger or more complex data chunks and reduce resources for less demanding tasks. This adaptability ensures the model's robust performance despite fluctuations in data volume and complexity.

12.4 EXPERIMENTAL RESULTS

The experiment was carried out in a robust computing environment, ensuring both speed and precision in our analysis. We utilized a system equipped with an Intel Core i5 processor, ensuring high performance and reliability. This was supplemented by a sizable 16.0GB of RAM, ensuring smooth data handling and efficient processing of complex computations. Our system was also reinforced with a 4GB NVIDIA GeForce RTX 3850 graphics card, a high-performing GPU that allowed us to leverage the power of parallel processing for tasks, crucially accelerating the computations and modeling involved in our experiment. All our procedures and processes were run on the Jupyter Notebook, an open-source web application that allowed us to create and share documents containing live code, visualizations, and narrative text. Moreover, our work was employed in a virtual environment using TensorFlow, an end-to-end open-source platform for machine learning, allowing us to better manage our project's dependencies and maintain the integrity and reliability of our codebase. The TensorFlow environment [50] was set up with CUDA [51], for a parallel computing platform created by Nvidia, enhancing our computational capacity by making use of the GPU's processing power.

Initially, to facilitate patient interactions and anomaly detection, the dataset was represented as a graph. Each patient record is represented as a node in the graph, and relationships between patients are represented as edges. The 9,990 nodes and 1,552,447 edges of this graph capture the complex network of patient interactions. The GraLSTM model was trained for 100 epochs, each epoch involving processing a batch of data and updating the model's parameters based on the calculated loss. The first factor to be observed was the loss values obtained by L1 loss [52] reported at the end of each epoch, which indicates how well the model is performing towards anomalies. Observing the epochs, the loss in Figure 12.7a. starts at a relative value of 8.697 and gradually decreases over the course of the training. After 100 epochs, the final reported total average loss was 4.518, suggesting that the model has made significant improvements in its predictive capability.

Our GraLSTM model was also compared over the span of 100 epochs with the GCN and GNN models. The GCN model reported a higher total average loss of about 12.68, with the GNN model reporting a total average loss of 5.3087 (see Figures 12.7b and c. These data points show that the GraLSTM model proved superior to the other two in terms of performance and anomaly detection. Losses for the three models are compared and plotted in Figure 12.8.

The second factor considered was the batch loading times [53]. The average loading time of GraLSTM is approximately 0.95 seconds per epoch, which is the most efficient, and remains fairly consistent throughout the training process. In contrast, GCN shows a slightly higher loading time, averaging around 2.36 seconds per epoch. The GNN model has an average loading time of 2.5 seconds per batch, showing relatively stable loading times across the 100 epochs. As Figure 12.9 shows, the GraLSTM model distinctly outperforms both the GCN and GNN models in terms of predictive accuracy, as indicated by lower loss, and computational efficiency, as demonstrated by lower batch loading times.

The GraLSTM model allows simultaneous data processing by workers—in this case IoT devices—each of which processes a portion of the total data independently and in parallel with others. In the IoT ecosystem, these workers can be individual sensors, devices, or systems that have computational capabilities. With this distributed learning approach, the workload is shared among multiple IoT devices, enhancing scalability and efficiency. Losses for 200 epochs were observed from every worker: worker 2 scored the lowest loss for 200 epochs, gradually reducing from 8.28 to 4.18, and the second lowest loss was obtained by worker 3 with 8.29 to 4.38 (see Figure 12.10).

Figure 12.10 shows that worker 2 outperforms the others with the lowest loss at the end of training, indicating the most efficient learning. Workers 1, 3, and 10 also show considerable learning, significantly reducing their losses

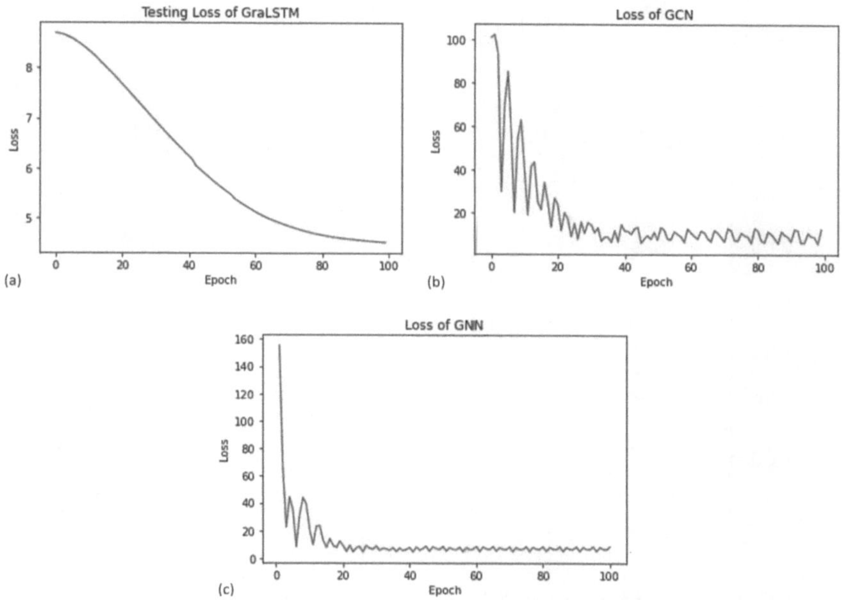

Figure 12.7 Loss of three models (a) GraLSTM loss, (b) GCN loss, (c) GNN loss.

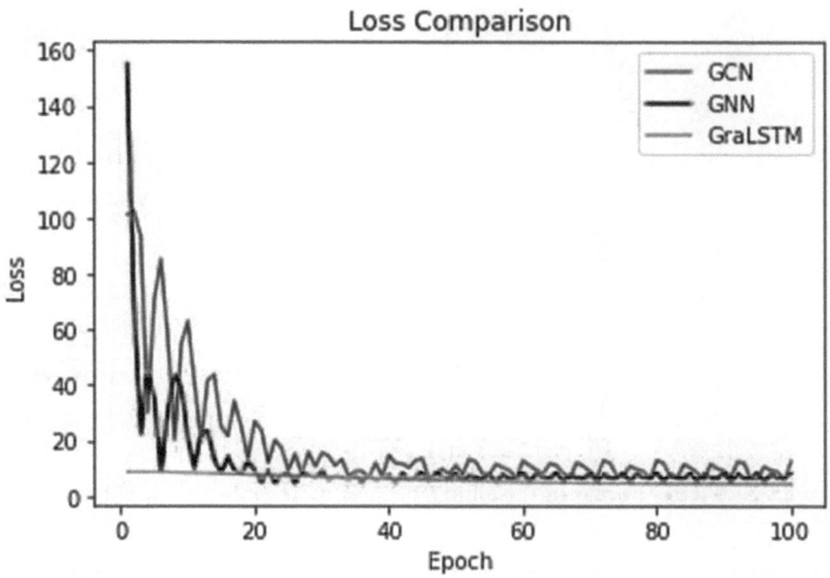

Figure 12.8 Comparison of loss of three models.

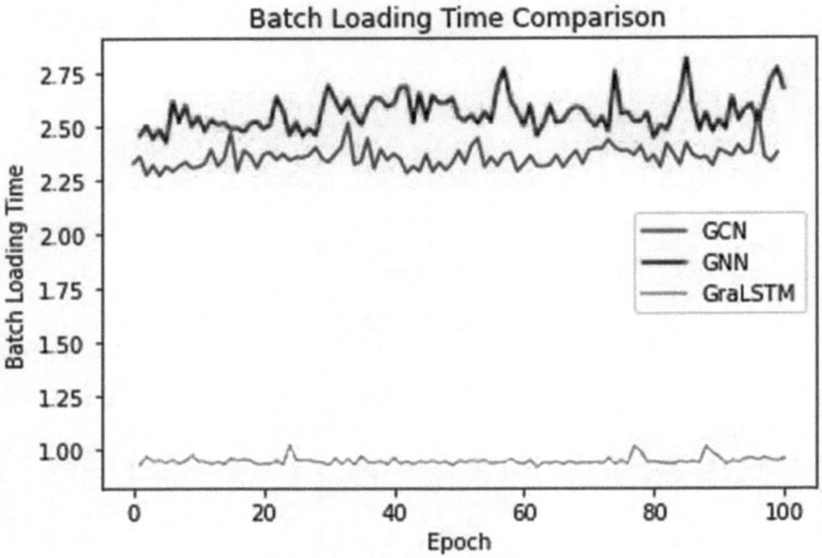

Figure 12.9 Comparison of batch loading time of three models.

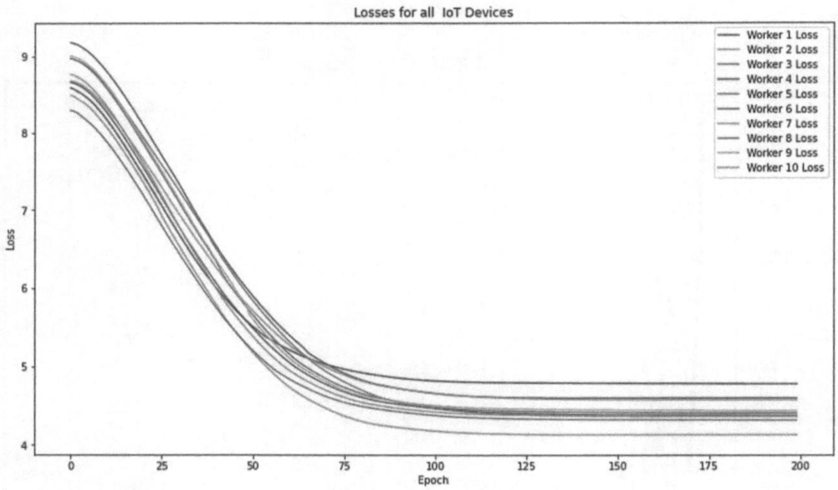

Figure 12.10 Loss observed for all IoT devices as activated as workers.

from the first to the final epoch. Workers 4, 5, 6, 7, and 8 show good progress too, but their final losses remain slightly higher. Despite having the highest initial loss, worker 9 demonstrates commendable progress, reaching a relatively low final loss. Proceeding with the batch loading time with total

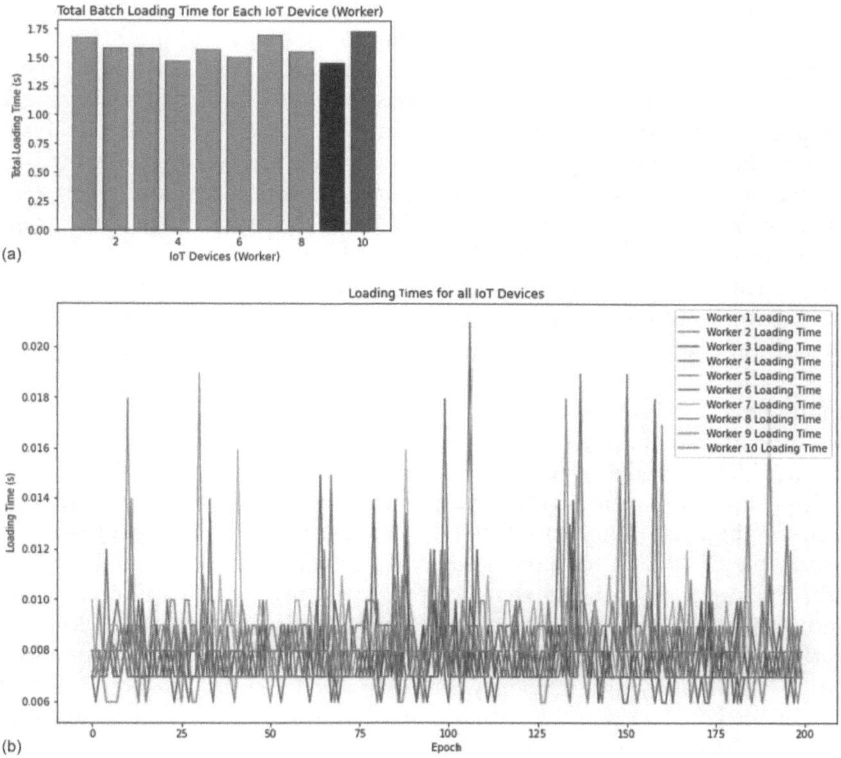

Figure 12.11 Batch loading times of workers (a) total loading time, (b) average loading time.

loading time, for all 200 epochs, worker 9 showed the lowest total loading time of 1.48 seconds, and the highest total loading time was taken by worker 10 with 1.73 seconds to complete the modeling of data, as shown in Figure 12.11a. Our work also computed the average batch loading time for each worker.

In an IoT real-time work environment, the process of summing up the total batch loading times and calculating the average batch loading time is crucial for several reasons: performance optimization, load balancing, bottleneck identification, and real-time responsiveness. Summing up the total for calculated batch loading times shows the total time a worker spends on this task across all epochs, giving an overall measure of the worker's efficiency. Averaging the batch loading time provides a typical overview of how long a worker takes to load a single batch. In a parallelized IoT setting, the average batch times for workers varied (see Figure 12.11b). Workers 0 and 1 were the fastest, averaging around 0.0076 seconds per batch. Worker 9 was close behind. Workers 2 and 8 were slightly slower, with times around

0.0078 seconds, while workers 3 and 7 hovered around 0.008 seconds. Workers 6 and 5 had times of approximately 0.0082 and 0.0081 seconds, respectively. Worker 4 was the slowest, averaging 0.0084 seconds per batch.

In this experiment, our model demonstrated the advantages of using an advanced computing environment and intelligent modeling when analyzing patient data for anomaly detection. This enables healthcare professionals to analyze a large volume of complex patient interaction data efficiently, leading to improved decision-making capabilities. The application of the graph-based approach in the GraLSTM model with the combination of GNN and LSTM enhanced anomaly detection. By transforming the complex patient interaction data into a structured network, the model provides healthcare practitioners with a clearer understanding of the relationships and patterns within the data. This also improves interpretation, can assist in identifying anomalies, such as irregularities in cardiovascular disease-related parameters, and aids in making more informed clinical decisions. The superiority of the GraLSTM model over other models, such as GCN and GNN, in terms of loss and computational efficiency, reinforces its potential for clinical decision-making. Its ability to effectively manage large datasets and maintain consistency in batch loading times was crucial for real-time analysis and decision support, ensuring that healthcare professionals can access timely and accurate information. Moreover, the application of the GraLSTM model in the IoT for distributed learning shows the importance of resource allocation for scalability and efficiency. Individual IoT devices (workers) demonstrated varying levels of efficiency in learning from the data, with worker 2 achieving the lowest average loss. This suggests that appropriate allocation of resources, such as processing power and speed, to specific devices can optimize their contributions to the learning process. This optimized resource allocation can further enhance the overall performance and scalability of the system, allowing for more effective clinical decision-making.

12.5 CONCLUSION

In this work, the potential integration of the internet of things (IoT) and AI with electronic health records (EHR) in the healthcare industry was focused on improving clinical decision-making. By developing the GraLSTM model, which combines graph neural networks (GNNs) with LSTM, processed data for developing the EHR ecosystem was achieved. The GraLSTM model effectively captured spatial and temporal data dependencies. The experiment involved training the GraLSTM model for 100 epochs and evaluating its performance in terms of loss and computational efficiency. The results show that the GraLSTM model achieved the lowest loss among the three models with a total L1 loss of training of 4.51, outperforming other models like GCN and GNN, indicating its superior anomaly detection capabilities. Additionally, the model exhibited consistent and efficient batch loading times throughout the training process, showing good computational efficiency. The research work further applied to the GraLSTM model, which demonstrates scalability by efficiently distributing the workload among multiple virtual IoT devices as workers. The analysis of the workers' performance revealed that worker 2 showed the lowest average loss, indicating its high efficiency in learning from the data. The low batch loading times observed from worker 9 in the experiment proved the model's ability to handle large datasets efficiently, contributing to faster analysis and decision support. By achieving scalability and fast processing speed, the GraLSTM shows efficacy and efficiency for clinical decision-making in the IoT-AI-based EHR ecosystem.

ACKNOWLEDGMENTS

The authors would like to acknowledge the support and encouragement received from VIT-AP University.

REFERENCES

[1] F. Alshehri and G. Muhammad, "A comprehensive survey of the Internet of Things (IoT) and AI-based smart healthcare," *IEEE Access*, vol. 9, pp. 3660–3678, 2021.

[2] J. D. Alvarado Moreno, L. C. Luis Garcia, W. C. Hernandez and A. M. Barrera Obando, "Embedded Systems for Internet of Things (IoT) Applications: A Review Study," *2018 Congreso Internacional de Innovación y Tendencias en Ingeniería (CONIITI)*, Bogota, Colombia, 2018, pp. 1–6, doi: 10.1109/CONIITI.2018.8587092

[3] IoT in Healthcare Market worth $289.2 billion by 2028. [Online]. Available: https://www.marketsandmarkets.com/PressReleases/iot-healthcare.asp. [Accessed: June, 2023].

[4] Artificial Intelligence (AI) segment in healthcare projected to reach USD 102.7 billion by 2028, with a CAGR of 47.6%. [Online]. Available: https://www.marketsandmarkets.com/Market-Reports/artificial-intelligence-healthcare-market-54679303.html. [Accessed: June, 2023].

[5] G. Devi and S. A. M. Rizvi, "Integration of Genomic Data with EHR Using IoT," *2020 2nd International Conference on Advances in Computing, Communication Control and Networking (ICACCCN)*, Greater Noida, India, 2020, pp. 545–549.

[6] S. Lee and H.S. Kim, "Prospect of artificial intelligence based on electronic medical record," *Journal of Lipid and Atherosclerosis*, vol. 10, no. 3, pp. 282, 2021.

[7] N. Han, S. Gao, J. Li, X. Zhang, and J. Guo, "Anomaly Detection in Health Data Based on Deep Learning," *2018 International Conference on Network Infrastructure and Digital Content (IC-NIDC)*, Guiyang, China, 2018, pp. 188–192.

[8] L. Zeng et al., "GNN at the edge: Cost-efficient graph neural network processing over distributed edge servers," *IEEE Journal on Selected Areas in Communications*, vol. 41, pp. 720–739, 2023.

[9] S. Godhbani, S. Elkosantini, W. Suh, and S. M. Lee, "ADL based Framework For Multimodal Data Fusion in Traffic Jam prediction," *2022 14th International Conference on Software, Knowledge, Information Management and Applications (SKIMA)*, Phnom Penh, Cambodia, 2022, pp. 126–132.

[10] L. Zhang et al., "An Adaptive Resource Allocation Approach Based on User Demand Forecasting for E-Healthcare Systems," *2022 IEEE International Conference on Communications Workshops (ICC Workshops)*, Seoul, South Korea, 2022, pp. 349–354.

[11] C. Ying, C. Ying, and C. Ban, "A performance optimization strategy based on degree of parallelism and allocation fitness," *EURASIP Journal on Wireless Communications and Networking*, vol. 2018, pp. 1–8, 2018.

[12] D. Warneke and O. Kao, "Exploiting dynamic resource allocation for efficient parallel data processing in the cloud," *IEEE Transactions on Parallel and Distributed Systems*, vol. 22, pp. 985–997, 2011.

[13] Y. Wu et al., "Intelligent resource allocation scheme for cloud-edge-end framework aided multi-source data stream," in *EURASIP Journal on Advances in Signal Processing*, vol. 2023, pp. 1–20, 2023.

[14] B. Chong et al., "The global syndemic of metabolic diseases in the young adult population: A consortium of trends and projections from the Global Burden of Disease 2000–2019," *Metabolism*, vol. 141, art. no. 155402, 2023.

[15] Y. Hong and M. L. Zeng, "International classification of diseases (ICD)," *KO Knowledge Organization*, vol. 49, pp. 496–528, 2023.

[16] Y. Liu et al., "Does the quality of street greenspace matter? Examining the associations between multiple greenspace exposures and chronic health conditions of urban residents in a rapidly urbanizing Chinese city," *Environmental Research*, art. no. 115344, 2023.

[17] MarketsandMarkets, "IoT in Healthcare Market by Component (Medical Device, Systems & Software, Services, and Connectivity Technology), Application (Telemedicine, Connected Imaging, and Inpatient Monitoring), End User, and Region-Global Forecast to 2025," 2019. [Online]. Available: https://www.marketsandmarkets.com/Market-Reports/iot-connectivity-market-94811752.html [Accessed: June, 2023].

[18] F. Hu, D. Xie et al., "On the Application of the Internet of Things in the Field of Medical and Health Care," *2013 IEEE International Conference on Green Computing and Communications and IEEE Internet of Things and IEEE Cyber, Physical and Social Computing*, Beijing, China, 2013, pp. 2053–2058.

[19] N. W. Lo, C. Wu, and Y. Chuang, "An authentication and authorization mechanism for long-term electronic health records management," *Procedia Computer Science*, vol. 111, pp. 145–153, 2017.

[20] G. Ma, J. Liu et al., "The Portable Personal Health Records: Storage on SD Card and Network, only for one's childhood," *International Conference on Electrical and Control Engineering*, Wuhan, China, 2010, pp. 4829–4833.

[21] M. Lubell, R. Guinta, and A. Moran, "Maintaining person's medical history in self-contained portable memory device," U.S. Patent 8 195 479B2, June 5, 2012.

[22] H. Hamidi, "An approach to develop the smart health using Internet of Things and authentication based on biometric technology," *Future Generation Computer Systems*, vol. 91, pp. 434–449, 2019.
[23] J. C. Mandel et al., "Smart on FHIR: A standards-based, interoperable apps platform for electronic health records," *Journal of the American Medical Informatics Association* (JAMIA), vol. 23, pp. 899–908, 2016.
[24] V. Chandola, A. Banerjee, and V. Kumar, "Anomaly detection: A survey," *ACM Computing Surveys* (CSUR), vol. 41, pp. 1–58, 2009.
[25] M. Reif, M. Goldstein, A. Stahl, and T. M. Breuel, "Anomaly detection by combining decision trees and parametric densities," *International Conference on Pattern Recognition*, Tampa, FL, pp. 1–4, 2008.
[26] R. Chalapathy, and S. Chawla, "Deep learning for anomaly detection: A survey," arXiv preprint arXiv:1901.03407, 2019.
[27] M. Sakurada, and T. Yairi, "Anomaly detection using autoencoders with nonlinear dimensionality reduction," *Proceedings of the MLSDA 2014 2nd workshop on machine learning for sensory data analysis, Gold Coast, Australia*, pp. 4–11, 2014.
[28] D. Xu et al., "Video anomaly detection based on a hierarchical activity discovery within spatio-temporal contexts," *Neurocomputing*, vol. 143, pp. 144–152, 2014.
[29] P. Malhotra et al., "LSTM-based encoder-decoder for multi-sensor anomaly detection," arXiv preprint arXiv:1607.00148, 2016.
[30] T. Schlegl et al., "Unsupervised anomaly detection with generative adversarial networks to guide marker discovery," *Information Processing in Medical Imaging: 25th International Conference*, Boone, NC, pp. 146–157, 2017.
[31] Y. Li et al., "Deep Structured Cross-Modal Anomaly Detection," *International Joint Conference on Neural Networks (IJCNN)*, Budapest, Hungary, pp. 1–8, 2019.
[32] K. Ding et al., "Deep anomaly detection on attributed networks," *Proceedings of SIAM International Conference on Data Mining*, Arizona, USA, pp. 594–602, 2019.
[33] I. Duta, A. Nicolicioiu, and M. Leordeanu, "Discovering dynamic salient regions for spatio-temporal graph neural networks," *Advances in Neural Information Processing Systems*, vol. 34, pp. 7111–7125, 2021.
[34] M. M. Bronstein et al., "Geometric deep learning: going beyond Euclidean data," *IEEE Signal Processing Magazine*, vol. 34, pp. 18–42, 2017.
[35] W. Hamilton, Z. Ying, and J. Leskovec, "Inductive representation learning on large graphs," *Advances in Neural Information Processing Systems*, vol. 30, 2017.
[36] V. Toporkov, D. Yemelyanov, and M. Grigorenko, "Optimization of Resources Allocation in High Performance Computing Under Utilization Uncertainty," *Computational Science–ICCS 2021: 21st International Conference*, Krakow, Poland, pp. 540–553, 2021.
[37] H. Gupta, A. Vahid Dastjerdi, S. K. Ghosh, and R. Buyya, "iFogSim: A toolkit for modeling and simulation of resource management techniques in the Internet of Things, edge and fog computing environments," *Software: Practice and Experience*, vol. 47, pp. 1275–1296, 2017.
[38] M. Chen et al., "Distributed learning in wireless networks: Recent progress and future challenges," *IEEE Journal on Selected Areas in Communications*, vol. 39, pp. 3579–3605, 2021.

[39] A. E. W. Johnson et al., "MIMIC-III, a freely accessible critical care database," *Scientific Data*, vol. 3, Article no. 160035, 2016.

[40] A. Goldberger et al., "PhysioBank, PhysioToolkit, and PhysioNet: Components of a new research resource for complex physiologic signals," *Circulation* [Online], vol. 101, no. 23, pp. e215–e220, 2000.

[41] A. Johnson, T. Pollard, and R. Mark, "MIMIC-III Clinical Database Demo (version 1.4)," *PhysioNet*, 2019.

[42] A. Grover and J. Leskovec, "node2vec: Scalable feature learning for networks," in *International Conference on Knowledge Discovery and Data Mining*, San Francisco, USA, pp. 855–864, 2016.

[43] J. Zhou et al., "Graph neural networks: A review of methods and applications," *AI Open*, vol. 1, pp. 57–81, 2020.

[44] A. Guo et al., "Predicting cardiovascular health trajectories in time-series electronic health records with LSTM models," *BMC Medical Informatics and Decision Making*, vol. 21, pp. 1–10, 2021.

[45] S. Zhang, H. Tong, J. Xu, and R. Maciejewski, "Graph convolutional networks: a comprehensive review," *Computational Social Networks*, vol. 6, pp. 1–23, 2019.

[46] M. Fey and J. E. Lenssen, "Fast graph representation learning with PyTorch Geometric," arXiv preprint arXiv:1903.02428, 2019.

[47] X. Wang et al., "Traffic flow prediction via spatial temporal graph neural network," in *Proceedings of the web conference*, Chongqing, China, pp. 1082–1092, 2020.

[48] R. Vinayakumar, K. P. Soman, and P. Poornachandran, "Long short-term memory based operation log anomaly detection," in *International Conference on Advances in Computing, Communications and Informatics*, Udupi, India, pp. 236–242, 2017.

[49] T. C. Chiu et al., "Semisupervised distributed learning with non-IID data for AIoT service platform," in *IEEE Internet of Things Journal*, vol. 7, pp. 9266–9277, 2020.

[50] B. Pang, E. Nijkamp, and Y. N. Wu, "Deep learning with TensorFlow: A review," in *Journal of Educational and Behavioral Statistics*, vol. 45, pp. 227–248, 2020.

[51] J. Sanders and E. Kandrot, "CUDA by example: an introduction to general-purpose GPU programming," *Scalable Computing: Practice and Experience*, vol. 11, p. 401, 2010.

[52] S. Pesme and N. Flammarion, "Online robust regression via sgd on the l1 loss," *Advances in Neural Information Processing Systems*, vol. 33, pp. 2540–2552, 2020.

[53] N. C. Taher, I. Mallat, N. Agoulmine, and N. El-Mawass, "An IoT-Cloud Based Solution for Real-Time and Batch Processing of Big Data: Application in Healthcare," in *Proceedigns of 3rd International Conference on Bio-engineering for Smart Technologies (BioSMART)*, Paris, Dec. 2019, pp. 1–8.

Index

Pages in *italics* refer to figures and pages in **bold** refer to tables.